Cascade Aerodynamics

J. P. GOSTELOW
B.Eng., M.A., Ph.D., M.A.S.M.E., M.R.Ae.S.
Professor, Head of School of Mechanical Engineering,
The New South Wales Institute of Technology,
Sydney, Australia

PERGAMON PRESS
OXFORD · NEW YORK · TORONTO · SYDNEY · PARIS · FRANKFURT

U.K.	Pergamon Press Ltd., Headington Hill Hall, Oxford OX3 0BW, England
U.S.A.	Pergamon Press Inc., Maxwell House, Fairview Park, Elmsford, New York 10523, U.S.A.
CANADA	Pergamon Press Canada Ltd., Suite 104, 150 Consumers Road, Willowdale, Ontario M2J 1P9, Canada
AUSTRALIA	Pergamon Press (Aust.) Pty. Ltd., P.O. Box 544, Potts Point, N.S.W. 2011, Australia
FRANCE	Pergamon Press SARL, 24 rue des Ecoles, 75240 Paris, Cedex 05, France
FEDERAL REPUBLIC OF GERMANY	Pergamon Press GmbH, Hammerweg 6, D-6242 Kronberg-Taunus, Federal Republic of Germany

First edition 1984

Library of Congress Cataloging in Publication Data
Gostelow, J. P.
Cascade aerodynamics.
(Thermodynamics and fluid mechanics for mechanical engineers series)
Includes bibliographical references and index.
1. Cascades (Fluid dynamics) 2. Aerodynamics.
I. Title. II. Series.
TJ267.5.C3G67 1984 621.406 83-23103

British Library Cataloguing in Publication Data
Gostelow, J. P.
Cascade aerodynamics.—(Thermodynamics and fluid mechanics for mechanical engineers series).—
(Pergamon international library)
1. Aerodynamics
I. Title II. Series
629.132'3 TL570

ISBN 0-08-020428-7 (Hardcover)
ISBN 0-08-020427-9 (Flexicover)

In order to make this volume available as economically and as rapidly as possible the typescript has been reproduced in its original form. This method unfortunately has its typographical limitations but it is hoped that they in no way distract the reader.

Printed and bound in Great Britain by
William Clowes Limited, Beccles and London

Editorial Foreword

The books in the Thermodynamics and Fluid Mechanics division of the Pergamon International Library were originally planned as a series for undergraduates, to cover those subjects taught in a three-year course for Mechanical Engineers. Subsequently, the aims of the series were broadened and several volumes were introduced which catered not only for undergraduates but, also, for postgraduate students and engineers in practice. These included new editions of books published earlier in the series.

The present work by Professor Gostelow is an advanced text which will be useful for designers in industry and it will help postgraduate students in the period of transition between first degree studies and research work. The treatment of the various topics is of sufficient depth to meet the needs of the experienced research worker. This book, therefore, represents another milestone in the development of the Thermodynamics and Fluid Mechanics division of the Pergamon International Library.

W.A.W.

Preface

Cascade Aerodynamics is intended to cover in as comprehensive a manner as is possible in a short reference work the extensive specialist field concerned with the flow around a two-dimensional row of aerofoils, usually termed a cascade. The cascade arrangement is commonly encountered in practical engineering situations when it is desired to turn a flow around a corner. Even more widespread is the use in turbomachinery design. Flows through compressors, fans, propellers, pumps and turbines are modelled by consideration of the performance of cascades of blades.

Increasing economic pressures to efficiently extract more energy from turbines, which are expected to be smaller and quieter than their predecessors, compel greater sophistication in aerodynamic design methods. The need for increased loading affects steam and gas turbine generating plant and aircraft gas turbines alike. These trends are equally evident in fan and compressor design. In all cases further progress is contingent on continuing refinement of our understanding of the flow in cascades.

It was difficult to decide whether the book should be a reference work, a literature review or a student text. This is the first specialist work to be written on the topic in the English-speaking world and the temptation was to attempt the former objective. I have been conscious from the outset that the need for a text for graduate and specialist students was at least as great.

This is an engineering work and the result is, characteristically, a compromise. I considered a 'combined operation' to be the best strategy because there is a continuing need to integrate the work of university researchers and of designers in industry. The ambitious objectives are therefore to provide a convenient reference source for the designer whilst giving adequate depth for the research worker and a sufficiently coherent account to introduce the topic to the graduate student.

A knowledge of fluid mechanics and thermodynamics to first degree level is presumed. An essential minimum of introductory material has been presented in Chapter 1. References are given to appropriate supporting texts. In Chapters 2, 3 and 4 low speed experimental work, the effects of three dimensionality and high speed experimental work are described. Chapter 5 covers basic low speed theory and provides benchmarks for the evaluation of design procedures. This discussion is extended to high speed and viscous flows in Chapters 6 and 7 respectively. Chapter 8 surveys developments in unsteady flows and, in Chapter 9, I have chosen to focus attention on three special applications of contemporary interest. In Chapter 10 the quality of available predictions is assessed. Calculation techniques have advanced in a relatively systematic fashion but much progress is also made by the accumulation of experimental evidence. Some of this experience is presented in Chapter 11 and conclusions are drawn concerning problem areas and likely advances.

In some chapters I have placed strong emphasis on the historical development of the field. I hope that this will provide useful perspective. Not only is the field of Cascade Aerodynamics one whose progress has been inadequately documented, it is also a field which is very lively at the present time. When the literature survey closed, late in 1982, there was much exciting activity; the field is barely mature and certainly not stagnant. At some time in the future the cascade model will probably be superseded by fully three-dimensional flow field investigations. That time has not yet arrived; I believe that further progress, especially for high speed flows, will require many careful theoretical and experimental studies in the cascade plane.

Although it is intended that most of the book should provide a balanced account of the field the reader may consider that undue emphasis is placed on the test case solutions of Chapter 5 and the attendant theory. This has provided a necessary basis for evaluative work in the field. Considerable research attention has been devoted to the provision of accurate calibration results for use by designers. An important aim of the research and of this book has been to establish confidence in good design procedures.

Security classifications and language difficulties make it impossible to do justice to such an extensive field as the aerodynamics of cascades. Some aspects, such as work on particle impingement and cascade testing in water, have not been covered. I apologise if important contributions have been neglected or injustices have been perpetrated.

The preparation of this work has been a lengthy project and has involved the willing support and co-operation of a number of people. Where a reference source is cited for a figure this constitutes my acknowledgement of the source and of my gratitude for the use of the material. In particular I am indebted to the following organisations for the provision of photographs:- Amalgamated Power Engineering, GEC Turbine Generators Ltd., General Electric, Mitsubishi Heavy Industries Ltd., ONERA, VKIFD, Liverpool University and Oxford University. Prof. J.H. Horlock was especially helpful during the formative years and Prof. R.I. Lewis, Dr. S.L. Dixon and Dr. G.J. Walker have commented helpfully on various aspects of the work. Good work was performed on the drawings by C.E. Evans and C. Kiriklidis and good photography by J.F. Pickering. My secretary, Mary Jackman, has provided invaluable support in the office and in typing the captions and tables.

The most important acknowledgement goes to my family who have given ungrudging support. My wife, Valerie, willingly took on the typing and preparation of camera-ready copy for the manuscript. I dedicate this book to my family.

Sydney, August 1983 J.P. Gostelow

Contents

List of Symbols

A	area
AVDR	axial velocity density ratio
AVR	axial velocity ratio
C_D, C_D	drag coefficient
C_L, C_L	lift coefficient
C_p	static pressure coefficient
D	drag force
<u>DW</u>	downwash perturbation velocity
E	energy
E	entrainment rate
F	log hodograph coordinate, $F = \lambda + i\theta$
H	stagnation enthalpy (Ch. 6)
H	head (Ch. 1)
H	boundary layer form factor
H'	energy thickness form factor
K	kernel of integral
K_1	constant in entrainment equation
L	lift force
L(m)	Thwaites parameter (Ch. 7)
M	Mach Number
2N	number of points on profile
P	total pressure
Q	heat supplied or transferred
R	gas constant
$\mathcal{R}$	auxiliary plane, $\mathcal{R} = \mathcal{P} + i\mathcal{Q}$
R	radius
Re	Reynolds' Number
R_r	reaction ratio
R_m	reaction ratio at mid-height
R_t	reaction ratio at tip
R_θ	momentum thickness Reynolds' No.
S	entropy
S*	Strouhal Number
S	total surface length (Ch. 10)
T	temperature
T_u	turbulence level
U	blade speed (Ch. 1)
U	free stream velocity
U	velocity component in ξ direction
V	velocity component in η direction
W	work done
W	circulatory velocity component
X	axial force
Y	tangential force
$\overline{Y}_c$	compressor loss coefficient
$\overline{Y}_t$	turbine loss coefficient
$\overline{Z}$	compressor loss coefficient
a	sonic velocity
a	Joukowsky transformation coefficient
a	longitudinal spacing
a^*	axial displacement of tandem blades
a	coefficient in transformation
a	crossflow crossover coefficient
b	bar diameter
b	lateral spacing
b	transformation coefficient
c	flow velocity
c	blade chord
c_f	skin friction coefficient
c_p	specific heat at constant pressure
c_v	specific heat at constant volume
d	diameter
f	frequency
g	acceleration due to gravity
g_t	staggered spacing
h	specific enthalpy
h_f	heat transfer coefficient with cooling
h_0	" " " without cooling
h	blade height
h^*	tangential displacement of tandem blades
i	incidence angle
k	slot width
l	coordinate representation of surface, $l = m + in$
l	parameter defined in Ch. 6
ℓ	transform plane coordinate, $\ell = m + in$
l	skin friction parameter (Ch. 7)

m	pressure gradient parameter (Ch. 7)
m	coefficient for deviation angle
m	mass flow rate
n	distance in normal direction
n	index for deviation angle
p	static pressure
q	source distribution
q	velocity
q_f	heat transfer rate with film cooling
q_0	" " " without film cooling
r	distance in radial direction
r	leading edge radius
s	blade spacing
s	distance in streamwise direction or along surface
t	time
t	thickness of blade
t_c	tip clearance
u	velocity component in x direction
$\bar{u}$	gust propagation velocity
v	velocity component in y direction
w	relative velocity, $w = u - iv$
w	velocity component in z direction
z	physical coordinates, $z = x + iy$
z	spanwise direction
z	axial direction (Ch. 1)
Γ	circulation
Π	Coles wake parameter
Φ	flow coefficient
Ψ	pressure rise coefficient
Ω	hodograph plane coordinates
Ω_s	streamwise vorticity component
α	flow angle, absolute flow angle
β	flow angle, relative flow angle
β, β'	oval semi-major axis (sect. 5.2)
γ	parameter (sect 5.2)
γ	specific heat ratio
γ^*	crossflow strength coefficient
γ	vorticity distribution
γ_o	bound vorticity - frozen gust
δ	deviation angle
δ'	absolute thickness
δ^*	displacement thickness
ε	deflection angle
ε	crossflow angle
$\bar{\zeta}$	enthalpy loss coefficient
ζ	vorticity
ζ	coordinates, $\zeta = \xi + i\eta$
η	efficiency
θ	circumferential direction
θ	camber angle
θ	momentum thickness
θ	coordinate direction, $\tan^{-1}(v/u)$
ϑ	stagnation point location
λ^*	sweep angle
λ, λ'	parameters (defined in Chs. 5, 6)
λ_θ	pressure gradient parameter
μ	viscosity
μ	wave angle (Ch. 6)

μ^*	dihedral
μ	inclination of streamlines to axial
ν	kinematic viscosity
ξ'	distance downstream from mid-chord
$\bar{\xi}$	enthalpy loss coefficient
ρ	density
σ	throat width
τ	shear stress
τ_e	trailing edge thickness
φ	parameter in Schlichting method
φ^*	inclination of normal to Bernoulli surface
χ	stagger angle
ω	complex potential function, $\omega = \varphi + i\psi$
$\omega(M)$	Prandtl-Meyer expansion angle
$\bar{\omega}$	total pressure loss coefficient
	SUBSCRIPTS
CR	critical or throat condition
DC	defect at centre
2D	at AVDR of unity
E	entrainment
LE	leading edge
MS	at mid-span
NS	normal shock
T	at trailing edge
TE	across trailing edge
x	at location x
ad	adiabatic
aw	adiabatic wall
ax	axial
b	at base
c	compressor
c	compressible
c	coolant
g	mainstream gas
h	hydraulic
i	incidence
inc	incompressible
iso	isothermal
l	lower
m	mean direction or value; mid-height
m	modified
max	maximum
min	minimum value
n	normal component
nom	nominal
opt	optimum
p	perturbation
p_{min}	minimum profile drag
q	source
r	re-compression
ref	reference condition
rev	reversible
s	isentropic
s	camber line
s	secondary component
s	streamwise direction
t	turbine

t	throat
t	thickness
t	transition inception
ts	total-to-static
tt	total-to-total
u	upper
w	at wall
x	component in x direction, axial component
y	component in y direction
δ	deviation
γ	vorticity
θ	circumferential or tangential component
o	stagnation condition
0	reference blade
1	upstream
2	downstream
10	at 10% thickness
∞	at infinity

SUPERSCRIPTS

—	average
'	blade or metal angle
'	perturbation due to singularity
*	choking condition
·	time rate of change
t	at time t

CHAPTER 1

The Cascade Model

For the purposes of this work a cascade is defined as an infinite row of equidistant similar bodies. These bodies, or blades, are usually aerofoil-shaped.

Such rows of blades are used directly in engineering practice when it is desired to divert a stream of fluid. The more common usage is in the design of turbomachines and the analysis of their performance. The central role played by turbomachinery in contemporary society has influenced the author to present the cascade model in that context. It would be possible to treat the subject of cascade aerodynamics with little reference to the practical problems which have spurred the growth of the discipline. Such a treatment would, however, lose much of the relevance to engineering practice which is an important objective of this book.

Although the main context is one of turbomachinery flows the subject matter is cascade aerodynamics. The reader in search of a more general text on the design of turbomachinery is referred to those by Dixon[1.1] and Horlock[1.2, 1.3].

It is important to emphasize at the outset that the cascade is merely a model. In fluid mechanics the process of modelling is one of physical and mathematical simplification and is intended to result in a rationalisation and deeper appreciation of the actual situation.

Axial flow steam turbines were developed in the late nineteenth century without recourse to the cascade model and often with little understanding of the fluid mechanics involved. Trial and error processes resulted in adequate performance for power generation and eventually propulsion. In the early part of this century the first primitive cascade wind tunnels, called nozzle testers, were developed to consolidate the advances and provide data for machines of higher rating and increased efficiency. Today's steam turbines for power generation are capable of delivering 600 MW of power or more. Their design is critically dependent upon advanced fluid mechanics and the cascade model is an essential tool in maximising efficiency at these high ratings. The interferogram on the cover of this book is taken from a steam turbine manufacturer's catalogue. Most of the electrical power generated in most countries of the world is produced by large, modern steam turbine plant. A small but consistent improvement in blading efficiency will have corresponding economic implications for the nation.

The flow in a turbine largely takes place under the influence of a favourable pressure gradient in the direction of flow. Turbines, at least while the flow is subsonic, therefore have reasonable tolerance to blade design and alignment errors. The same cannot be said of the axial flow compressor, fan or pump. These operate under adverse pressure gradients. In such a diffusing flow the danger of flow separation is much more pronounced. Most early axial flow compressors were badly designed because the necessary aerodynamic knowledge was simply not available. Parsons patented the use of a reversed turbine as an axial flow compressor in 1884 and later built a number of them for blast furnace use[1.4]. Although the pressure rating was low, efficiencies of only around 55% were recorded.

It was about this time (in 1903) that the Wright brothers tested aerofoil sections in a home-made wind tunnel prior to the first manned and powered flight. The aerodynamic technology was barely available for isolated aerofoils, let alone the much more complicated rows of blades needed for the axial flow compressor.

During the next three decades various separate developments were pursued which appeared unrelated at the time but which were all essential to the eventual appearance of the jet engine as we know it today.

The wind tunnel pioneers began extending the success of the Wright brothers to biplane arrangements and to elementary cascade representations of propeller blading. In a parallel mathematical development the theory of isolated wings was gradually (and with considerable tedious computation) extended to cascades. These early developments are described in chapters 2 and 5. Those were the 'technology push' factors. The 'demand pull' factors were associated with the forward thinking of the gas turbine designers. Early work by the Brown Boveri Company and subsequent work by Griffith, Von Ohain and Whittle was prominent here. It was quite clear from their cycle calculations that a compressor efficiency of at least 70% (dependent upon pressure ratio, turbine inlet temperature and turbine efficiency) was essential before self sustaining operation of a gas turbine with the production of a net thrust would become possible. However if this could be achieved many limitations on aircraft propulsion and in the marine and industrial fields would be removed.

Whittle's decision was that the centrifugal compressor offered the best opportunity to demonstrate the efficacy of the jet engine. Under the immediate pressures of wartime this was to prove the correct assessment. Griffith had considered the axial compressor to have more potential and this view was shared by the German and Italian designers. In the long term although the centrifugal compressor has been ideal for smaller engines an efficient axial compressor has proved quite essential for most fan, jet and gas turbine engines.

As a result of the systematic improvement of the aerodynamic art, and the more urgent compilation of cascade data under wartime pressures, it became possible to design an efficient axial flow compressor. The foundations were laid for the modern gas turbine and virtually all design work on axial flow machines since has built upon that foundation. The present work is firmly rooted in that tradition. Computational developments may one day make it possible to transcend the cascade model. In the author's view that day has not yet arrived.

Ironically it may well prove to be the more difficult problem of the radial flow machine which is solved first by fully three dimensional viscous flow computer programs. Cascade data can be applied to centrifugal machines and have often been applied successfully to the diffusers of compressors and pumps and to the impellers of fans. However the additional complexities of such three dimensional flows make agreement between theory and experiment tenuous in all but the simplest of cases. Whilst the reader is encouraged to apply cascade data to radial and mixed flow situations the author considers that an accurate representation of such flows is not yet susceptible to routine solution.

1.1 The Meridional and Cascade Planes

As indicated in Fig. 1.1 the flow is fully three dimensional and is conveniently represented in radial, r, circumferential, θ, and axial, z, coordinates.

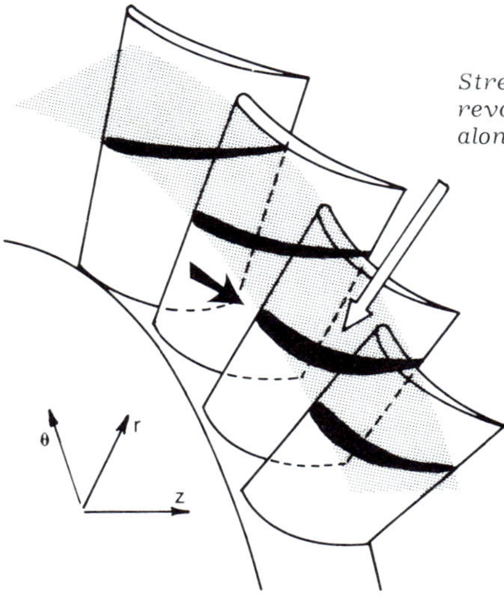

Fig. 1.1. The cascade plane for an axial compressor blade row.

The flow field is the result of various influences. Adjacent blade rows have rotational motion with respect to each other and blades generate fluid disturbances, such as wakes, shock waves and potential interactions. Accordingly all blades are in an unsteady environment. Furthermore the blades themselves may vibrate, rendering the physical boundary conditions time-variant. Acoustic signals are also generated as a result of aerodynamic interactions and in turn have an aerodynamic effect.

Possibly the most important and intractable feature of the flowfield is its viscosity. Only a three dimensional fully viscous solution can be expected to correctly represent the real flow phenomena.

Although progress is being made towards this end [1.5] the problem cannot yet be considered solved. The current state of the art of axial turbomachinery design follows the approach which has been most successful for a period of forty years. This book is in the classical tradition of turbomachinery design and the cascade model is a key element in that classical tradition. The author anticipates that in the next two decades a fully three dimensional, viscous and unsteady approach to machine design will begin to emerge especially for radial-flow machines. Such a 'modern' approach would probably not discard the cascade model entirely since there would be a continuing need for accuracy checks and the physical understanding that the cascade model allows.

The essence of the classical approach is the splitting of the three-dimensional problem into two separate sub-problems the solutions of which are then obtained and superposed iteratively.

1.1.1 <u>The meridional plane</u>. The first sub-problem is that of the flow in the meridional plane. In this model a plane having only the r, z coordinates of Fig. 1.1 is examined. The principal assumption is that the flow is axisymmetric, i.e. that variations in the flow properties in a circumferential direction are neglected. It is assumed that such effects, the variations in flow from one blade to an adjacent one in the same row, can be represented correctly by averaged data resulting from consideration of the second sub-problem.

The experimental elucidation of meridional plane data is generally by means of radial traverses taken between blade rows. The radial variation of stagnation temperature and the distribution of axial velocity are two important properties which need to be calculated for a design but which also need experimental verification. The axial flow compressor chapter of reference 1.1 gives a good example of such data. A conclusion which could be drawn from such data is that as aerodynamic designs improve the quality of representation by the classical division of the flow field into two complementary planes also improves - a self-reinforcing tendency.

Most design offices today have structured the design process upon the basis of a computer program for flow computation in the meridional plane. It forms a framework for both rapid preliminary evaluations and more detailed final designs. The simple radial equilibrium method is widely used for this purpose. In addition to the axisymmetric flow assumption it is assumed that any radial flows occur within the blade row, the flow outside the blade row being in radial equilibrium. It is demonstrated[1.1] that the equation of motion in a radial direction under these conditions is

$$\frac{1}{\rho}\frac{dp}{dr} = \frac{c_\theta^2}{r} \tag{1.1}$$

and that for a machine having radially constant stagnation enthalpy and entropy

$$c_x \frac{dc_x}{dr} + \frac{c_\theta}{r}\frac{d}{dr}(r\, c_\theta) = 0 \,. \tag{1.2}$$

Equation (1.2) holds for the flow between blade rows in an adiabatic, reversible turbomachine in which equal work is delivered by the rotor rows at all radii.

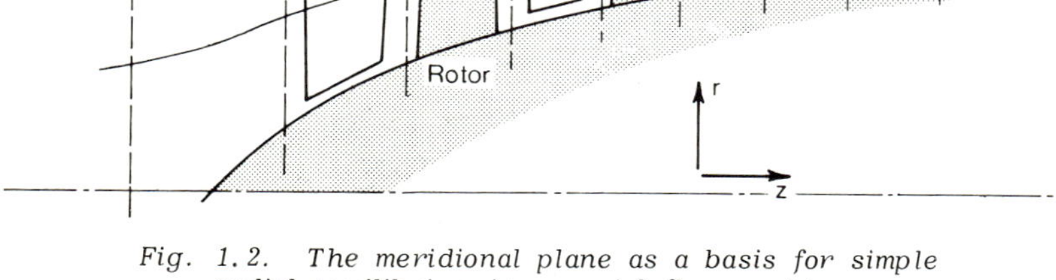

Fig. 1.2. The meridional plane as a basis for simple radial equilibrium in an axial-flow machine.

Simple radial equilibrium theory would be applied on a 'hub, pitch and tip' basis or possibly the flowpath could be divided by five streamlines as indicated in Fig. 1.2. Axial stations would be placed between each row of blades. This stage of complexity is not demanding in computer time and many early designs were performed in this way using manual computation and are still flying, pumping or otherwise fulfilling their assigned commercial duty. Contemporary usage of these simplified calculations would be in preliminary design work on an interactive visual display basis. Flowpath changes are typically input by cursor and their effects on velocity distribution assessed in a matter of seconds.

For most detailed design work simple radial equilibrium theory is not sufficiently precise. It is then necessary to implement radial equilibrium theory in full as described by Smith [1.6]. A description of the theory is beyond the scope of this book but it can safely be said that without such theories transonic fans and hence today's wide-bodied airliners would not have been developed. In computer program form the theory is set up for solution of the flow through a large number of streamtubes (eight would be a typical number actually used) and a large number of stations, many of which would be placed within a single blade-row.

Whereas calculations and traverses in the meridional plane provide an essential structural framework for design process they do not of themselves provide the input of the most important physical data. The essential quantitative information and qualitative understandings of the flow phenomena involved are obtained by turning attention to the complementary 'blade-to-blade' or cascade plane.

1.1.2 The cascade plane. The cascade plane is obtained by viewing along the blade axis a projection having the z, θ coordinates of Fig. 1.1. If the radial equilibrium analysis has indicated that in the region of interest the streamlines have a radius which varies through the blade-row then the 'cut' is made along the streamline and then viewed in a direction parallel to the blade axis or stacking line [1.7]. In the simplest case of a machine having a cylindrical hub, casing and intermediate streamlines and blades stacked radially the cylindrical cut is projected radially. In these circumstances the blade row thus obtained is unrolled from a cylinder by the simple transformation

$$x = z, \qquad y = r\theta .$$

The errors implicit in this unrolling process are small for a reasonable number of blades and hub-tip ratio [1.8].

A more precise definition of the cascade plane was given by Wu [1.9]. Wu considered S1 and S2 planes illustrated in Fig. 1.3. Through-flow equations were derived for these

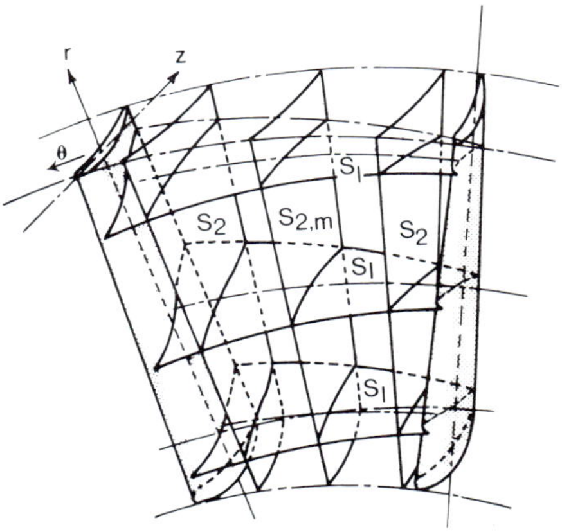

Fig. 1. 3. Schematic showing Wu's S1 and S2 surfaces.

planes. Until adequate computer power became available it had not proved possible to solve these equations. In 1968 Marsh [1.10] achieved this for both planes in what became known as the 'matrix through-flow' analysis.

The above discussion has mostly concerned the mathematical basis of solutions to the turbomachinery design problem. The breakthroughs which led to the development of successful axial flow fans, pumps and most importantly compressors for gas turbine engines resulted from the generation of accurate experimental data on the setting of blades and the corresponding flow turning and losses. Furthermore the effective development of large steam turbines has only been possible as a result of increased information on high speed flows through turbine blading.

1.2 Cascade Notation and Definitions

Cascade projections taken at a certain radius for stages from an axial flow compressor and an axial flow turbine are illustrated in Fig. 1.4. In each case a complete stage is shown. For a compressor this consists of a rotor row followed by a stator row; for a turbine the stage has a nozzle row followed by a rotor row. Cascade projections could be drawn in an essentially similar manner for mixed and radial flow machines.

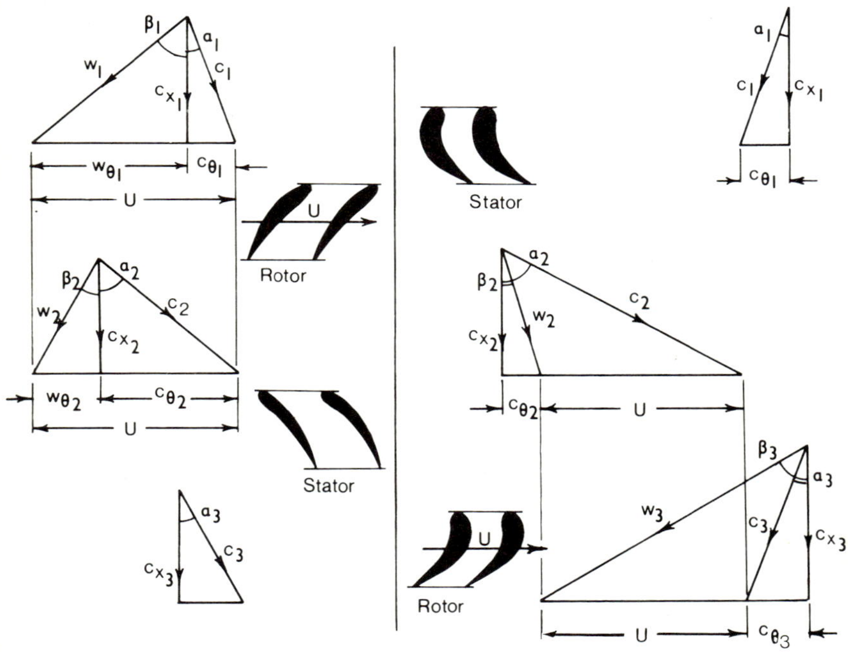

(a) Compressor stage *(b) Turbine stage*

Fig. 1.4. Velocity triangles for compressor and turbine stages.

Also indicated in Fig. 1.4 are the corresponding velocity vectors. Each velocity triangle is composed of absolute, relative and blade velocities. The important axial and tangential velocity components are also indicated. Absolute flow angles are designated α and relative angles β. The sense in which these angles are drawn is important. In some texts, especially German works and those dealing with radial machines, the angles α and β represent the flow angle with respect to the tangential direction. In this work the more usual convention for axial machines of measuring these angles with respect to the axial direction will be followed. Throughout this work flow angles following the sense of Fig. 1.4 will be designated positive flow angles.

A more detailed cascade projection for representative compressor and turbine blade rows is presented in Fig. 1.5. The purpose of Fig. 1.5 is to represent the notation which is commonly used for cascade geometrical parameters and which will be followed in this work.

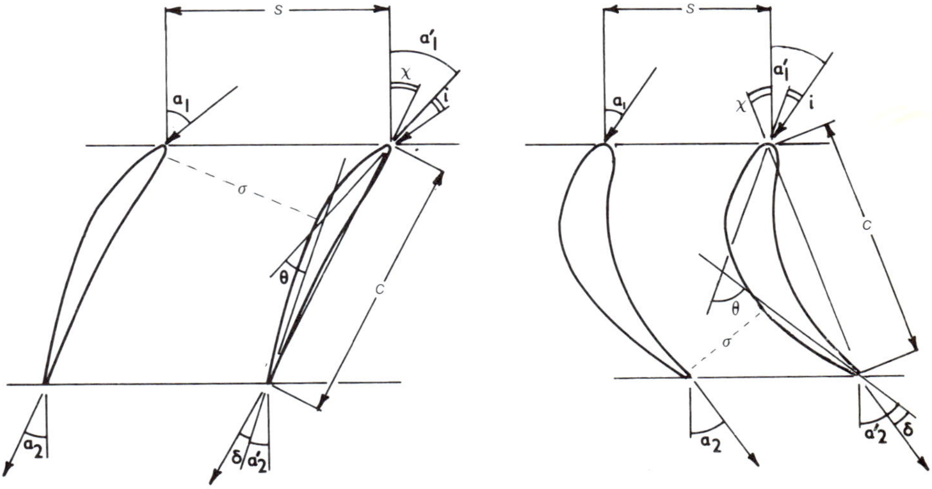

(a) *Compressor cascade* (b) *Turbine cascade*

Fig. 1.5. Cascade geometry and notation.

Figs. 1.5(a) and 1.5(b) are drawn for stationary blade rows with α representing the flow angle. Rotating rows having β as the flow angle could as readily have been used. With regard to an individual blade row, as opposed to a stage, the blade inlet is designated station 1 and the outlet, station 2. The superscript α' represents the blade 'metal' angle generally measured between the axial direction and the tangent to the camber line at leading and trailing edges. The camber line is obtained by bisecting the thickness of the aerofoil. This process may be carried out on a normal to the aerofoil chord. The chord is defined as the straight line joining the leading and trailing edges. Although the leading and trailing edge points are usually located at the points of maximum curvature this is not always the case.

The aerofoil chord makes a certain angle with respect to the axial direction. This angle is designated the blade stagger angle, χ. If the tangents to the camber line at the leading and trailing edges are extrapolated the angle included between these lines is termed the camber angle, θ. The distance between blade leading and trailing edges is the chord length, c, and that between the edges of two adjacent blades is the spacing, s. The British cascade practice followed in this work uses the space-chord ratio, s/c, whereas American practice uses the solidity, c/s.

The incidence angle, i, is defined as the difference between the flow and blade angles at inlet, i.e.

$$i = \alpha_1 - \alpha_1' .$$

In general there will be an underturning of the fluid by the blade row; accordingly a deviation angle, δ, is defined for discharge angles. For a compressor blade this is

$$\delta = \alpha_2 - \alpha_2' .$$

For a turbine blade

$$\delta = \alpha_2' - \alpha_2 ;$$

each of these definitions gives a positive value of deviation angle most of the time.

A fluid deflection angle, ϵ, is also sometimes used. For compressor blading this is defined as

$$\epsilon = \alpha_1 - \alpha_2 ,$$

whereas for turbine blading

$$\epsilon = \alpha_1 + \alpha_2 .$$

1.3 The Equations of Motion

The concept of a control volume and the control surface surrounding the control volume are helpful ones in deriving the equations of motion of a fluid. The size and shape of the control volume may be chosen arbitrarily but are often drawn to coincide with the physical boundaries of a body or to identify flows normal to a particular station in a duct or free stream. The surface of the control volume is always closed and is referred to as a control surface.

The principal equations of motion, as discussed in reference 1.2 or in earlier works of this series, are the continuity equation, the First Law of Thermodynamics, Newton's Second Law of Motion and the Second Law of Thermodynamics.

1.3.1 The continuity equation. It is convenient to consider a small volume element as depicted in Fig. 1.6. The total mass flowing into the element is

$$\dot{m}_{in} = \rho u\, \delta y\, \delta z + \rho v\, \delta x\, \delta z + \rho w\, \delta x\, \delta y .$$

The total mass flowing out is

$$\dot{m}_{out} = (\rho u + \frac{\partial(\rho u)}{\partial x}\delta x)\, \delta y\, \delta z + (\rho v + \frac{\partial(\rho v)}{\partial y}\delta y)\delta x\, \delta z + (\rho w + \frac{\partial(\rho w)}{\partial z}\delta y)\delta x\, \delta y .$$

The difference between these is the rate at which mass is stored in the volume. The only way in which the stored mass can change is by a change in density. The rate of change of mass stored is then

$$\frac{\partial \rho}{\partial t}(\delta x\, \delta y\, \delta z) .$$

The continuity equation is an expression of conservation of mass. The mass entering is therefore equal to that discharged plus that stored.

$$\therefore\ \rho u(\delta y\, \delta z) + \rho v(\delta x\, \delta z) + \rho w(\delta x\, \delta y) = (\rho u + \frac{\partial(\rho u)}{\partial x}\delta x)\delta y\, \delta z + (\rho v + \frac{\partial(\rho v)}{\partial y}\delta y)\delta x\, \delta z + (\rho w + \frac{\partial(\rho w)}{\partial z}\delta z)\delta x\, \delta y + \frac{\partial \rho}{\partial t}\delta x\, \delta y\, \delta z .$$

This simplifies to

$$\frac{\partial \rho}{\partial t} + \frac{\partial(\rho u)}{\partial x} + \frac{\partial(\rho v)}{\partial y} + \frac{\partial(\rho w)}{\partial z} = 0 . \qquad (1.3)$$

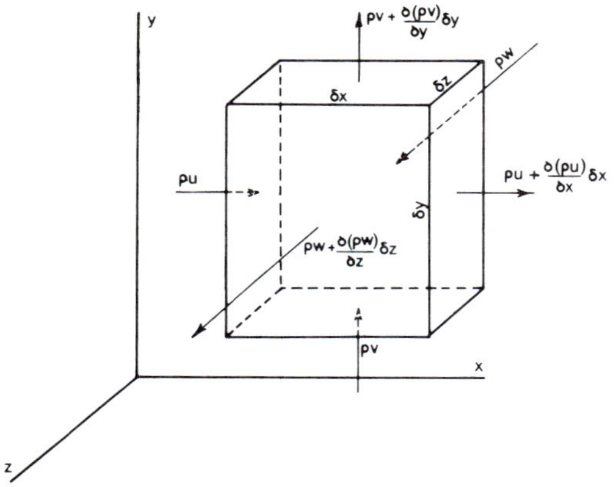

Fig. 1.6. Control volume considered for continuity equation.

For steady flow this becomes

$$\frac{\partial(\rho u)}{\partial x} + \frac{\partial(\rho v)}{\partial y} + \frac{\partial(\rho w)}{\partial z} = 0 \tag{1.4}$$

and for incompressible steady, two dimensional flow

$$\frac{\partial u}{\partial x} + \frac{\partial v}{\partial y} = 0\,. \tag{1.5}$$

If the flow is irrotational, then the vorticity, ζ , is given by

$$\zeta = \frac{\partial v}{\partial x} - \frac{\partial u}{\partial y} = 0$$

and a potential function φ may be defined

$$u = \frac{\partial \varphi}{\partial x}\,, \qquad v = \frac{\partial \varphi}{\partial y}$$

and substitution into equation (1.5) gives

$$\frac{\partial^2 \varphi}{\partial x^2} + \frac{\partial^2 \varphi}{\partial y^2} = 0\,. \tag{1.6}$$

This, the Laplace equation, may be applied to the problem of calculating the two-dimensional, steady, incompressible irrotational flow through a cascade.

1.3.2 <u>The First Law of Thermodynamics.</u> For a system undergoing a cyclical process

$$\oint dQ = \oint dW \tag{1.7}$$

where Q represents the heat supplied to the system and W the work done by the system during a cycle.

For a process in which a change of state from condition 1 to condition 2 is involved it can be shown that the change in total energy, E , of the system is given by

$$E_2 - E_1 = \int_1^2 (dQ - dW)\,. \tag{1.8}$$

This is true regardless of the route taken between conditions 1 and 2 and therefore E is a property of the system. It is often convenient to subdivide the total energy into internal energy, kinetic energy and potential energy components.

For the steady flow of a fluid through a control volume the difference between the heat transfer rate from the surroundings to the control volume $\dot{Q}$ and the work produced by the system via a shaft, $\dot{W}$ is

$$\dot{Q} - \dot{W} = \dot{m}[(h_2 - h_1) + \tfrac{1}{2}(c_2^2 - c_1^2) + g(H_2 - H_1)] \,. \qquad (1.9)$$

This representation of the First Law for flow situations is termed the Steady Flow Energy Equation.

The potential energy term is usually only significant for hydraulic machines and is otherwise ignored. Similarly the heat transfer rate from the surroundings may be taken to be zero. It is convenient to combine the static enthalpy and kinetic energy terms to form the stagnation enthalpy, $h_o = h + \tfrac{1}{2}c^2$. Therefore, for a work-producing non-hydraulic machine,

$$\dot{W}_t = \dot{m}(h_{o1} - h_{o2})$$

and, for a work absorbing non-hydraulic machine (note sign convention),

$$\dot{W}_c = \dot{m}(h_{o2} - h_{o1}) \,.$$

1.3.3 Newton's Second Law of Motion. The algebraic sums of the body forces on the matter within the control volume and the externally imposed forces on the control surface, in a certain direction, are equal to the net outgoing momentum flux in that direction.

For the steady flow of a fluid through an elementary control volume and in the absence of friction or shaft work, Euler's equation of motion for one dimensional flow

$$\frac{dp}{\rho} + c\,dc + g\,dH = 0 \qquad (1.10)$$

is obtained.

If the above equation is applied to a control volume, bounded by a streamtube and normal stations 1 and 2, then Bernoulli's equation is obtained

$$\int_1^2 \frac{dp}{\rho} + \tfrac{1}{2}(c_2^2 - c_1^2) + g(H_2 - H_1) = 0. \qquad (1.11)$$

If the influences of density change and altitude change are absent the Bernoulli equation reduces to its simplest form

$$p_0 = p + \tfrac{1}{2}\rho c^2 = \text{constant}. \qquad (1.12)$$

In considering the energy transfer processes of turbomachines it is convenient to apply Newton's Second Law to the moments of forces about an axis.

This process results in Euler's pump and turbine equations. For a work absorbing machine the rate of work done on the fluid is given by

$$\dot{W}_c = \dot{m}(U_2 c_{\theta_2} - U_1 c_{\theta_1}) > 0 \,. \qquad (1.13)$$

For work producing machines the sign of $\dot{W}$ is reversed and the Euler turbine equation is obtained

$$\dot{W}_t = \dot{m}(U_1 c_{\theta_1} - U_2 c_{\theta_2}) > 0 \,. \qquad (1.14)$$

1.3.4 The Second Law of Thermodynamics. Inefficiencies occur in a real machine and the Second Law enables these to be quantified. The Second Law states that it is impossible to construct a heat engine which operates in a continuous cycle, receiving heat from a single source and doing work.

If a reversible cycle is postulated then

$$\oint \frac{dQ_{rev}}{T} = 0 , \tag{1.15}$$

where dQ is an element of heat transferred to the system at an absolute temperature T. If the cycle is a real one the inequality of Clausius states that

$$\oint \frac{dQ}{T} \leq 0 . \tag{1.16}$$

Entropy is defined as a property such that

$$dS = \frac{dQ_{rev}}{T}$$

in a reversible process. For a change from state 1 to state 2 by any reversible path

$$S_2 - S_1 = \int_1^2 \frac{dQ_{rev}}{T} . \tag{1.17}$$

For an incremental change of state

$$dS = \frac{dQ_{rev}}{T} . \tag{1.18}$$

If a fluid flows through a control volume and experiences a change of state

$$\int_1^2 \frac{dQ}{T} \leq S_2 - S_1 . \tag{1.19}$$

Many turbomachines are virtually adiabatic. In an adiabatic process $d\dot{Q} = 0$ and $S_2 \geq S_1$. If the process is reversible as well $S_2 = S_1$. Thus for an adiabatic machine the ideal process is one in which the entropy remains unchanged during the process. The performance of a real machine may be evaluated with reference to the ideal reversible process.

1.4 Definitions of Loss and Efficiency

The compression and expansion processes are conventionally represented on an enthalpy-entropy diagram (Fig. 1.7). The efficiency of a machine is often most readily defined in relation to these diagrams.

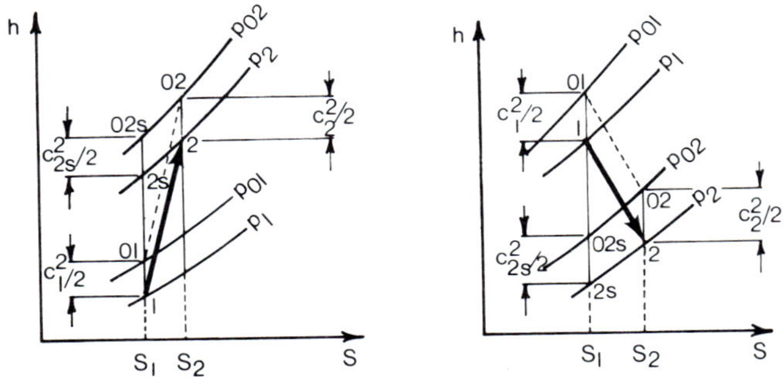

(a) Compressor *(b) Turbine*

Fig. 1.7. Enthalpy-entropy diagrams for a compressor and a turbine.

For a compressor the minimum possible (isentropic) work input to the machine is divided by the actual work input to give the isentropic or adiabatic efficiency,

$$\eta_c = \frac{h_{02s} - h_{01}}{h_{02} - h_{01}} . \tag{1.20}$$

The hidden thermodynamic effect of 'small stage' compression leads to the alternative definition of a polytropic efficiency [1.1]. Although the difference between polytropic and isentropic efficiencies is most marked for high pressure ratios it can be significant even for pressure ratios of two or less. For a compressor the polytropic efficiency is higher than the isentropic efficiency whereas for a turbine it is lower.

For a turbine the actual work delivered by the turbine is divided by the maximum possible (isentropic) work available from the machine to give a total-to-total isentropic or adiabatic efficiency,

$$\eta_{tt} = \frac{h_{01} - h_{02}}{h_{01} - h_{02s}} . \tag{1.21}$$

However in some circumstances the discharge kinetic energy cannot be recovered. In this case the total-to-static efficiency,

$$\eta_{ts} = \frac{h_{01} - h_{02}}{h_{01} - h_{2s}} \tag{1.22}$$

is employed.

The above definitions are not readily applicable when the working fluid is a liquid. For a pump the hydraulic efficiency is defined as

$$\eta_h = \frac{\Delta W_{p\ min}}{\Delta W_p} = \frac{g(H_2 - H_1)}{\Delta W_p} \tag{1.23}$$

and for a hydraulic turbine

$$\eta_h = \frac{\Delta W_t}{\Delta W_{t\ max}} = \frac{\Delta W_t}{g(H_1 - H_2)} . \tag{1.24}$$

Although the above definitions encompass the performance of fluid machines and their stages, the incorporation of losses into the design process will be dependent upon the result of detailed cascade investigations. A principal end result of a cascade test will be in the form of a loss coefficient. Different investigators have used a variety of forms of loss coefficient. Some loss coefficients used for compressors are as follows:

$$\bar{Z} = \frac{p_{01} - p_{02}}{\frac{1}{2}\rho c_1^2} \tag{1.25}$$

$$\bar{\omega}_c = \frac{p_{01} - p_{02}}{p_{01} - p_1} \tag{1.26}$$

$$\bar{\xi} = \frac{h_2 - h_{2s}}{\frac{1}{2}c_1^2} = \bar{\zeta} . \tag{1.27}$$

The equivalent loss coefficients used for turbines are :

$$\bar{Y}_t = \frac{p_{01} - p_{02}}{\frac{1}{2}\rho c_2^2} \tag{1.28}$$

$$\bar{\omega}_t = \frac{p_{01} - p_{02}}{p_{02} - p_2} \tag{1.29}$$

$$\bar{\xi} = \frac{h_2 - h_{2s}}{\frac{1}{2}c_2^2} = \bar{\zeta}\,. \tag{1.30}$$

The above turbine loss coefficients and their origins are described in reference [1.11]. Brown[1.12] has performed a useful comparative analysis of the above coefficients. It was shown that for incompressible flow the above loss coefficients are equal and interchangeable. Any low speed data correlated in terms of one loss coefficient may be directly applied to others.

Brown considered typical compressor and turbine stage velocity triangles and plotted adiabatic and polytropic efficiencies predicted by each of the loss coefficients as a function of Mach number. As the Mach number was raised the discrepancies increased. For compressor blades it was concluded that coefficients $\bar{Z}$ and $\bar{\omega}_c$ predicted too high an efficiency whereas that obtained from $\bar{\xi}$ was too low. For turbine blades $\bar{Y}_t$ and $\bar{\omega}_t$ predicted too high an efficiency whereas $\bar{\xi}$ was reasonably reliable. The efficiency discrepancy between the various loss coefficients at the sonic condition is of the order of 4%.

It is necessary to be precise about the application of loss coefficients obtained from two-dimensional cascade tests to a three-dimensional machine. Flow conditions at the radial location under consideration must correspond to those of the cascade data used. Since the flow in the hub and casing regions will be influenced by wall boundary layers and secondary flows it is important to make allowances for these effects by the application of appropriate cascade data or corrections in these regions. The cascade data must be extensive and applied with diligence if the intention is to integrate cascade losses in a spanwise direction to obtain the efficiency of a blade row or machine. Accurate allowances for three-dimensionality and compressibility effects are therefore quite essential.

1.5 Cascade Force Analysis

If the flow is steady, incompressible and reversible the Bernoulli equation (1.12) gives

$$p_1 + \tfrac{1}{2}\rho c_1^2 = p_2 + \tfrac{1}{2}\rho c_2^2\,. \tag{1.31}$$

If, furthermore, the flow is two dimensional, the continuity equation (1.5) gives

$$c_1 s \cos\alpha_1 = c_2 s \cos\alpha_2$$

$$\text{or}\quad c_1 \cos\alpha_1 = c_{x_1} = c_2 \cos\alpha_2 = c_{x_2} = c_x\,. \tag{1.32}$$

Combining these the static pressure rise (across a compressor cascade) may be expressed in terms of the inlet dynamic head and the inlet and discharge flow angles,

$$p_2 - p_1 = \tfrac{1}{2}\rho(c_1^2 - c_2^2) = \tfrac{1}{2}\rho c_1^2\left\{1 - \frac{\cos^2\alpha_1}{\cos^2\alpha_2}\right\}.$$

A static pressure rise coefficient for the cascade may be defined as

$$\Delta C_p = \frac{p_2 - p_1}{\frac{1}{2}\rho c_1^2}$$

$$\text{or}\quad \Delta C_p = \left\{1 - \frac{\cos^2\alpha_1}{\cos^2\alpha_2}\right\}. \tag{1.33}$$

If a total pressure loss through the cascade, Δp_0, is introduced, then

$$\Delta C_p = \left[1 - \frac{\cos^2 \alpha_1}{\cos^2 \alpha_2}\right] - \overline{\omega}_c \qquad (1.34)$$

where the total pressure loss coefficient, $\overline{\omega}_c$ is given by

$$\overline{\omega}_c = \Delta p_0 / \tfrac{1}{2} \rho c_1^2 . \qquad (1.35)$$

Application of the momentum equation to a control surface enclosing a blade is illustrated in Fig. 1.8. The control surface extends sufficiently far upstream and downstream for flow conditions at those boundaries to be considered uniform.

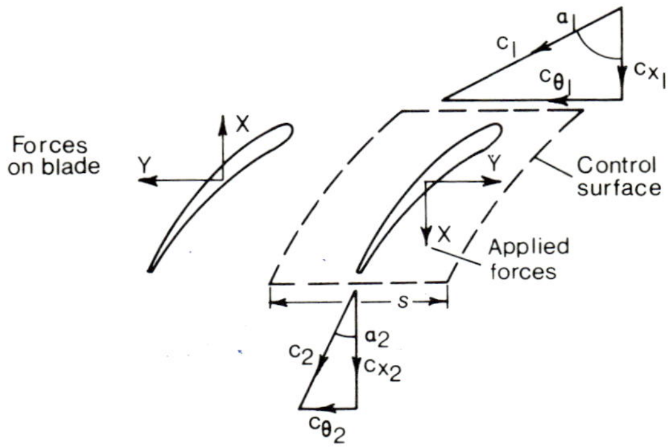

Fig. 1.8. Control surface around blade in cascade.

Axial and tangential forces, X and Y respectively, are exerted upon the fluid by unit depth of blade.

The force in the axial direction is given by

$$X = s(p_2 - p_1)$$

and from the Bernoulli equation

$$X = \tfrac{1}{2} \rho s (c_1^2 - c_2^2) - s \Delta p_0$$

$$\text{or} \quad X = \tfrac{1}{2} \rho s c_x^2 (\tan^2 \alpha_1 - \tan^2 \alpha_2) - s \Delta p_0 . \qquad (1.36)$$

The force in the tangential direction is given by

$$Y = \rho s c_{x_1} (c_{\theta_1} - c_{\theta_2})$$

$$\text{or} \quad Y = \rho s c_x^2 (\tan \alpha_1 - \tan \alpha_2) . \qquad (1.37)$$

If a mean flow direction is defined as

$$\tan \alpha_m = \frac{\tan \alpha_1 + \tan \alpha_2}{2}$$

then the lift force on the cascade perpendicular to that direction is

$$L = X \sin \alpha_m + Y \cos \alpha_m$$

$$\therefore \quad L = \rho s c_x^2 \sec \alpha_m (\tan \alpha_1 - \tan \alpha_2) - s \Delta p_0 \times \sin \alpha_m . \qquad (1.38)$$

The lift coefficient, C_L , is defined as

$$C_L = \frac{L}{\frac{1}{2}\rho c_m^2 c}$$

and is given by

$$C_L = 2(\tfrac{s}{c}) \cos \alpha_m (\tan \alpha_1 - \tan \alpha_2) - \frac{\Delta p_0 (\tfrac{s}{c}) \sin \alpha_m}{\frac{1}{2}\rho c_m^2} . \tag{1.39}$$

The drag force, D , in the mean flow direction is

$$D = Y \sin \alpha_m - X \cos \alpha_m = \Delta p_0 \, s \cos \alpha_m .$$

Defining a drag coefficient, C_D , by

$$C_D = \frac{D}{\frac{1}{2}\rho c_m^2 c} \tag{1.40}$$

$$C_D = \bar{\omega}_c (\tfrac{s}{c}) \left[\frac{\cos^3 \alpha_m}{\cos^2 \alpha_1}\right]. \tag{1.41}$$

The lift coefficient may now be conveniently written as

$$C_L = 2(\tfrac{s}{c}) \cos \alpha_m (\tan \alpha_1 - \tan \alpha_2) - C_D \tan \alpha_m . \tag{1.42}$$

Situations are quite common in which the axial velocity at discharge is different from that at inlet. It is relatively straightforward to repeat the above analysis taking such differences into account.

Although the static pressure rise and loss coefficient parameters in the analysis have been based on inlet velocity the discharge velocity may be used as a basis and hence parameters produced which are more relevant to turbine cascade operation.

1.6 Degree of Reaction

The concepts of 'impulse' and 'reaction' are important ones which apply to both work-absorbing and work-producing machines.

A simple example of an impulse machine is a paddle wheel rotated by the impingement of a jet of water onto its vanes. The jet is formed in a stationary nozzle whose only function is to transform the pressure energy of the working fluid into kinetic energy. This kinetic energy is transferred to the spinning wheel by pure impulse action. The jet and wheel are unbounded and may be open to the atmosphere.

A simple example of a reaction machine is the type of lawn sprinkler in which water is transferred internally to a rotor and emerges tangentially from a spinning nozzle. The water enters the sprinkler with high pressure energy which is transformed to kinetic energy in the rotating nozzle. This nozzle, being part of the rotor, is free to move in a circular path. The force of reaction caused by the jet's change of momentum sustains the rotation.

Obviously a turbine could be constructed combining the above principles, having a stationary nozzle and a downstream rotor. In both of these devices some of the pressure energy would be converted to kinetic energy. Such a combination opens up a wide range of design possibilities for both work-absorbing and work-producing machines. To classify these further it is necessary to formally state a "degree of reaction".

A general definition of the degree of reaction is the static enthalpy change in the rotor divided by that in the entire stage.

For a turbine this would then have the form

$$R_r = \frac{h_2 - h_3}{h_1 - h_3} . \tag{1.43}$$

If the absolute velocity c_3 is identical to c_1 then the stagnation enthalpies may be used in the denominator

$$R_r = \frac{h_2 - h_3}{h_{01} - h_{03}} \tag{1.44}$$

and this is the most common form for an axial flow turbine.

Although a corresponding definition may be used for a compressor, the different functions of the machines lead to different definitions for reaction. The function of a compressor is to raise the static pressure of a fluid whereas that of a turbine is to produce work. Accordingly it is more useful to define the degree of reaction for a compressor as the ratio of static pressure rise in the rotor to that in the entire stage,

$$R_r = \frac{p_2 - p_1}{p_3 - p_1} . \tag{1.45}$$

For compressible and irreversible flows this definition differs from those based on enthalpy. Nevertheless the closer relationship of the pressure basis (1.45) to the functional objective leads to its widespread usage for compressors.

Since frictional losses generally rise as the square of the velocity, highest efficiency is obtained by having moderate velocities in both rotor and stator, rather than high in one row and low in the other. There is therefore a tendency for a degree of reaction around 50% to be preferred if efficiency is an important criterion.

However other criteria might favour an impulse machine. Leakage losses would be one such consideration, since it is much easier to seal a stator row than a rotor row.

Similarly, where compactness is a consideration for turbines, the work output is limited by wheel speed which is limited by blade stresses. The higher fluid velocities possible, for a given wheel speed, in an impulse turbine, make this an attractive choice for auxiliary units etc. Other reasons for impulse turbines to be considered are their minimisation of end thrust and their more predictable creep behaviour.

It can be readily shown [1.1] that if an enthalpy basis is used for the definition of stage reaction then, if the absolute entry and discharge velocities are equal and the axial velocity is constant through the stage, two simple special cases result. Referring to Fig. 1.4 (b) if $\alpha_2 = \beta_3$ the reaction is 50%. If $\beta_2 = \beta_3$ the reaction is zero.

An important distinction that Dixon makes is that between a zero reaction turbine stage (defined on an enthalpy basis) and an impulse stage. In a zero reaction stage, because of irreversibilities, there is a small drop in static pressure through the rotor row. An impulse stage is one which is defined as having no static pressure drop through the rotor row; it will therefore have a slight degree of negative reaction.

1.7 Compressible Flows

It is beyond the scope of this work to provide a comprehensive survey of the relevant gas dynamics. Texts by Shapiro [1.13], Liepmann and Roshko [1.14], and Daneshyar [1.15], give detailed treatments of high speed flows. Compressible flow data are usually provided in tabular form and the tables of Keenan and Kaye [1.16] are recommended for this purpose.

1.7.1 One dimensional isentropic flow. For steady, adiabatic flow of a perfect gas

$$\frac{T}{T_0} = \left[\frac{c}{c_0}\right]^2 = \left[1 + \left(\frac{\gamma - 1}{2}\right) M^2\right]^{-1} . \qquad (1.46)$$

Using the isentropic law pressure, density and temperature variations can be expressed in terms of their isentropic values

$$\left[\frac{p}{p_0}\right] = \left[\frac{\rho}{\rho_0}\right]^{\gamma} = \left[\frac{T}{T_0}\right]^{\frac{\gamma}{\gamma-1}} . \qquad (1.47)$$

If equations (1.46) and (1.47) are combined then

$$\left[\frac{p}{p_0}\right] = \left[1 + \left(\frac{\gamma - 1}{2}\right) M^2\right]^{-\frac{\gamma}{\gamma-1}} \qquad (1.48)$$

and

$$\left[\frac{\rho}{\rho_0}\right] = \left[1 + \left(\frac{\gamma - 1}{2}\right) M^2\right]^{-\frac{1}{\gamma-1}} . \qquad (1.49)$$

Equations (1.46), (1.48) and (1.49) are commonly tabulated for practical values of γ. Air, being a diatomic gas, has a γ of 1.4.

An expression for a mass flow function may be obtained in terms of the above relationships and it can be shown[1.2] that this flow function reaches a maximum, termed the choking condition, when the local Mach Number reaches unity. Sonic conditions are marked with an asterisk.

$$M = 1; \qquad c^* = \sqrt{\gamma R T^*} .$$

For air the critical pressure ratio at which the sonic condition is reached is 0.528.

The maximum possible mass flow which may be passed may be expressed in terms of throat area, A^*.

$$\dot{m}_{max} = \rho^* A^* c^* = \frac{A^* p_0}{\sqrt{T_0}} \sqrt{\frac{\gamma}{R} \cdot \left(\frac{2}{\gamma + 1}\right)^{\frac{\gamma+1}{\gamma-1}}} . \qquad (1.50)$$

For $\gamma = 1.4$ this reduces to

$$\dot{m}_{max} = 0.686 \frac{A^* p_0}{\sqrt{RT_0}} . \qquad (1.51)$$

1.7.2 Shock waves. If the continuity, momentum and energy equations are applied to the constant area, frictionless, adiabatic flow of a perfect gas, certain relationships between properties at an upstream station (1) and a downstream station (2) may be derived.

These relationships are known as the Rankine - Hugoniot equations and define the strength of a normal shock wave :

$$\frac{p_2}{p_1} = \frac{\left(\frac{\gamma + 1}{\gamma - 1}\right)\left(\frac{\rho_2}{\rho_1}\right) - 1}{\left(\frac{\gamma + 1}{\gamma - 1}\right) - \frac{\rho_2}{\rho_1}} \qquad (1.52)$$

and similarly

$$\frac{\rho_2}{\rho_1} = \frac{c_1}{c_2} = \frac{1 + \left(\frac{\gamma + 1}{\gamma - 1}\right)\left(\frac{p_2}{p_1}\right)}{\left(\frac{\gamma + 1}{\gamma - 1}\right) + \frac{p_2}{p_1}} . \qquad (1.53)$$

The flow properties are usually required in terms of the Mach Number upstream of the shock. This Mach Number may be expressed in terms of pressure and density ratios and the relationship

$$M_1^2 = \left(\frac{\gamma + 1}{2\gamma}\right)\frac{p_1}{p_2} + \left(\frac{\gamma + 1}{2\gamma}\right) \tag{1.54}$$

derived. A similar approach yields an expression for M_2.

Normal shock tables usually provide, for given values of M_1 and γ, the values of M_2, static and stagnation pressure ratios and static temperature ratio.

It is important in cascade aerodynamics to note that if a pitot tube is placed into a supersonic stream the measured pressure will be affected by the presence of a curved shock which stands off ahead of the tube (Fig. 1.9). A correction for this may be readily derived by assuming that the stagnation streamline passes through a normal shock. In addition the compression between the shock and stagnation point may be regarded as isentropic.

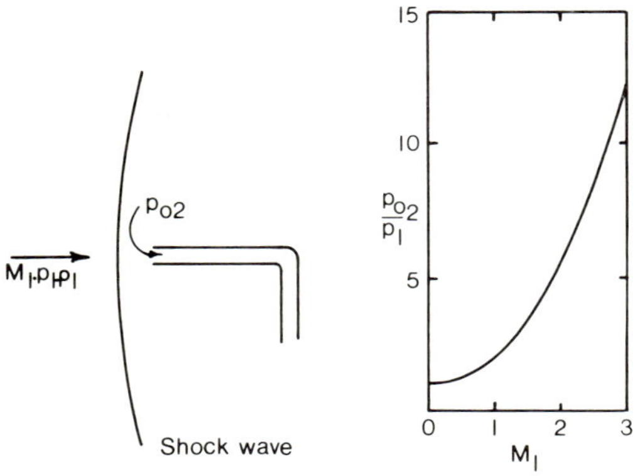

Fig. 1.9. Pitot tube in supersonic flow.

The correction is the Rayleigh supersonic pitot formula and is given by

$$\frac{p_1}{p_{02}} = \left[\frac{p_1}{p_{01}}\right]\left[\frac{p_{01}}{p_{02}}\right] = \frac{\left\{\left(\frac{2\gamma}{\gamma+1}\right) M_1^2 - \left(\frac{\gamma-1}{\gamma+1}\right)\right\}^{\frac{1}{\gamma-1}}}{\left\{\frac{\gamma+1}{2} M_1^2\right\}^{\frac{\gamma}{\gamma-1}}} \tag{1.55}$$

More complex boundary conditions than the pitot tube case give rise to oblique shock waves. The treatment of these is more difficult than the normal shock and for boundary conditions such as those described in Chapter 6, involve quite sophisticated flow computations.

REFERENCES

1.1 DIXON, S.L. Fluid Mechanics, Thermodynamics of Turbomachinery. Pergamon, Oxford, 3rd ed. (1978)

1.2 HORLOCK, J.H. Axial Flow Compressors. Butterworths, London (1958)

1.3 HORLOCK, J.H. Axial Flow Turbines. Butterworths, London (1966)

1.4 STONEY, G. Scientific activities of the late Hon Sir Charles A. Parsons, FRS. Engineering 144 (1937)

1.5 DODGE, P.R. A non-orthogonal numerical method for solving transonic cascade flows. ASME Paper 76-GT-73 (1976)
1.6 SMITH, L.H.Jr. The radial equilibrium equation of turbomachinery. Trans ASME, J. Eng. Power 88 pg. 1 (1966)
1.7 SMITH, L.H.Jr. and YEH, H. Sweep and dihedral effects in axial-flow turbomachinery. Trans. ASME, J. Basic Eng. pg. 401 (1963)
1.8 ROSSOW, V.J. An analysis of the error involved in unrolling the flow field in turbine problems. Mitt. Inst. Aero. (Zurich) No. 23 (1957)
1.9 WU, CHUNG HUA. A general theory of three-dimensional flow in subsonic and supersonic turbomachines of axial, radial and mixed flow types. NACA TN 2604 (1952)
1.10 MARSH, H. A digital computer program for the through-flow fluid mechanics on an arbitrary turbomachine using a matrix method. ARC R&M 3509 (1968)
1.11 HORLOCK, J.H. Review of losses and efficiencies in axial-flow turbines. Int. J. Mech. Sci., Vol 2, pg. 48 (1960)
1.12 BROWN, L.E. Axial flow compressor and turbine loss coefficients: A comparison of several parameters. Trans. ASME, J. Eng. Power (1972)
1.13 SHAPIRO, A.H. The Dynamics and Thermodynamics of Compressible Fluid Flow. Vols. I and II. Ronald Press (1953-54)
1.14 LIEPMANN, H.W. and ROSHKO, A. Elements of Gas Dynamics. Wiley, N.Y. (1957)
1.15 DANESHYAR, H. One-Dimensional Compressible Flow. Pergamon, Oxford (1976)
1.16 KEENAN, J.H. and KAYE, J. Gas Tables. Wiley, N.Y. (1948)

CHAPTER 2

Low Speed Cascade Testing

The basic information required by the designer of turbomachines takes, for a given blade shape, the form of graphs showing the variation of outlet angle and total pressure loss with inlet angle while such parameters as stagger, space-chord ratio, Reynolds' Number and free stream turbulence level are also varied. Since the variables are so numerous it is not surprising that this information has been most adequately provided in the form of loose correlations of all existing results.

In general there are too many variables for such design information to be obtained from the systematic testing of turbine and compressor rigs. Any attempt to do so usually introduces a wide range of uncertainty resulting from other spurious effects present in a three dimensional rotating blade row. For this reason successful designs were not forthcoming until suitable data became available from the testing of two-dimensional cascades.

In this chapter some of the early cascade work which led to the axial compressor for gas turbine propulsion will be described. Incorporation of the data into design correlations and the design process will then be discussed. Details will be given of effects of Reynolds' Number and Turbulence Level and the evolution of the side wall suction and variable incidence mechanisms. Finally information will be provided on high quality facilities for low speed cascade testing in air.

Some good cascade testing has been performed using water as a working fluid. It is not possible to adequately describe this work here and the reader is referred to original papers. In an extensive programme of blade profile development Numachi has designed and tested a range of hydrofoil and propeller profiles with a view to minimising cavitation effects [2.1]. One particular cascade was tested by Wade and Acosta [2.2] over a wide range of conditions between fully wetted and fully cavitating.

At low speeds better flow visualisation is possible in water than in air. Small free surface tunnels have been used for visualisation and teaching work. In elegant quantitative work in a water channel Clayton and Massey [2.3] used the hydrogen bubble technique to study flow separation in a fan cascade at a Reynolds' Number of 3000. The results were applied to cooling fan behaviour in high flying aircraft.

2.1 Early Cascade Experiments

Although Parsons had built an axial flow compressor in 1901 its efficiency was poor because of incorrect blade setting angles. Such deficiencies had been of little hinderance to his previous successful development of the steam turbine and much useful turbine design information is still obtained from relatively crude nozzle passages in which little attention is paid to periodicity of flow from one passage to another.

The fundamental difference between the flows encountered around compressor blading and those encountered in turbines is that the former is a diffusing situation, involving adverse pressure gradients, whereas turbine flows are accelerating, keeping the boundary layers thin and stable. Obviously in compressor blading a small alignment error could tend to increase any local adverse pressure gradient and cause premature boundary layer separation.

During the first two decades of this century the need for an aerodynamic approach to the design of turbines and compressors was gradually realised. The design of wings and isolated aerofoils was well understood and attempts had been made to apply the isolated aerofoil approach to turbomachines. However this had failed even in the, then urgent, design of airscrews.

Wood, Bradfield and Barker [2.4], in 1919, realised that the interference of neighbouring blades should be applied to airscrew theory and obtained modified lift coefficients in the first known cascade tests. They found that the cascade effect accounted for by far the largest part of the unexplained discrepancy between simple aerofoil theory and airscrew experiment. A 7' (2.1m) low speed wind tunnel was used with an ingeious mechanism for variation of incidence angle.

In 1928 extensive cascade investigations were reported by Christiani [2.5] in Germany and by Harris and Fairthorne [2.6] in England. Christiani tested compressor cascades having five or more blades of 200mm chord. The aspect ratio was 3.05 and incidence variation was by simple rotation of the central blade.

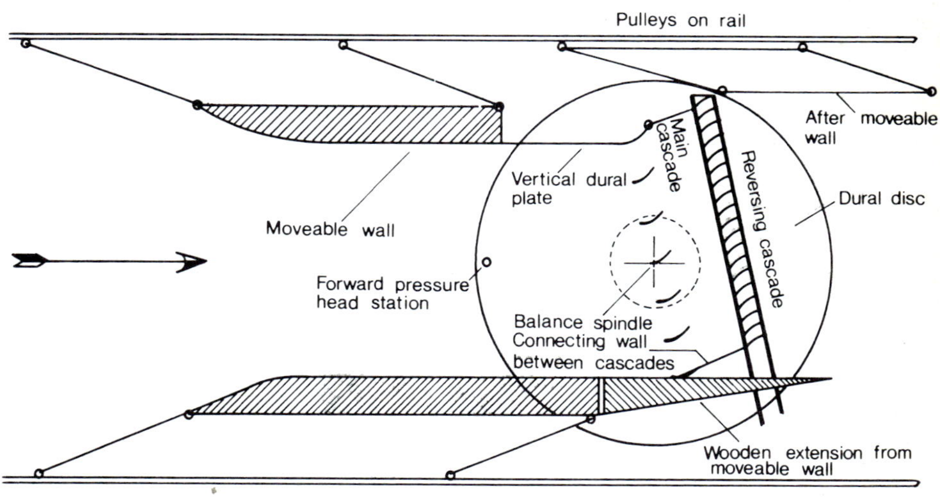

Fig. 2.1. Plan of Harris and Fairthorne tunnel.

The tunnel used by Harris and Fairthorne is shown in Fig. 2.1. It was desired to mount the cascade in an existing 4' x 2' (1.2m x 0.6m) low speed working section. "To avoid the use of a bent tunnel " Harris and Fairthorne produced an arrangement whereby any deflection imparted by the test cascade was removed by a second cascade two chords downstream.

The test configurations chosen had stagger angles of 0°, $22\frac{1}{2}^{o}$ and 45° and space/chord ratios of 0.5, 1.0 and 1.5 and an R.A.F. 27 profile with 45.2^{o} camber was used throughout. Although the chord was only 3" (76mm) this did mean that an aspect ratio of 7:1 could be used. This feature almost certainly gave the results more lasting value than might otherwise have been the case.

The other noteworthy feature is Harris and Fairthorne's ingenious means of inlet angle variation. The test cascade was mounted on a 40" (1.02m) diameter dural disc. Harris and Fairthorne recognised the importance of having representative flow conditions in the end passage and this generally entailed, for any change of inlet angle, a variable height between top and bottom walls. They were the first investigators to mount 'dummy' walls in a parallelogram arrangement such that for any inlet angle change the walls remained parallel.

The major shortcoming, however, was that the tests were all performed in the critical Reynolds' Number region between 10^{5} and 1.5×10^{5}. Laminar separation is almost inevitable under these circumstances, resulting in lower flow deflections than would attain in an operational compressor. Results were presented for deflection, lift and drag and downstream traverses over the range of incidence, stagger and space/chord ratio provided a useful basis for subsequent design despite the low Reynolds' Number.

British axial compressor designs of the late 1930's used the Harris and Fairthorne data, which were adequate for the development of the first axial machines having reasonable efficiency, although possibly biasing the design to rotor blading of relatively low stagger and high camber. As an example of a compressor design using these results, the C.A. Parsons Number 2 compressor [2.7] had a brake efficiency approaching 80% whereas that of the Number 1 compressor, having high stagger, low camber rotor blading without the benefit of the Harris and Fairthorne cascade data, had been 10% lower. The experience of Metropolitan-Vickers, in producing axial compressors for jet propulsion engines, was similar and again the design of the first compressors giving acceptable efficiencies for viable gas turbine operation utilised the Harris and Fairthorne cascade data.

The equivalent early cascade results which formed the basis of successful Swiss axial compressor designs were those of Keller [2.8].

A series of tests on vanes having 90° turning was undertaken by Collar [2.9] which resulted in identification of a profile well suited to application in the turning vanes of wind tunnels.

The last known pre-war cascade tests were performed in Japan by Shimoyama [2.10] in 1937. These very thorough tests on Göttingen 549 blades were carried out for space/chord ratios between 0.75 and 2 and stagger angles between 60° and 80°. The tunnel used had a working section 530mm square and the blade chord was 100mm. In addition to a comprehensive range of curves for lift and drag as a function of incidence, an interesting range of pressure distributions was also recorded. Typical results are shown in Fig. 2.2 in which these early observations of typical loss-incidence 'buckets' and the effect of close spacing of the pressure surface pressure distributions are particularly noteworthy.

In concluding our survey of pre-war cascade experiments it could be argued that the effort on steam turbine nozzle and bucket passages, which was begun by Rateau [2.11],

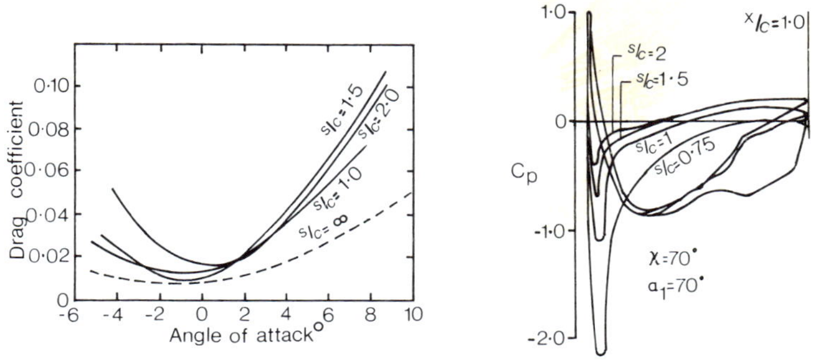

Fig. 2.2. Some cascade results of Shimoyama.

constituted the first cascade experiments. Such work was generally performed in steam with up to nine blade passages and the axial and tangential forces were measured using a balance. The most systematic testing of this type was carried out in the 1920's by the Institution of Mechanical Engineers, reviewed by Guy [2.12].

Although the steam turbine nozzle tests provided useful design data, the flows investigated were undoubtedly very three dimensional and the measurements made no attempt to eliminate such effects. The major contribution of Harris and Fairthorne and the new aerodynamics had been to focus attention on the specifically two-dimensional effects.

2.2 Production of British Compressor Design Data

Because of the lower pressure ratios involved, compressor designers have been able to make greater use of low speed data than have turbine designers. The collection of turbine cascade data will therefore be discussed in later chapters.

The earliest correlation of compressor cascade data was by Howell [2.13] and has been widely used in axial compressor design. As a basis upon which to relate the available data, Howell postulated a "nominal" deflection as being 80 per cent of the maximum deflection. Finding this nominal deflection to be a function of outlet angle, space/chord ratio and Reynolds' Number, Howell used a rule originally discovered by Constant [2.14] to relate the nominal deviation to the camber and space/chord ratio,

$$\delta = m\theta(s/c)^n \quad . \tag{2.1}$$

Howell gave suitable values of n for various types of blading and Carter [2.15] gave a relationship between m and the outlet angle.

Instead of using the nominal conditions as a basis for his correlation, Carter [2.15] started at the point of maximum lift /drag ratio, which he termed the "optimum" condition. For any given cascade he showed that it was possible to plot contours of constant lift /drag ratio on a graph giving the relationship between deflection and outlet angle. Carter also produced a useful design curve of optimum incidence against optimum camber and space/chord ratio (see Chapter 11).

The basic cascade data upon which the Howell correlation was founded were limited in extent. The major works at that time had been those of Constant [2.14] and those mentioned in the previous section. In addition to this cascade testing had been performed during the second world war at various British establishments including Fraser and Chalmers, C.A. Parsons, Metropolitian-Vickers, Power Jets and the R.A.E.

A representative low-speed cascade tunnel from this period is the N.G.T.E. No. 2 low-speed cascade tunnel (Fig. 2.3). Atmospheric air was forced through the tunnel by an axial flow blower driven by a variable speed 18.6kw motor. Velocities of 61m/sec were obtained in the 0.31sq.m working section. Air entered the settling chamber via a diffuser and then passed through a honeycomb flow straightener and two fine screens followed by a gradual contraction. Thirteen blades of 102mm chord, giving an aspect ratio of four, were usually used. End wall boundary layers were drawn off through suction slots but no suction was provided on the side walls. The usual technique was to vary the quantity of end-wall bleed until side wall static pressures one chord up-stream were constant.

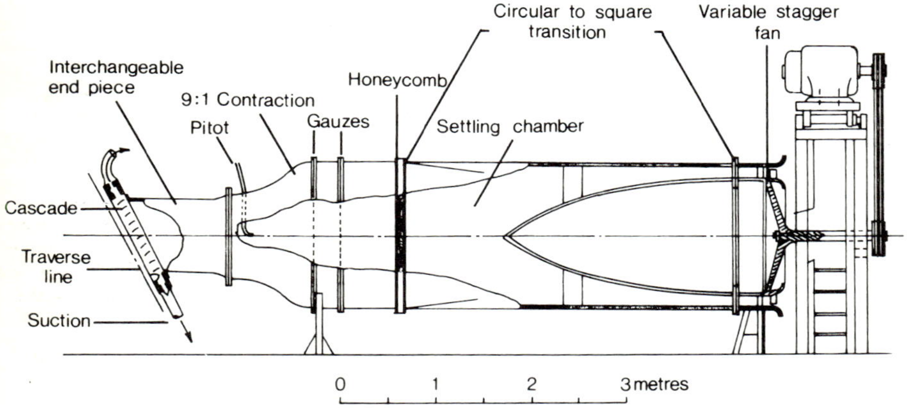

Fig. 2.3. The N.G.T.E. No. 2 low speed cascade tunnel.

This low speed tunnel, dating back to around 1940, had fully automatic traverse gear for surveying flow angles and downstream total pressures. The whole emphasis was on ensuring good operating conditions for a wide range of turbine and compressor cascades and a large test throughput. Unfortunately, however, this particular tunnel also represents the very features which led, for many years, to consistent discrepancies between British and American compressor design practice. Before elaborating on this question it is necessary to trace the development of American low speed cascade tunnels.

2.3 American Work on Low Speed Cascades

In the United States work had already begun on cascades of NACA 65 series blading. The early results of Bogdonoff [2.16] were found to be unreliable since a large degree of contraction was taking place in the flow through the cascade. It had been argued that such a flow was acceptable since the flow in a compressor was also subject to three dimensional end effects. Erwin and Emery [2.17] showed that the similarity was only superficial and that the flow within a compressor blade row was usually more nearly two-dimensional than that in early cascade tunnels. They then proceeded to develop a cascade tunnel, having porous side walls, which permitted the realisation of two dimensional conditions.

Erwin and Emery gave the following definition of two-dimensional flow in a cascade tunnel :-

1. Equal pressures, velocities and directions exist at different spanwise locations.

2. The static pressure rise across the cascade equals the value associated with the measured turning angle and wake.

3. No regions of low energy flow, other than blade wakes, exist. These wakes are constant in the spanwise direction.

4. The measured force on the blades equals that associated with the measured momentum and pressure change across the cascade.

5. The various performance values do not change with aspect ratio, number of blades or other physical factors of the tunnel configuration.

In practice conditions (1) and (3) were not obtained since it was not possible to bleed away the whole of the boundary layer. It was, however, concluded that whereas conventional cascade tests with solid side walls did not give two-dimensional conditions, the use of the porous side wall technique made it possible to approach the ideal. For all practical purposes two-dimensional flow was attained.

The thorough preliminary work of Erwin and Emery made possible a series of systematic cascade tests by Herrig, Emery and Erwin [2.18] on NACA 65 blading. For selected values of blade camber and space/chord ratio, the deflection and loss coefficient were obtained for different stagger angles. This method of presentation was considered inconvenient and the results were replotted by Mellor [2.19] in the form of graphs of α_2 against α_1 for lines of constant stagger. Companies using 65 series blading have gone a step further and correlated the data in a manner similar to that of the Howell correlation.

The tunnel which embodied the above philosophy and produced the NACA results was the Langley 5" (127mm) tunnel, now at the von Karman Institute in Belgium. Commissioned in 1944, the tunnel has a working section of up to 0.093 sq.m and a maximum flow velocity of 46m/sec. The flow discharges as a free jet to the atmosphere a few inches beyond the blade row. Settling chambers of 1.37m x 2.03m contain three fine screens. The distinctive features of the tunnel are the side and end wall treatments. The end walls allow boundary-layer suction over the entire flow surface including a flexible portion in the blade region. This flexible portion replaces the end blades (see Fig. 2.4) and its shape is adjusted by the operator until good periodicity from passage to passage is obtained. Side wall boundary layers are firstly removed by 4.76mm protruding slots, about 127mm upstream of the blades. In addition to this, continuous suction is provided through porous side walls.

With this degree of control over wall boundary layers, flow three-dimensionality is no problem, regardless of aspect ratio. It was therefore possible to use a span of only 127mm giving an aspect ratio of unity.

After the work on the 65 series blading and work leading to the formulation of Lieblein's diffusion factor [2.20], research at NACA on cascades was temporarily discontinued. Fortunately, however, one very important discovery had been made which led to renewed British interest in the American results. Felix and Emery [2.21] had tested an identical 10C4/30C50 section to one which had previously been tested in England. The disturbing results of Felix and Emery, from tests carried out at an aspect ratio of unity with porous side walls, gave values of outlet angle between 1° and 3° higher than earlier results. This discrepancy turned out to represent a consistent disparity between the American and British tests. It was not fully explained until more complete studies had been performed upon the effects of axial velocity variation.

2.4 Axial Velocity Variation through Cascades

In tests on cascades having solid side-walls it is generally observed that for a given cascade configuration and inlet flow and with a given incoming side wall boundary layer the secondary flow influences are independent of aspect ratio. More precisely the dominating three-dimensional flow feature is the commonly-observed corner stall which emanates from the first region of reasonably unfavourable pressure gradient in the end wall region. It is this corner stall which causes considerable blockage and therefore a higher axial velocity in the mid-span section of the passage (see Fig. 2.5).

The axial velocity ratio of a cascade is defined as

$$AVR = \int_0^S (c_{x2})_{MS}\,dy \Big/ \int_0^S (c_{x1})_{MS}\,dy \qquad (2.2)$$

Because the physical area of corner stall is unaffected by changes in aspect ratio (unless the aspect ratio is so low that the stalled regions from the corners actually meet at the mid-span) the axial velocity ratio in solid side-wall tests is strongly dependent upon the aspect ratio. For most of the NGTE tests upon which the British compressor design data were based the axial velocity ratio was around 1.1, occasionally rising to 1.2.

As a result of the discovery by Felix and Emery of discrepancies between American and British results extensive investigations were carried out on one particular cascade. Pollard[2.22] tested a cascade of 10C4/30C50 blades at a stagger of 36°, an inlet angle of 52° 50' and a space/chord ratio of unity, in order to understand the effect of porous side-wall suction more thoroughly. Besides the Liverpool University No. 1 and No. 2 low speed tunnels, the former NACA 5" (127mm) tunnel was also used.

Fig. 2.4. Former NACA 5" (127mm) tunnel now at VKIFD, Belgium.

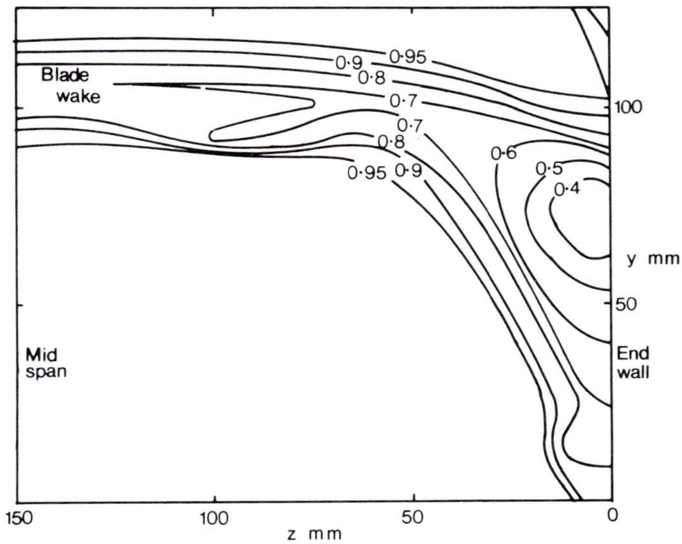

Fig. 2.5. Contours of P_2/P_1 downstream of compressor cascade.

For tests on this particular cascade in porous side-wall tunnels the effect of axial velocity ratio on deviation angle could be summarised by the relationship:

$$\delta = \delta_{2D} - 10(AVR-1) \tag{2.3}$$

where δ_{2D} is the deviation at AVR = 1.0. The results for AVR > 1.0 were obtained by allowing the end-wall loss region to build up and the results include a correlation for the ensuing secondary flow. Taking the axial velocity ratio of the Howell and Carter correlation results for deviation angle to have been over 1.1, and that for American data as typified by those of Lieblein [2.20] to have been unity, equation (2.3), gives the basis for an explanation of differences between the British and American data.

Although there are other small differences such as those of turbulence level and Reynolds' Number, it appears that British cascade data were obtained in flows having at least 10% axial velocity increase through the cascade, whereas the American tests were conducted under nominally two-dimensional conditions. As a result the British tests will have resulted in deviation angles lower by at least a degree than the corresponding values from American tests.

2.5 Influence of Reynolds' Number

The performance of axial turbines and compressors generally deteriorates as Reynolds' Numbers are decreased. Low Reynolds' Numbers are most commonly encountered in the last stages of steam turbines and in high-flying aircraft or small-engine components. Figure 2.6 illustrates the decrease in inlet Reynolds' Number with altitude for a typical aircraft engine. As the Reynolds' Number of axial compressor blading drops below a critical value, which is generally around 10^5, the falls in efficiency, flow and pressure ratio are particularly pronounced.

For tests upon a given cascade in an open circuit low speed cascade tunnel the Reynolds' Number is simply proportional to flow velocity, apart from second order temperature corrections. There should therefore be no difficulty in acquiring data over a wide range of Reynolds' Numbers. There is, however, a shortage of such data because in practice their acquisition is not really convenient.

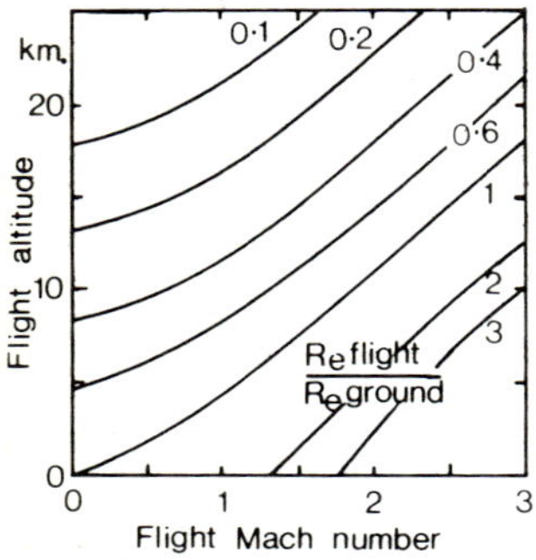

Fig. 2.6. Effect of altitude and Mach number on Reynolds' Number.

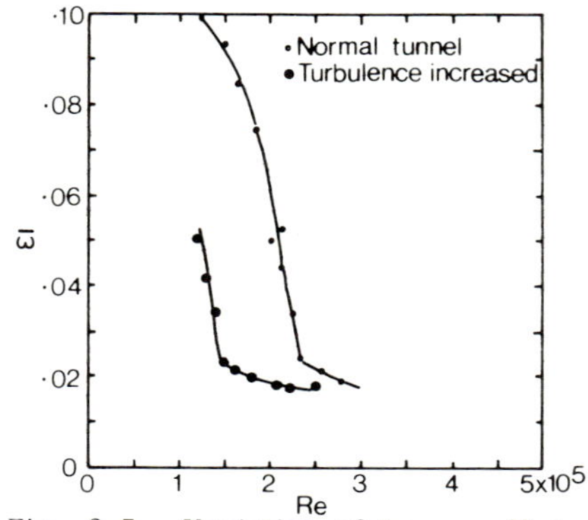

Fig. 2.7. Variation of loss coefficient with Reynolds' Number.

For a 150mm chord cascade in an atmospheric density open jet tunnel the attainable Reynolds' Number range will be between 0.25 x 10^5 and 6.6 x 10^5. Below this range velocities are only of the order of a metre per second and are thus very difficult to measure. Above the range compressibility effects begin to confuse the situation.

The most comprehensive series of experiments on Reynolds' Number effects in compressor cascades has been performed by Rhoden[2.23] in a low turbulence wind tunnel. For negative incidences at Reynolds' Numbers below 10^5 there was complete laminar separation from the suction surface. At slightly higher Reynolds' Numbers the laminar separation was bounded by a turbulent reattachment. As the Reynolds' Number increased above 2 x 10^5 and at positive incidences transition occurred before the laminar boundary layer could separate but the ensuing turbulent layer tended to separate near the trailing edge.

A useful correlation of Reynolds' Number data for turbine cascades has been produced by Soderberg and is fully described in Horlock's book [1.3].

Figure 2.7 gives some results from Blight and Howard's tests [2.24] on a compressor cascade designed by Lighthill. The critical nature of a reduction in Reynolds' Number below 2 x 10^5 on loss coefficient is clearly seen. The other curve is for a case in which the tunnel turbulence level was artificially increased. Clearly the critical Reynolds' Number has been almost halved by the introduction of turbulence. The point is that the effects of Reynolds' Number and turbulence level are closely linked. An increase in turbulence level has a similar effect on the boundary layer to an increase in Reynolds' Number, both leading to a high rate of exchange of momentum between free stream and boundary layer.

2.6 Effect of Free Stream Turbulence

Free-stream turbulence has a major effect on the flows that occur in turbomachinery. This is mainly due to its influence on the transition from laminar to turbulent boundary layer, its further development and hence its separation. Furthermore, the boundary layer displacement thickness affects the free-stream flow and thus alters the conditions that would be expected from potential flow theory.

The nature of the turbulence in a wind tunnel of conventional design, i.e. with proper settling chamber containing honeycombs and gauzes plus an aerodynamically-designed contraction, can be expected to be homogeneous and nearly isotropic (Schlichting[2.25]). Townsend[2.26] states that the flow downstream of a uniform grid placed across the entrance to a wind tunnel working section, which is the experimental approximation to isotropic homogeneous turbulence, is not a homogeneous flow but a stationary one with an intensity gradient in the direction of mean flow. Theoretical isotropic turbulence, on the other hand, is homogeneous but not stationary in time. However under ordinary conditions the differences between the flows are slight. Therefore, at a certain distance from the screens, mean square velocity fluctuations in the three coordinate directions are assumed to be equal to each other.

$$\overline{u}^2 = \overline{v}^2 = \overline{w}^2 .$$

In such cases the turbulence level can be described by the quantity

$$T_u = \frac{\sqrt{\overline{u}^2}}{U} = \frac{\sqrt{\frac{1}{3}(\overline{u}^2 + \overline{v}^2 + \overline{w}^2)}}{U} \times 100\ (\%) \qquad (2.4)$$

In any series of tests in which the turbulence level is to be systematically varied the most convenient procedure is to ensure that the basic turbulence level of the 'clean' tunnel is equal to or lower than the lowest desired test value. It is then necessary to insert suitable devices which will give the required levels of near-isotropic turbulence.

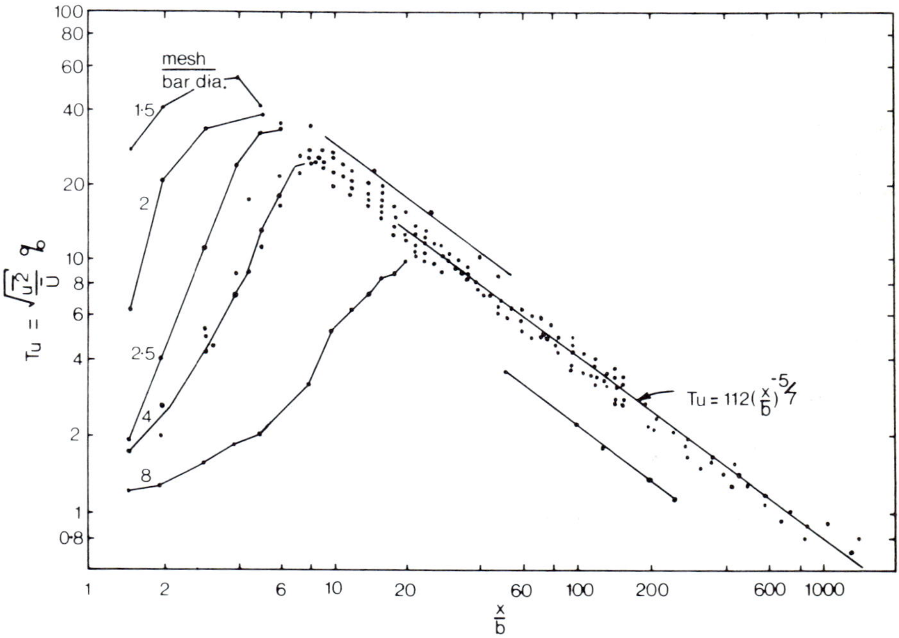

Fig. 2. 8. Turbulence attenuation downstream of bi-planar grid.

Various types of device have been used for this purpose. These include turbulence-generating fans mounted in the settling chamber, rows of small flags oscillating in the inflow and continuous chains of transverse wires passing upstream of the cascade[2.27]. However the most common and reliable device is the simple bi-planar grid of cylindrical bars mounted at the upstream end of the working section. It is the bar diameter which is the significant parameter for the strength of shed vorticity and hence the turbulence level at downstream locations. A correlation, based upon the work of references [2.28] to [2.31], presents the intensity of turbulence at different downstream locations in Fig. 2.8.

Figure 2.9 indicates some typical turbulence levels encountered in multistage compressors [2.27] and Fig. 2.10 shows how the turbulence level varies over the characteristic

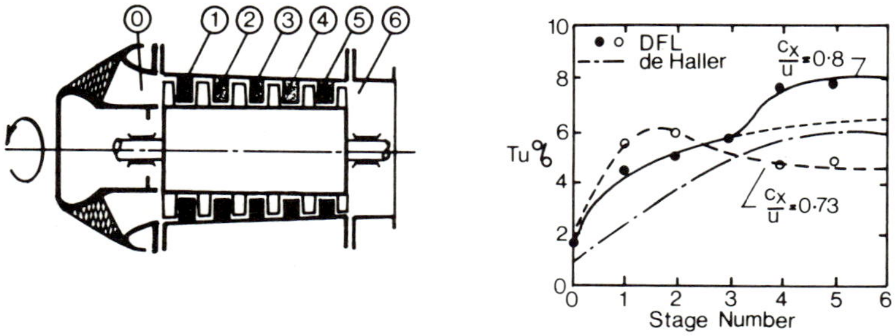

Fig. 2.9. Typical turbulence levels in multistage compressors.

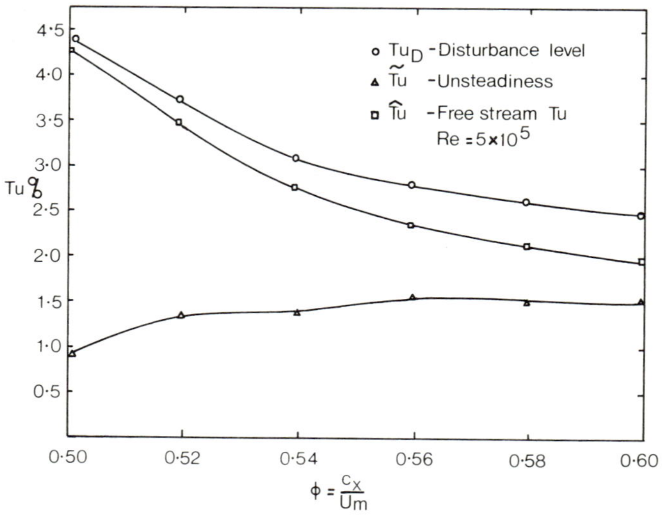

Fig. 2.10. Variation of turbulence level with flow rate for single stage compressor (Evans[2.32]).

for a single stage compressor[2.32]. This figure distinguishes between the random and periodic disturbance levels and shows that the random turbulence increases near stall. Disturbance levels in turbines are generally higher, reaching 18% in some measurements [2.33]. Koschel and Kühl [2.34] discriminated between random and periodic constituents behind a turbine stage. Of the peak 9.9% root mean square disturbance level, 8.9% was random and 4.2% periodic. Rotor blade passage through the stator wake was accompanied by a 65° variation in relative inlet angle.

Because the effects of turbulence level and Reynolds' Number are closely related it has been suggested by some workers that turbulence effects may be accounted for by postulating an "effective Reynolds' Number" - a multiplier on the actual Reynolds' Number. The work of Martlew[2.35] showed the inadequacy of this approach. The relationship is not a linear one - as the turbulence level was raised from 0.18% to 1.5% the separation point moved rearwards but a further increase to 3.0% had little effect. Although closely related, the Reynolds' Number and turbulence level must be treated as separate experimental variables.

2.7 Some Design Features of Low Speed Cascade Tunnels

Some of the desirable features for low speed cascade tunnels have emerged from the previous discussion. Any general purpose cascade tunnel should offer as wide a range of free-stream velocities, and hence Reynolds' Numbers, as possible. The flow should have a low turbulence level but it should be possible to increase this by the insertion of suitable grids or other means.

A further requirement, as with any high quality tunnel, is that variations in flow velocity and angle should be very small over the entire working section. It is usual to perform calibration traverses, resulting in contours of velocity and flow angle, to ensure that this is the case.

Periodicity of flow from passage to passage is the essential requirement of a good cascade test. This should be evident not only in traverses taken downstream of the blade row but also those taken upstream. These traverses should show minimal variations in velocity and flow angle profiles from passage to passage.

The upstream traverse is most conveniently performed by means of a tongued bar which forms a part of one side wall and may be translated tangentially to the cascade at a distance of one or two chords upstream. Figure 2.11 shows such an inlet traverse bar installed in the No. 1 cascade tunnel at Liverpool University. A suitable probe would be carried in this bar, probably a claw probe having one total pressure tube and two side tubes which may be nulled for yaw angle. Static pressures are more conveniently measured from a row of tappings placed in the side-wall upstream of the blade row.

It will be usual when a cascade is first tested for the above instrumentation to indicate marked variations in the tangential direction of either side-wall static pressures or flow angle or both of these. This is true of both turbine and compressor cascades and is generally caused by thick end-wall boundary layers stalling the end blade.

The usual solution to this problem is to deploy boundary layer bleed slots just upstream of the end blade, thus removing the offending end-wall boundary layer. Obviously the quantity of bled flow may be varied and it is usually found that by controlling the suction rate through these end passage slots, a strong effect on the variations in static pressure and yaw angle may be obtained. The setting may then be optimised to give good upstream periodicity from passage to passage.

Having set up the cascade test in this way the function of the downstream traverse is merely to record what effect the cascade has had on the carefully controlled entry

Fig. 2.11. Inlet traverse arrangement on Liverpool University No. 1 Cascade Tunnel.

flow. The question of how far downstream of the blade trailing edge this traverse should be made is, however, an open one. Most low speed tunnels have the cascade discharging an unrestricted jet to the atmosphere. There are no mechanical constraints on the downstream location of the traverse, which could be anywhere between the trailing edge plane and downstream infinity. The latter location is theoretically desirable but fortunately not necessary.

In the trailing edge plane the variations in total pressure will amount to a dynamic head and the corresponding angle variations will be at least 10° as shown in Fig. 2.12. This will lead to uncertainties in deviation angle and loss prediction. Measurement close to the trailing edge is therefore only recommended for special purposes. By a distance of one chord downstream most of the mixing has taken place and pitchwise flow angle variation will be only small. It is therefore a common practice to perform the outlet traverse at a distance of one chord downstream.

Measurement of total pressure in this plane presents no problems, a small pitot tube of perhaps 1.6mm diameter is usual. Since the cascade discharges to the atmosphere the static pressure can safely be assumed to be the atmospheric ambient value. The method of flow angle measurement is, however, debatable. The principle difficulty is the use of a yawmeter in a region of total pressure gradient, i.e. the blade wake. Different types of probe, such as the claw, the wedge and the cobra will perform differently because the side holes pick up an element of total pressure. Each will thus respond differently. The angle distributions of Fig. 2.12 were measured using a wedge probe and represent typical features of a true angle distribution. The use of differing types of probe will not usually have any significant effect upon the average angle across a blade passage, but the extent of any indicated peaks and troughs in a blade wake will vary according to the nature of the instrument. Provided this potential source of local discrepancy is borne in mind it is therefore reasonably safe to assume that integrated values of flow angle and loss coefficient will not be greatly affected by the choice of instrumentation.

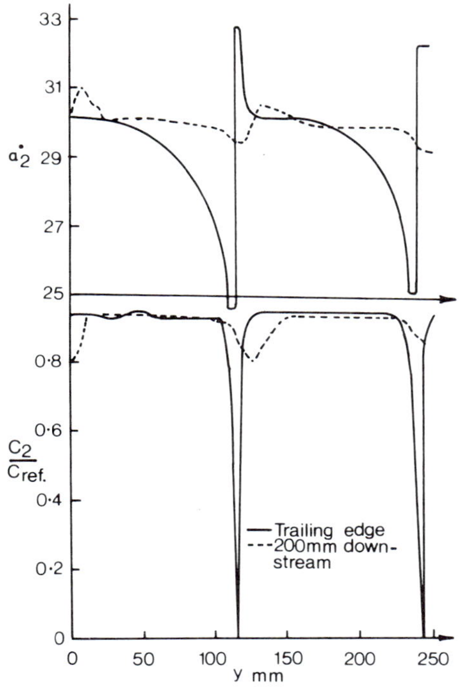

Fig. 2.12. Angle and velocity distributions 200mm downstream and in trailing edge plane.

Calibration for yaw angle is often conveniently performed in situ but with the cascade removed. Because of manufacturing errors yawmeters are not perfectly symmetrical about the pitot tube. They must therefore be calibrated so that the direction of flow giving balance of pressure in the two side orifices is related to some fixed datum on the probe. Such a datum can be conveniently provided by a miniature spirit level. Such levels are obtainable with an accuracy of ± 3'. On a large claw probe the level may even rest directly on the pitot tube.

For calibration of the yaw probe in situ the cascade is removed and the flow angle is measured at the same location introducing the probe from firstly one side and then the other. In this way any probe angle error may be distinguished from the tunnel flow inclination and a suitable correction made.

Cascade traverses can be tedious, as can the reading of many manometer pressures and the subsequent reduction and integration of data. The whole process lends itself well to automation and many contemporary tunnels have not only remote traverse gear but also computer-based data logging systems. In routine running it is convenient for a computer program to make transducer calibration and temperature drift corrections, apply numerical integration procedures and produce all final output values. For parameters such as axial velocity ratio it is especially helpful if this computation can be carried out by an on-line computer.

These procedures are even more economically justifiable for high-speed testing. Measurement of pressure is also simpler in high speed work because of the wide range of available pressure transducers for higher pressures.

Instead of a downstream traverse some workers prefer to use a 'comb' of total pressure tubes covering one or two passages. Total pressure tubes are insensitive to angle variations and as long as the tubes are aligned approximately with the flow angle no significant errors, other than from probe interference, are anticipated. The principal difficulty with this method is that the pitot orifices should all be located at the same distance downstream from the blade row. For any given comb this condition can only be satisfied for one particular flow angle. To cover a suitable working range it is therefore necessary to stock a 'family' of combs. This procedure is less accurate and less convenient than a good automated traverse with one combination pitot-yaw probe.

The derivation of good passage periodicity, the basic criterion for a good cascade test, has been associated with the use of end-wall boundary layer bleed. Another fundamental criterion, the two-dimensionality of the flow, is associated with the aspect ratio of the blading and the provision of side-wall boundary layer bleed. This side-wall bleed can take two forms:

(i) provision of a porous wall or slots upstream of the blade row.

(ii) provision of a porous wall through the blade row (bleed slots at the blade-wall juncture have also been tried with some success[2.36]).

For either compressor or turbine cascades careful removal of the upstream side-wall boundary layer will be beneficial. The author's experience is that a porous wall makes a cleaner job of this removal then protruding slots.

The remaining questions are whether to provide continuous side-wall suction in the immediate vicinity of the cascade and what aspect ratio to choose for the cascade. The answer will be different for decelerating and accelerating cascades.

For a turbine cascade, with the possible exception of an impulse cascade having a substantial region of adverse pressure gradient, the favourable pressure gradients will tend to preclude corner stall. The discharge axial velocity will therefore not substantially exceed the inlet value and there is little point in careful control of the side wall boundary layer. Provided the aspect ratio is adequate the two-dimensionality should be reasonable.

Compressor cascades are much more susceptible to the effects of separating side wall boundary layers. In early tests aspect ratios of four and higher were not sufficient to preclude large three dimensionality effects on deviation angle. The lesson is that porous side walls are essential for compressor cascades, giving proper control over axial velocity ratio, a parameter which should be recorded in any cascade test. Fortunately, however, if side wall suction is used the aspect ratio is no longer of significance. Under these conditions an aspect ratio of unity is acceptable.

For these reasons a tunnel designed for the testing of turbine cascades might tend to look rather different than one for compressor cascades. The former might have a reasonably high aspect ratio (dependent upon the blade shape) but will not need complicated porous side wall arrangements. For the latter a low aspect ratio and therefore a narrow tunnel would be acceptable but extensive boundary layer bleed arrangements will be needed. In practice the tunnel might be needed for testing all kinds of cascades and some suitable design compromise is made.

2.8 Some Contemporary Low Speed Tunnels

Table 2.1 lists some low speed tunnels presently in use and outlines their features. Since for many of the tunnels there is no published record of turbulence level this parameter has not been listed. None of the tunnels for which the turbulence level was available has a value above 0.4%. The term 'variable incidence mechanism' means any arrangement whereby the inlet angle may be continuously varied without resort to

Table 2.1 Some Low Speed Linear Cascade Tunnels

ORGANISATION	CITY	CONTACT OR REFERENCE	SIZE	FEATURES
JAPAN				
Kyushu Univ.	Kyushu	Inoue	305 x 533	abc
Nagoya Univ.	Nagoya	Otsuka	300 x 420	-
Ship Research Inst.	Tokyo	Hiraoka	200 x 1200	
Univ. of Tokyo	Tokyo	Takata	180 x 420	abcd
Univ. of Tokyo	Tokyo	Tanaka	180 x 420	
Univ. of Tokyo	Tokyo	Tanaka	250 x 600	bd
U.S.A.				
MIT	Cambridge	Sotak	508 x 864	acd
Univ. of Minnesota	Minneapolis	Eckert	60 x 469	-
NASA Lewis	Cleveland	-	254 x 762	ab
UTRL	Hartford	Dring	278 x 685	a
Westinghouse	Philadelphia		203 x 610	ac
US Naval Postgrad.Sch.	Monterey	Shreeve	254 x 1525	-
U.K.				
Liverpool University	Liverpool	Norbury	305 x 737	abcd
Liverpool University	Liverpool	Norbury	737 x 737	ac
Whittle Laboratory	Cambridge	Whitehead	610 x 1168	abc
C.A. Parsons	Newcastle	Gibson	110 x 300	-
NGTE	Pyestock	Calvert	400 x 750	a
INDIA				
BHEL (R & D)	Hyderabad	Srivastava	450 x 800	c
IIS	Bangalore	Soundranayagam	300 x 750	ac
IIT	Madras	Prithvi Raj	335 x 400	c
Nat. Aero. Lab.	Bangalore	Paranjpej	140 x 300	c
WEST GERMANY				
Tech.Univ. of Braunschweig	Braunschweig	Stark	300 x 500	a
Tech.Univ. of Braunschweig	Braunschweig	Stark	600 x 900	a
BELGIUM				
VKIFD	Brussels	Breugelmans	127 x 762	ab
FRANCE				
Ecole Centrale	Lyons	-	254 x 1346	abc

Key to FEATURES:

a. End passage bleed
b. Side wall bleed
c. Variable incidence mechanism
d. Upstream suction slot

the addition of fixed wedges. The ideal such mechanism is that invented by Montgomery[2.37] as incorporated into tunnels at M.I.T., Liverpool, Cambridge and Bangalore. As an example of a high quality cascade tunnel which is in daily use for research and undergraduate teaching some features of the Cambridge tunnel are now described.

2.8.1 The Cambridge University No. 1 Cascade Tunnel. Now located in the Whittle Laboratory, the tunnel has evolved from an original design by Rhoden [2.38]. The 112 kw variable speed induction motor drives a centrifugal fan which discharges into a wide-angled diffuser containing a honeycomb. In a 'trial and error' development process splitter plates were inserted into the diffuser and suitable screens and perforated plates were placed in the settling chamber for constancy of flow angle across the working section to within $\pm \frac{1}{3}^{\circ}$ and constancy of dynamic head to $\pm \frac{1}{2}\%$.

A new contraction for this tunnel reducing the 1.98m x 1.07m settling chamber to a working section of 1.17m x 0.61m was designed by Evans[2.28]. Design methods for contractions are discussed by Gibbings [2.39], Mills [2.40] and Mikhail[2.41]. The problem with many contraction theories is that the "fairing in" occurs asymptotically at $\pm\infty$. In practice to keep the axial distance short severe truncations are needed.

A prime consideration in contraction design is the avoidance of regions of adverse pressure gradient which could cause separation. Regardless of the precise contour or shape the essential criterion is that the boundary curve should be continuous in slope and curvature. Separation regions are avoided by the maintenance of a fair contour rather than by dimensional accuracy.

A further consideration in contraction design is the avoidance of Görtler vorticity which may persist into the working section. There are strong arguments for the upstream portion of the contraction to be effectively a 'bellmouth' within the settling chamber with no concave surfaces presented to the flow.

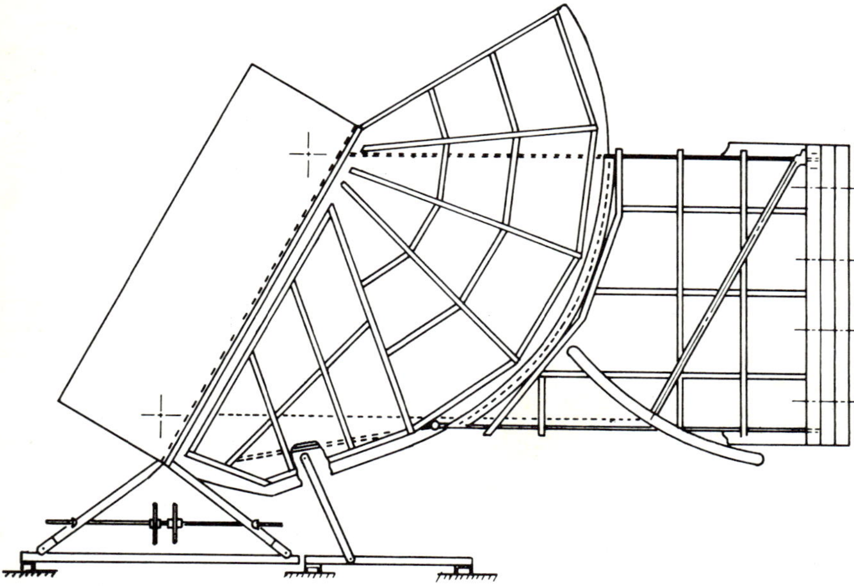

Fig. 2.13. Variable incidence mechanism on the Cambridge University No. 1 Cascade Tunnel.

The variable incidence mechanism in use is shown in elevation in Fig. 2.13. Whereas the M.I.T. device was hinged about a vertical axis, the air stream being deflected in a horizontal plane, in the Liverpool, Cambridge and Bangalore tunnels the circular arc side-wall of the 'drum' axis is horizontal, coincident with the leading edge of the end blade and the air stream is deflected upwards by the cascade.

The problem which Montgomery's design overcomes is that, as inlet angle is varied, in order to preserve the same integral number of blade passages at constant spacing the working section height must vary. To preserve good quality inflow all tunnel walls must remain parallel. To achieve this a false floor is provided. The downstream end of this floor is hinged to the bottom end blade and by a parallelogram link mechanism the false floor is free to rise and fall as incidence changes, always remaining parallel to the top wall of the working section.

As the false floor moves so a variable quantity of air flow must be by-passed and in such a way as not to disturb the main flow. To achieve this the false floor leading edge has a sharp splitter lip and is provided with a yawmeter to ensure that the flow is parallel to the false floor. To control this condition the resistance of the by-pass path must be the same as that through the test cascade. The pressure change across the cascade will be variable and it is therefore necessary to provide a variable angle diffuser which can be set to give the same by-pass diffusion as is present for the main flow.

In addition to a variable incidence mechanism the Cambridge tunnel also has provision for boundary layer bleed through the cascade side-walls. The suction fan employed is powered by a 15kw motor. The bleed collection plenum chambers contain the blade mounting framework and coarse perforated metal plates for support of the side-walls. Adjacent to the perforated plate is a coarse metal gauze and the final smooth interface with the flow is a 2mm thick layer of 'Vyon'. This plastic has the porosity characteristics of a fine callendered gauze but has a smoother surface.

Blades for low speed cascade testing may be formed from an alloy on a numerically controlled machine or made from a soft wood having good warpage and shrinkage characteristics finished by applying a coating of shellac. For low speed testing at Cambridge the blades, which are 610mm span and 152mm or 305mm chord, are constructed using a glass-reinforced epoxy resin technique described by Gregory-Smith and Marsh[2.42]. Considerable precision and reliability are obtained from this technique which lends itself well to the embedding of static pressure instrumentation inside the blades.

2.9 Future Low Speed Cascade Tunnels

Low speed cascade testing has served the aerodynamicist well in the past by providing systematic design data and enabling him to look more closely at fundamental questions which could not be answered in a rotating machine.

It may well be that there are further basic questions which can best be answered by low speed cascade investigations. Until perfection is attained in theoretical flow predictions it will always be necessary to check the accuracy of these by comparison with experimental data.

It seems very likely that low speed cascade testing will play a role in the evaluation of blading for special purpose applications. One such purpose is surprisingly basic. Very few data are available on the performance of cascades at high space-chord ratios. The designer of propellors for air and marine use, low pressure fans or 'propulsors', windmills and other devices often finds himself in uncharted territory between applicable isolated aerofoil data and cascade data. Good tunnels capable of taking long cascades are needed for this work.

Other special types of low speed cascade testing may include investigation of unsteady cascade response, generation of boundary layer data and the study of secondary flow effects. These are all pressing design problems to which answers are needed firstly in two-dimensional cascades, then in the three dimensional cascade situations which will be discussed in the next chapter, and finally in the actual machine.

REFERENCES

2.1 NUMACHI, F. Cavitation tests on hydrofoils designed for accelerating flow cascade: Report 3 - Five profiles generated from prescribed pressure configurations of Types 1 and 3. Trans. ASME, J. Basic Eng. pg. 543 (1964)

2.2 WADE, R.B and ACOSTA, A.J. Investigation of cavitating cascades. Trans. ASME, J. Basic Eng. (1967)

2.3 CLAYTON, B.R. and MASSEY, B.S. Flow separation in an aerofoil cascade. J.R.Ae.S. pg.559 (1967)

2.4 WOOD, R.McK., BRADFIELD, F.B. and BARKER, M. Multiplane interference applied to airscrew theory. ARC R&M 639 (1919)

2.5 CHRISTIANI, K. Experimentelle Untersuchung eines Tragflugelprofils bei Gitteranordnung. Luchtfahrtfers 2 (1948)

2.6 HARRIS, R.S. and FAIRTHORNE, R.A. Wind tunnel experiments with infinite cascades of aerofoils. ARC R&M 1206 (1928)

2.7 BOWDEN, A.T. and JEFFERSON, J.L. The design and operation of the Parsons Experimental Gas Turbine. Proc. I. Mech. E., 160, pg. 454 (1949)

2.8 KELLER, C. Axialgeblase vom Standpunkt der Tragflugeltheorie. Leeman, Zurich (1934)

2.9 COLLAR, A.R. Some experiments with cascades of aerofoils. ARC R&M 1768 (1937)

2.10 SHIMOYAMA, Y. Experiments on rows of aerofoils for retarded flow. Mem. Fac. Eng. Kyushu Univ., 8(4) (1938), NACA TM 1190 (1947)

2.11 RATEAU, A. Mode du Calcul des Turbines a Vapeur. Assoc. Tech. Mar., Paris (1909)

2.12 GUY, H.L. Some research on steam turbine nozzle efficiency. J.I.Civ.E. 13, 91 (1939)

2.13 HOWELL, A.R. The present basis of axial flow compressor design. Part 1 - Cascade theory and performance. ARC R&M 2095 (1942)

2.14 CONSTANT, H. Performance of cascades of aerofoils. ARC R&M 1768 (1937)

2.15 CARTER, A.D.S. The low speed performance of related aerofoils in cascade. ARC CP 29 (1949)

2.16 BOGDONOFF, S.M. NACA cascade data for the blade design of high performance axial flow compressors. J. Aero. Sci. (Feb. 1948)

2.17 ERWIN, J.R. and EMERY, J.C. Effect of tunnel configuration and testing technique on cascade performance. NACA Rep. No. 1016 (1951)

2.18 HERRIG, L.J., EMERY, J.A. and ERWIN, J.R. Systematic two dimensional tests of NACA 65 series compressor blades at low speeds. NACA TN 3916 (1957)

2.19 MELLOR, G.L. The 65 - series cascade data. MIT, GTL (unpublished) (1956)

2.20 LIEBLEIN, S. Loss and stall analysis of compressor cascades. Trans. ASME, J. Basic Eng. 81, pg. 387 (1959)

2.21 FELIX, A.R. and EMERY, J.C. A comparison of typical NGTE and NACA axial flow compressor blade sections in cascades at low speeds. NACA TN 3937 (1957)

2.22 POLLARD, D. Low speed performance of two dimensional cascades of aerofoils. Ph.D thesis, Liverpool Univ. (1964)

2.23 RHODEN, H.G. Effects of Reynolds' Number on the flow of air through a cascade of compressor blades. ARC R&M 2919 (1956)

2.24 BLIGHT, F.G. and HOWARD, W. Tests on four aerofoil cascades. Part 1 - Deflection, drag and velocity distribution. Part 2 - Boundary layer characteristics Reps. E.74 and E.75 ARL, Australia (1952)

2.25 SCHLICHTING, H. Boundary Layer Theory. 4th ed. McGraw Hill, N.Y. (1960)

2.26 TOWNSEND, A.A. The Structure of Turbulent Shear Flow. Cambridge Univ. Press (1956)

2.27 SCHLICHTING, H. and DAS, A. On the influence of turbulence level on the aerodynamic losses of axial turbomachines. in Dzung, L.S. (ed) "Flow Research on Blading". Elsevier (1970). See also discussion by W.H. Gibson (pg. 270)

2.28 EVANS, B.J. Effects of free-stream turbulence on blade performance in a compressor cascade. Ph. D. thesis, Cambridge Univ. (1971)

2.29 BAINES, W.D. and PETERSON, E.G. An investigation of flow through screens. Trans. ASME, 73, pg. 467 (1951)

2.30 FRENKIEL, F.N. The decay of isotropic turbulence. Trans. ASME, 70, pg. 311 (1948)

2.31 VICKERY, B.J. On the flow behind a coarse grid and its use as a model of atmospheric turbulence in studies related to wind loads on buildings. NPL Rep. 1143 (1965)

2.32 EVANS, R.L. Turbulent boundary layers on axial flow compressor blades. Ph.D thesis, Cambridge Univ. (1973)

2.33 WOOD, N.B. Flow unsteadiness and turbulence measurements in the low-pressure cylinder of a 500MW steam turbine. I. Mech. E. Conf. Publ. 3, pg. 115 (1973)

2.34 KOSCHEL, W. and KUHL, W. Turbine flow phenomena and their effect on the rotor blade heat transfer. Proc. 6th Conf. Fluid Mech., Budapest (1979)

2.35 MARTLEW, D.L. Measurements of the performance of a cascade of compressor blades under various degrees of intensity of turbulence in the incident air stream. ARC 15,249 (1952)

2.36 PEACOCK, R.E. Flow control in the corners of cascades. ARC 27,291 (1965)

2.37 MONTGOMERY, S.R. The design of a low speed cascade tunnel. MIT GTL Rep. 43 (1958)

2.38 RHODEN, H.G. The development of a wind tunnel. Engineering, 171 pg. 677 (1951)

2.39 GIBBINGS, J.C. A note on the combination of a contraction with a supersonic nozzle for a wind tunnel. RAE TN 65,066 (1965)

2.40 MILLS, R.D. Some finite two dimensional contractions. Aero. Quart. 19 (1968)

2.41 MIKHAIL, M.N. Optimum design of wind tunnel contractions. AIAA J. 17,5, pg. 471 (1979)

2.42 GREGORY-SMITH, D.G. and MARSH, H. The manufacture of fibreglass rotor blades with pressure tappings. Aero J. 75;723 pg. 213 (1971)

CHAPTER 3

Three-Dimensional Flows and Non-Rectilinear Cascades

No real flow is ever truly two-dimensional. In the cascade model we are therefore pursuing an unattainable ideal. This is why the application of cascade data should never be considered to be more than the exploitation of a model.

However this is not always a limitation or disadvantage. In this chapter we shall consider some of the effects (axial velocity variation, aspect ratio, secondary flows, end wall boundary layers, tip clearance) that occur in real machines. We shall also explore the capacity of the cascade model for generalisation to two-dimensional situations in other planes, such as occur in radial and mixed flow machines, and the incorporation of cascade data into fully three-dimensional schemes of design and analysis. Finally the experimental bridge between linear cascades and machines - the increasing use of annular and rotating cascades - will be discussed.

3.1 Pseudo Three-Dimensional and Axial-Velocity-Ratio Effects

As discussed in Chapter 2 axial velocity variations resulting from differences in cascade testing technique can have a substantial effect on downstream flow. Furthermore, the resulting errors will tend to be systematic ones between countries and design groups adopting the different methods.

In any event the meridional streamlines of turbomachines do not generally form cylindrical surfaces and it becomes necessary to allow for axial velocities which vary across a blade row in some prescribed or assumed manner. If this is not done correctly the tip section of an axial compressor rotor or a stator hub section might be designed to give excessive axial flow diffusion with a catastrophic effect on efficiency and stall margin.

Cascade data should be deployed in the design process in an iterative manner. A first approximate solution of the radial equilibrium equation will be obtained. This will give information on the axial velocity ratio across the cascade which should be used to modify the cascade deviation angle and loss coefficient.

When compressibility effects are significant the axial velocity density ratio

$$\text{AVDR} = \int_0^s (\rho_2 c_{x2})_{MS}\,dy \Bigg/ \int_0^s (\rho_1 c_{x1})_{MS}\,dy \qquad (3.1)$$

is used.

Cascade data on the effects of axial velocity variation cover only a limited operational range and theoretical treatments are not yet sufficiently reliable to cover the whole range. Accordingly this discussion is necessarily superficial. The effects on deviation angle are treated first, followed by the effects on loss coefficient and finally on pressure distribution.

3.1.1 Axial velocity ratio effects on deviation angle. Schulze et al [3.1] attempted to rationalise the effect of axial velocity variation on discharge angle by basing an equivalent 'two-dimensional' discharge angle on inlet axial velocity (c_{x1}) and mean axial velocity ($(c_{x1} + c_{x2})/2$) respectively.

For the two cases the equations

$$\tan\alpha_2 = \tan\alpha_{2.2D}/ \text{AVR} \tag{3.2}$$

and

$$\tan\alpha_2 = \frac{\tan\alpha_{2.2D}}{2(1+ \text{AVR})/\text{AVR}} \tag{3.3}$$

respectively, were obtained.

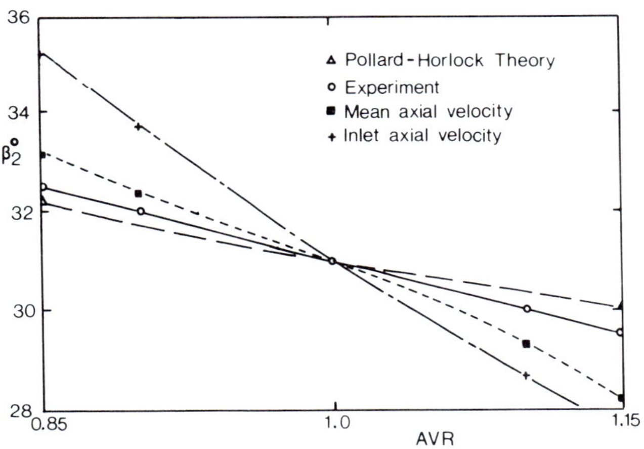

Fig. 3.1. Comparison of theoretical and experimental results for effect of axial velocity ratio on outlet angle.

In Fig. 3.1 the experimental results described in section 2.4 are compared with a theoretical method devised by Pollard and Horlock [3.2] and equations (3.2) and (3.3). It appears from this particular comparison that the Pollard and Horlock theory and the mean axial velocity approximation of equation (3.3) agree reasonably well with experimental data but not the inlet axial velocity approximation of equation (3.2).

Equation (3.3) is not dependent on inlet angle. It agrees reasonably well with test data for design conditions but does not represent the more scattered data at high incidence angles.

In an attempt to derive a correlation for the variation of deviation with AVDR over a wide range of incidence the available cascade data are plotted in Fig. 3.2.

The parameter chosen to represent deviation angle effects is

$$K = (\delta_{2D} - \delta)/(\text{AVDR} - 1) \tag{3.4}$$

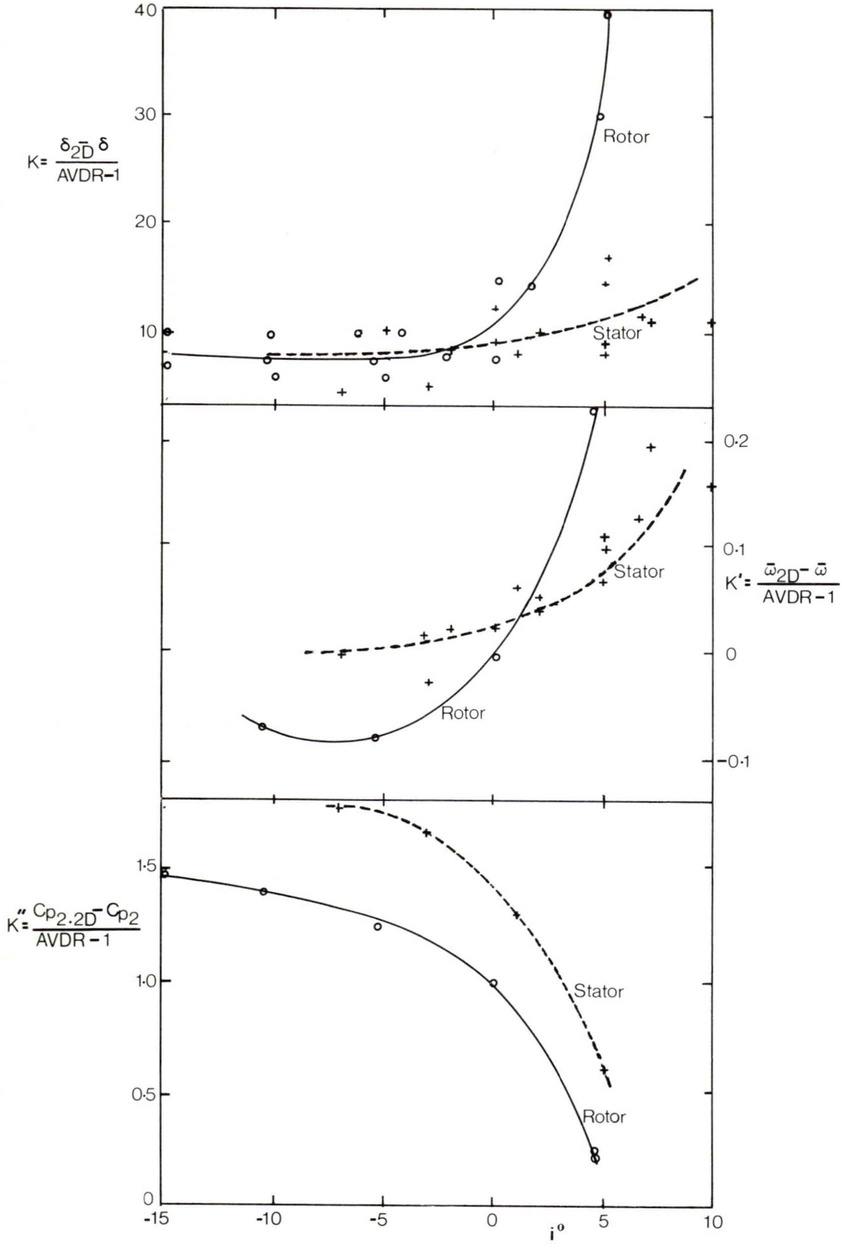

Fig. 3.2. Effects of axial density velocity ratio on deviation angle, loss coefficient and discharge pressure coefficient as a function of incidence angle.

This parameter reflects the observed linear dependence of deviation on AVDR[3.3, 3.4]. For incompressible flows AVDR reduces to AVR. Attempts to normalise incidence and deviation by nominal or optimal values did not reduce the scatter and were abandoned.

The results for deviation angle represent a wide range of testing from Liverpool University (refs. 3.4, 3.5) and German work (refs. 3.3, 3.6, 3.7).

The stagger angle was found to be a very significant variable[3.4, 3.7] and data are given for cascades having stagger greater than 30°, typical of compressor rotors and those having stagger of 20° and lower, typical of compressor stators. Median lines have been drawn for each. It is clear from the plot that axial velocity variations have a particularly marked effect on deviation angle for high incidence tests on high stagger or compressor rotor blading.

3.1.2 Axial velocity ratio effects on loss coefficient. The variation of total pressure loss coefficient may also be represented as a linear function of axial velocity ratio. For some data there is a tendency for the results to reach a minimum at an intermediate value of axial velocity ratio but the tendency is not sufficiently strong to preclude the drawing of a straight line through the data.

The parameter plotted for loss coefficient is

$$K' = (\bar{\omega}_{2D} - \bar{\omega})/(AVDR - 1) \quad . \qquad (3.5)$$

The data represent the results of five series of tests from refs. 3.3, 3.6 and 3.7 (stator cascades) and 3.5 (rotor cascade). The agreement between the four series of stator cascade data is quite good. That from the rotor cascade follows a somewhat different trend and it is not yet clear whether this discrepancy is a particular result of those rotor blade tests or represents a more systematic difference in trend of the type revealed in the deviation angle plots of Fig. 3.2. Because of the differentiable relationship between turning angle, loss coefficient and static pressure rise coefficient,

$$Cp = (1 - (\cos^2\alpha_1 / \cos^2\alpha_2)) - \bar{\omega} \quad , \qquad (3.6)$$

it is presumed that the results do show a consistently different trend for rotors and stators.

3.1.3 Axial velocity ratio effects on pressure distribution. As the axial velocity ratio is raised the contraction effects of thickening side-wall boundary layers are manifested as a gradual decrease in suction surface pressure. The pressure rise attained at the trailing edge by the diffusing suction surface flow governs the pressure levels on the pressure surface, which tend to fall uniformly with increase in axial velocity ratio. The dependence of the trailing edge pressure rise coefficient on axial velocity ratio may also be plotted as a function of incidence. The results available are ref. 3.3 (stator cascade) and refs. 3.4, 3.5 (rotor cascades). In Fig. 3.2 the parameter

$$K'' = (\Delta Cp_1 - \Delta Cp_2)/(AVDR - 1) \qquad (3.7)$$

is plotted and it can be seen that the results from the stator cascade are not in absolute agreement with those from the rotor cascade but do exhibit a similar trend.

3.2 Aspect-Ratio Effects

Because of the local effects of secondary flows and corner-stall regions the axial velocity ratio can be a misleading indicator of convergence or divergence effects on blade boundary layers. The axial velocity ratio is a parameter which averages these effects from blade to blade. However, as Fig. 3.3 from ref. 3.8 shows, although the pressure surface boundary layers may be virtually two-dimensional, the same blade may exhibit a degree of suction surface corner stall, which imparts extreme contraction and three-dimensionality to the suction surface boundary layer.

The flow at the cascade mid-span in this case is symmetrical and therefore the boundary layer is collateral. However the high level of mass addition due to convergence renders the boundary layer far from two-dimensional. Dunham [3.9] has performed boundary layer calculations using an entrainment method modified to incorporate a correction for stream convergence or divergence. Figure 3.4 compares Dunham's theory with Shaalan's measurements on a compressor cascade of aspect ratio 4.83. Whilst the measured axial velocity ratio was 1.43 it is necessary to assume a value of four to force agreement between theory and experiment. In the light of the suction surface patterns of Fig. 3.3 this is not too surprising. However it does lead one to exercise caution in replacing aspect ratio by axial velocity ratio as a parameter for cascade testing.

In many cases it is not convenient to employ side-wall bleed and in turbine cascade testing this is generally unnecessary. Furthermore, the question of aspect ratio effects on blading performance is of very real importance in its own right for direct use in machine design.

In theoretical and experimental investigations on the combined effect of axial velocity density ratio and aspect ratio Stark and Hoheisel [3.6] have shown that the aspect ratio has the effect of defining the distribution of axial velocity variation for a given wall shape and contraction ratio. In turn the axial velocity distribution decisively determines the pressure distributions, deviation angles and loss coefficients.

For axial flow compressors operating at or near design point, losses appear to increase proportionally as the reciprocal of the aspect ratio increases. Thus there is a considerable potential advantage in efficiency if aspect ratio can be maximised.

However stall margin is often at least as important as efficiency. An increase in aspect ratio (blade height divided by chord) has been experimentally observed to have a severe adverse effect on stall margin. The works of Fligg[3.10] and Fahmi [3.11] highlight this feature very clearly.

For turbine cascades the effect of aspect ratio is dependent upon the amount of turning imposed and upon whether substantial regions of adverse pressure gradient are present. This relates directly to the degree of reaction of the stage. Impulse rotor blading gives rise to stronger secondary flows than turbine nozzle guide vane blading and must therefore be tested at a higher aspect ratio to avoid strong contraction effects. Although the data are not conclusive, the use of aspect ratios as low as unity may be permissible for some nozzle blading. As a general rule, however, it would be prudent to use aspect ratios of four or higher.

3.3 Application of Cascades to Mixed and Radial Flows

For any given duty the specific speed concept [3.12] indicates whether an axial flow machine or a radial flow type is best suited for the job. The general case is the intermediate one of the mixed flow machine.

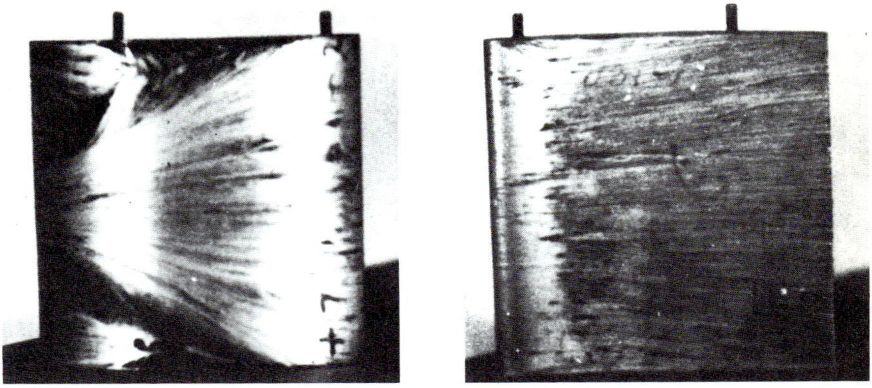

Suction surface Pressure surface

Fig. 3.3. Flow visualization on a compressor blade of unity aspect ratio.

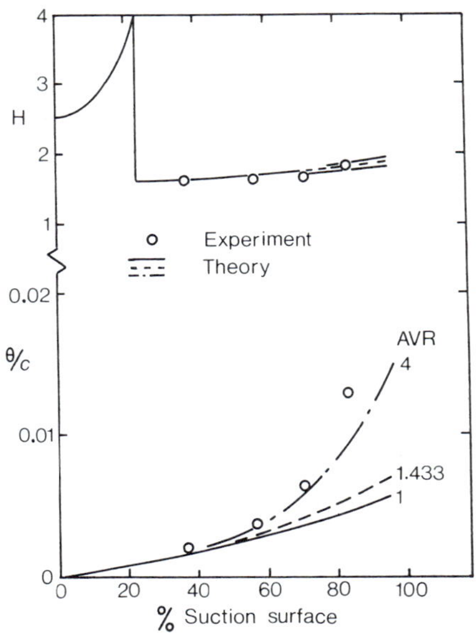

Fig. 3.4. Dunham's boundary layer calculations showing effect of axial velocity ratio on suction surface boundary layer.

In this general case flow variations in the axial, radial and circumferential directions must be considered. Before considering the more complicated mixed flow machine we shall first discuss the simplified case of a pure radial turbomachine - one in which the flow can be assumed to have no axial velocity components in the blading region.

3.3.1 Radial flow machines. The Ljungstrom turbine [1.1] is a pure radial flow machine and early Francis turbines were largely radial in flow. Today, however, pure radial flows tend to be limited to the stator vanes of radial flow machines and to some fan and pump impellers.

Cascade techniques may be applied to either impeller blades or diffuser vanes provided the axial velocity components are small. Flow data obtained from conventional axial cascades are applied by conformal transformation from the radial $z = re^{i\theta}$ plane to the axial $\zeta = \xi + i\eta$ plane. The transformation is simply

$$\zeta = \ln z,$$

giving

$$\xi = \ln r, \qquad \eta = \theta;$$

or upon differentiation

$$d\xi = dr/r, \qquad d\eta = d\theta .$$

The process is illustrated in Fig. 3.5.

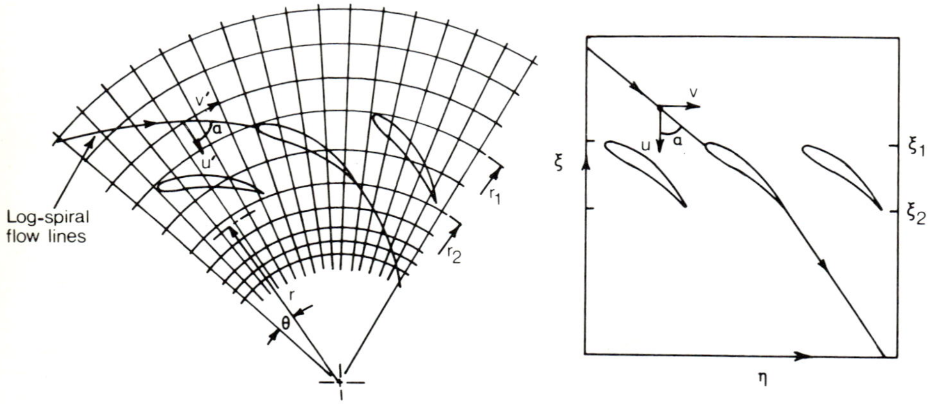

Fig. 3. 5. Conformal transformation of radial cascade into straight cascade.

It is worth noting that a straight streamline in the cascade plane transforms to a 'log spiral' streamline in the radial plane. The unique and favourable properties of log spiral lines in the radial plane were noted long ago and have been used by Busemann to give an accurate solution for slip factor [3.12].

Wiesner [3.13], in reviewing existing approached to slip factor estimation, has presented a generalization of the Busemann slip factor. Yedidiah [3.14] has subsequently questioned the validity of the classical relative eddy model and has presented an alternative approach which still requires experimental verification.

An interesting application of the Stanitz [3.15] subsonic compressible cascade design method to centrifugal diffuser vanes was described by Emerson and Horlock [3.16]. Although the design pressure rise was substantially achieved the blading was found to have high velocities in the leading edge region and was therefore excessively sensitive to incidence variation. Various authors have applied two-dimensional cascades, and the corresponding experimental data, to radial pump and compressor design. However, the conditions are often outside the usual range of cascade data and furthermore, the inlet conditions to such a diffuser are often severely non-uniform. As a result of radial pressure gradients the viscous effects might be expected to depart significantly from those of corresponding axial cascades.

Nevertheless Pampreen [3.17] has demonstrated that at subsonic speeds diffusers designed using cascade technology can be more efficient, have a larger stall-free range, and give a smaller compressor, than the corresponding conventional passage-type diffusers. Pampreen compared experimentally the performance of a compressor having a passage diffuser with that of the same machine but having the passage diffuser replaced by a tandem cascade of double circular arc and NACA 65 series aerofoils. The results showed superior performance for the cascade diffuser and also demonstrated that cascade data could be used directly and with confidence in the design of centrifugal compressor diffusers.

Although use of cascade technology for diffuser design appears promising at subsonic speeds, most success in high pressure ratio centrifugal compressors, and the corresponding high inlet Mach Number diffusers, appears to have been achieved with conventional vane-island, channel and pipe diffusers [3.18, 3.19]. Even in the design of these diffusers cascade technology may give valuable information.

The application of cascade data to centrifugal impeller blading is more difficult. The author knows of no convincing comparisons between cascade theory and impeller measurements for such a case. At first appearance the design of suitable impellers from cascade theory does not present a problem. However when the cascade results used are for unseparated flow, the agreement is invariably poor because in the majority of cases the flow within a centrifugal impeller will be separated. This was recognised in 1902 by Smith[3.20] and in 1923 by Carrard [3.21]. However recognition of the 'jet and separated zone' nature of the flow within an impeller passage has not deterred many sophisticated attempts to apply potential flow theory to mixed and radial wheel design.

This may, indeed, be justified for radial impellers having backward-leaning vanes typical of many pumps. Traverses have shown that such configurations may have no strong wakes or separated regions [3.22]. However such cases are rare and more usually a 'wake' of stagnant fluid occupies a considerable proportion of the passage. Dean [3.23] has argued that because of Coriolis forces this wake does not mix easily and does not therefore contribute significantly to the loss. The approach of many designers is therefore to design for high, but efficient, diffusion in the inducer and diffuser regions and even to deliberately provoke separation within the radial impeller passage in the knowledge that it would occur anyway. Obviously such an approach does not lend itself well to representation of impeller flows by the cascade model.

3.3.2 <u>Mixed flow machines.</u> The mixed flow machine is a more general case than either the radial or the axial machine. The qualifying remarks on separated flow within radial impellers might also be expected to apply to mixed flow pumps and compressors. However the radial diffusion will be less severe and hence numerical methods might be applied more successfully.

Most radial inflow turbines, whether large Francis units for hydraulic applications or small turbocharger turbines, are also mixed flow machines. Good numerical techniques should certainly be applicable to the design of such units.

Pollard [3.24] has presented a useful development of the Schlichting[3.25] cascade theory for application to mixed flow machines. Using a conformal transformation the meridional stream surface was mapped onto the cascade plane. The relative vorticity of the flow with respect to the rotor was also included in the transformation. The method was used to predict slip factors in addition to the flow field and gave good agreement with exact solutions. In a similar development Lewis et al[3.26] have extended the Martensen[3.27] theory for application to mixed flow machines.

Although the above extensions of cascade technology will be applicable to many configurations they will not be adequate for the typical centrifugal compressor impeller in which the flow is induced in an axial plane and discharged in a radial plane from blades which have little or no backward lean. This 'typical centrifugal compressor' impeller is precisely the type of machine which exhibits extensive separation and strong secondary flows [3.28].

Sophisticated matrix-through-flow[3.29] and finite element[3.30,3.31] techniques have been developed for such cases but any agreement between existing theories and experiment for centrifugal compressor impeller flows should only be regarded as fortuitous at present.

Comparisons between inviscid computations and flow measurements[3.32,3.33] have demonstrated that such predictions can be used with confidence only when the flow is not separated. There are, however, two classes of theory which appear promising.

In the approach of Sturge and Cumpsty[3.34] the flow was treated essentially as a potential flow having a separated region on one boundary. This is essentially a two-dimensional incompressible inviscid solution in which the flow is allowed to separate from the suction surface when the velocity has fallen to a specified proportion of the maximum value. This method has been applied by Hajek and Gopalakrishnan[3.35] to test data from Mizuki et al[3.36] and found to give satisfactory agreement between theory and experiment, subject to the arbitrary nature of the separation criterion.

The other new theoretical approach is that of three-dimensional viscous flow solution by a partially parabolic solution. Moore and Moore[3.37] have developed the method and have demonstrated that it has considerable potential for impeller flows in which leading edge effects do not dominate. Using a similar approach Malin et al[3.38] have also analysed the test data of ref.3.36 and have obtained reasonable agreement. Other three-dimensional viscous flow theories for impellers have been reported by Dodge[3.39], Fraser et al[3.40] and Swanson[3.41].

3.4 Secondary Flows and Losses

When applied to flows in blade cascades the term 'secondary flow' is restricted to considerations of the differential deflection imparted by the cascade to a sheared free stream. Such a free stream has a spanwise total pressure gradient which may or may not result from an incoming wall boundary layer. Use of the term to refer to corner shear stress asymmetries in a turbulent boundary layer[3.42] is excluded.

3.4.1 Qualitative explanation. Secondary flows of interest usually result from entry to the cascade of a flow having small shear; both small and large deflections may be analysed. Consideration of the case in which the inlet shear is large is a somewhat specialised, highly non-linear problem for which few solutions exist.

Obviously small-shear secondary flows are encountered in various situations, often far removed from cascade flows; such cases are the meandering of rivers, the flow in pipes, and the end wall flow around cylindrical bodies. In general, however, secondary flows will be most marked in situations where the end wall boundary layers are particularly thick, such as in axial compressors. In certain cases these boundary layers may be sufficiently thick to render the "large shear" analyses most appropriate. However this situation would be unusual and for most cases the classical small shear potential flow analyses of Squire and Winter[3.43] and Hawthorne[3.44] are adequate.

An understanding of the nature of secondary flow in a cascade may be derived from consideration of a streamline which enters the cascade from an "inviscid" side wall boundary layer. Assuming that this boundary layer is thin and that no separation is present the static pressure distributions outside the end wall region will tend to prevail right up to the wall. What may have been an acceptable degree of loading in the midspan region of a blade section subjected to the full free stream velocity, will have to be borne also by the side wall regions under a much reduced incoming velocity. The static pressure gradient normal to the stream line is given by

$$dp/dn \simeq \rho U^2/R .$$

Since this pressure gradient is the same in the side wall region as in the mid-span, whereas the velocity U is lower in the side wall region, the radius of curvature of the streamline R in the side wall region must also be lower, by a factor of $(U/U_{MS})^2$, to preserve the balance. The result, as may be seen in Fig. 3.6, is that whereas the

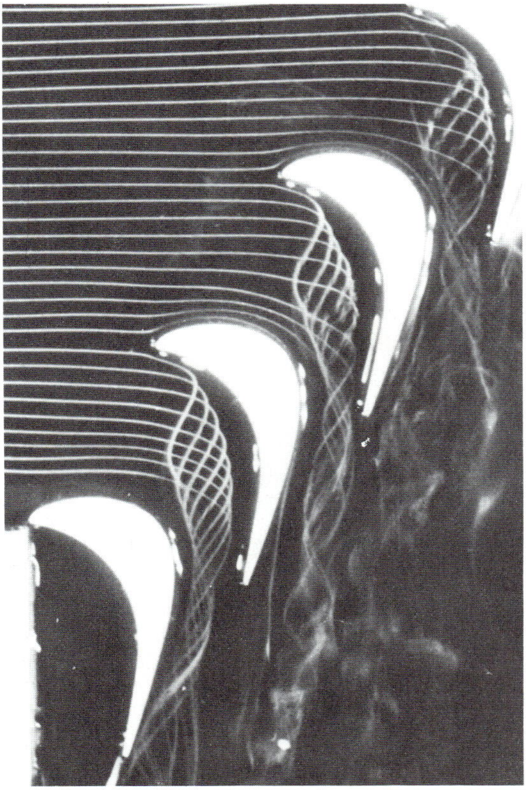

Fig. 3.6. Visualization of secondary flows near the end wall in a turbine blade cascade. (Courtesy of J. Fabri)

streamline might have followed the blade surface curvature in the mid-span region, the path it takes in the side-wall region has a much higher curvature. There results a velocity component perpendicular to that of the free stream velocities directed across the channel from the high pressure side. Since there can be no flow through the blade a circulatory motion of the type indicated in Fig. 3.6 is set up. This is the classical secondary flow situation and the "passage vortex" which is set up is often quite strong and causes severe performance deterioration, especially in multistage machines.

'Overturning' results near the side wall whereas some distance from the wall the flow is 'underturned'. It is this overturning which may result in the stalling of an axial flow compressor being initiated in the corner region.

The above is a simplified description of a complex phenomenon. Langston et al [3.45] have shown that in a turbine cascade the approaching wall boundary layer tends to roll up into a horseshoe vortex around the leading edge. Whereas one branch moves along the suction surface of the blade under consideration the other branch crosses the passage toward the suction surface of the adjacent blade as shown in Fig. 3.6. As it does so it rolls up most of the wall boundary layer which is discharged as a loss core adjacent to the suction surface - end wall corner. Majka[3.46] has observed similar behaviour in a compressor cascade.

Substantial passage vorticity results, however, from the imposition of cascade deflection upon an existing side wall boundary layer and is not dependent upon the postulation of any coupling mechanism between the vorticity and the wall boundary layer through the cascade. As a result some progress has been made in analysing secondary flow patterns using inviscid theory.

3.4.2 Prediction of passage vorticity and angle variation. In order to predict the change in outlet angle resulting from secondary flows it is firstly necessary to calculate the secondary flow vorticity in a plane (the Trefftz plane) downstream of the cascade exit and then, from this, the secondary flow stream function.

This is not intractable if the following simplifications are made:-

1. No body forces are imposed.
2. The flow is steady, inviscid and incompressible.
3. Only spanwise variations in inlet velocity are considered.
4. Trailing shed and filament vorticities are ignored.
5. Surfaces of constant total pressure remain plane and parallel to the end wall as they pass through the cascade.

On the basis of these assumptions Hawthorne[3.44] found that the change in streamwise vorticity between stations 1 and 2 could be expressed as

$$\left(\frac{\Omega_{s2}}{U}\right) - \left(\frac{\Omega_{s1}}{U}\right) = -2\int_1^2 \frac{|\text{grad } p_0/\rho|}{U^2} \sin\varphi^* d\varepsilon \tag{3.8}$$

where ε is the turning angle through the cascade and φ^* the angle between the normals to the streamline and the Bernoulli surface.

If the velocity, U , is constant along the streamline this expression reduces to

$$\Omega_{s2} - \Omega_{s1} = -2\varepsilon \cdot \partial U/\partial z \tag{3.9}$$

which was first derived in a different way and checked experimentally by Squire and Winter[3.43] . Hawthorne has derived a more complicated function which includes the effects of trailing shed and filament vorticity in addition to the passage vorticity. These three components may cancel if there is no turning of the flow, or if the flow is turned back to axial, or in the impulse blade situation. An improved theory giving a different expression for trailing shed vorticity has been given by Came and Marsh[3.47]. Glynn et al [3.48] have proposed an iterative scheme to allow for spanwise variations in discharge angle.

The exit air angles may be calculated by treating the expression for secondary vorticity in equation (3.8) as the right hand side of a Poisson equation for secondary stream function. Numerical solutions may be readily derived giving predictions of spanwise variation in outlet angle such as those of Fig. 3.7. The superposed experimental results reported in ref.[3.49] exhibit reasonable agreement with classical theory for turbine nozzle guide vanes. In fact Adkins and Smith[3.50] demonstrate that for compressors operating

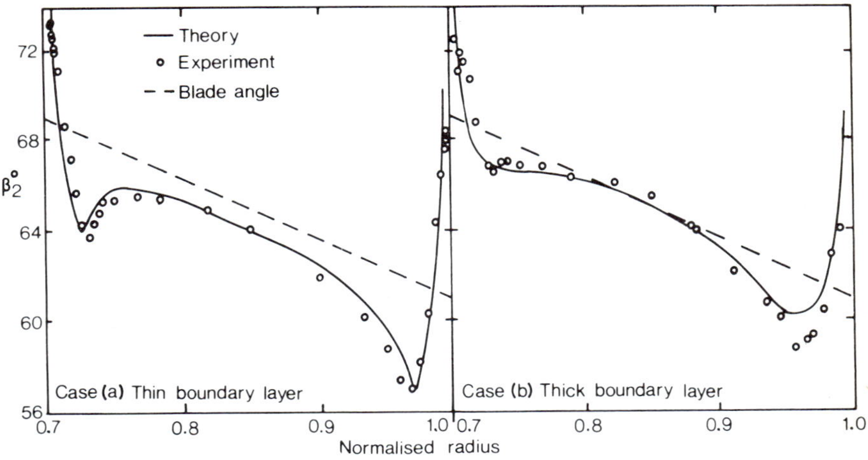

Fig. 3.7. Outlet angles from nozzle guide vane as measured by Hunter and predicted by Gregory-Smith [3.49].

near design point much of the difference between measured exit angles and two-dimensional cascade data can be attributed to secondary flows calculable by inviscid theory.

3.4.3 Secondary flow losses. Having obtained a description of the flow field resulting from secondary vorticity it would be logical to suppose that this information could be used directly in a subsequent loss calculation. Unfortunately such a procedure has proved illusory.

Kinetic energy is lost from the main stream into the secondary vorticity and, knowing the secondary flow field, this loss may be calculated. An alternative approach is the use of a triangular polar plot by Johnston [3.51] in which the form of equation (3.8) is fitted outside the boundary layer. However, the losses obtained by such inviscid approaches are many times less than observed secondary flow losses.

Louis [3.52] produced an elegant demonstration of this by generating a simulated wall boundary layer (actually half of a wake) at inlet to a cascade. When a dividing wall was placed downstream of this, corner stall was observed with a considerable increase in loss. It became clear that it was not the dissipation of induced secondary vorticities which was the main constituent of secondary losses but rather the stagnation pressure losses in a weak and often separated downstream wall boundary layer.

Since many cases do include a corner stall or separation region it is very difficult to pursue the logical path of firstly calculating the flow field and then performing a viscous loss calculation. Research on three-dimensional boundary layers and separated flows may make this route more feasible in the future. However the need for design data has led most workers to concentrate on empirical predictions.

Those predictions have been lucidly reviewed by Horlock [1.3], Dunham [3.53] and Salvage [3.54]. The secondary loss predictions all take the form $\bar{\omega}_S = L.\bar{\omega}_L$ where $\bar{\omega}_L$ is a loss coefficient due to loading and L is a length scale multiplier.

Over twenty authors have produced such correlations. Controversy over the correct loading parameter and length scale multiplier and differences in accounting for other factors leads to widespread disagreement between the various correlations. Chauvin [3.55] has summarised the situation for some of the correlations by calculating losses for turbine nozzle vanes turning from axial for a range of practical outlet angles. The results of Fig. 3.8 show widespread disagreement between the various methods. Similar comparisons were made for an impulse turbine blade and the disparities were even greater.

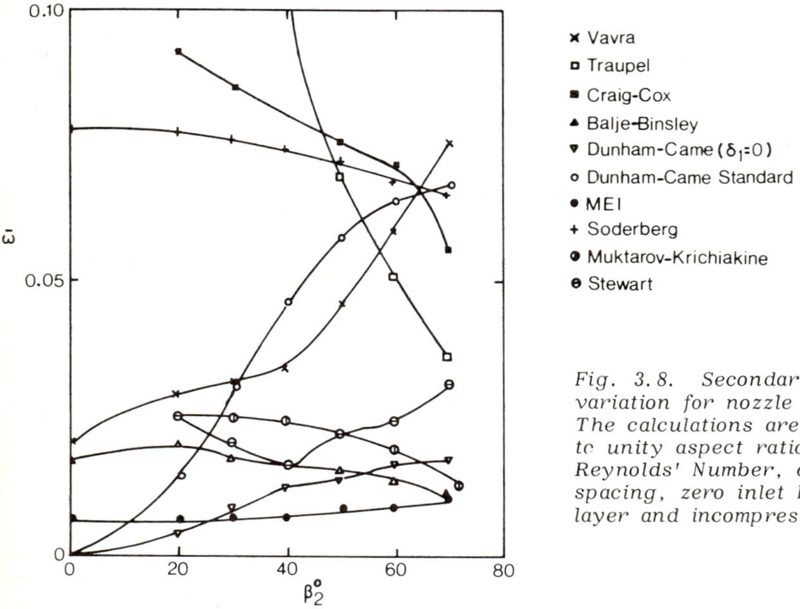

Fig. 3.8. Secondary loss variation for nozzle cascades. The calculations are referenced to unity aspect ratio, reference Reynolds' Number, optimum spacing, zero inlet boundary layer and incompressible flow.

Clearly the existing correlations are not very useful and further work needs to be performed before trustworthy predictions can be made. Three things are needed:-

(i) A clearer understanding of the physics of the situation and of the relevant non-dimensionalising parameters.

(ii) Better cascade tests, low speed at first, with careful control over inlet boundary layers and precise statement of test conditions.

(iii) Advances in the theory of boundary layers with cross-flow with the eventual objective of direct prediction of losses.

Griepentrog [3.56] has made a semi-empirical inviscid attempt to predict secondary losses taking account of the spacing between secondary vortices and this appears to give reasonable loss predictions for closely-spaced blades. However no method is sufficiently reliable for the author to recommend design use. Should the use of a rough correlation be essential that of Came [3.57] is probably safer than most:-

$$\bar{\omega}_S = \left(\frac{C_L}{s/c}\right)^2 \frac{\cos^3\alpha_2}{\cos\alpha_1' \cos^3\alpha_m}\left(0.008 + 0.64\frac{\delta_1'}{c}\right). \tag{3.10}$$

In this expression for the gross secondary-loss coefficient, which includes inlet and exit boundary-layer losses, and all other blade losses apart from the profile loss, the first term is the downstream wall boundary layer component and the second is the loss constituent of the vortex core.

3.5 End-Wall Boundary Layers

Information on end-wall boundary layers is needed for two main reasons. Firstly, further progress in research on secondary flows is dependent on a deeper understanding of the physics of end-wall boundary layers and the prediction of secondary flow losses is constrained by the available techniques for predicting cross-flows in skewed wall boundary layers. Any such progress is hard-won and can be expected to be applied firstly to flows in bends, then to cascades and eventually to rotating blade rows.

On a second level is the immediate need of the designer for information on annulus wall boundary layer behaviour in multi-stage machines. In a multi-stage axial compressor, for example, the stall characteristics, stage-matching and design flow will be governed largely by wall boundary layer growth. The matching of compressor stages is particularly difficult. Each stage must have its flow capacity matched precisely to those of other stages if loading and efficiency are not to be impaired. The flow capacity is directly dependent on the blockage effect of the annulus wall boundary layers, which must therefore be accurately predictable. In low aspect ratio turbines the wall boundary layers may provide a greater loss contribution than all other sources. Obviously the greatest contributions to wall boundary layer predictions are empirical, and will continue to be so for some time. In particular many recent measurements have indicated a strong coupling between annulus wall boundary layer growth and tip-clearance levels.

3.5.1 Research on end wall boundary layers in cascades. The use of boundary-layer theory is complementary to secondary flow theory, the end wall flow being treated as a viscous flow problem. A velocity component normal to the streamwise boundary layer may be incorporated, this indirectly introducing secondary flow concepts. Measurements such as those of Papailiou et al[3.58] have shown that any theory which neglects the existence of boundary layer crossflows within the passage can have only limited validity in turbomachinery situations where secondary flow effects are often very strong.

Two alternative types of boundary layer theory are available. Integral methods are relatively simple to use and give good physical understanding. For this reason the following discussion will be confined to integral methods. However it is questionable whether integral methods possess sufficient generality for prediction of many turbomachine flows[3.59]. For this reason some workers are now turning to the use of differential methods.

The circumferentially-averaged integral method for calculation of swirling boundary layers proceeds by selection of a suitable crossflow velocity profile, to complement the Coles profile usually applied to the streamwise component in the manner depicted in Fig. 3.9.

The simplest such profile is that due to Mager[3.60]

$$(v/u) = \tan \varepsilon_w (1 - (z/\delta'))^2 . \qquad (3.11)$$

The Squire and Winter expression of equation (3.9) reduces to

$$2\varepsilon(\partial U/\partial z) = (\partial v/\partial z) \qquad (3.12)$$

when the blades are widely spaced, a term in $(\partial w/\partial n)$ vanishing. If a boundary condition v= 0 at u = U is imposed then integration of equation (3.12) gives an alternative crossflow model

$$(v/u) = -2\varepsilon(1-(u/U)) . \qquad (3.13)$$

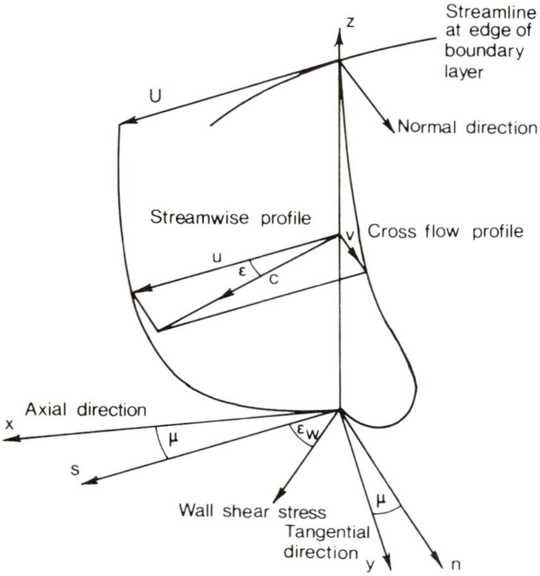

Fig. 3.9. Streamwise and crossflow velocity profiles.

This result was obtained by Johnston[3.51] for the outer portion of the boundary layer. For the inner portion Johnston suggested the collateral (constant angle) flow condition.

$$v/u = \tan \varepsilon_w \ .$$

Combining equations (3.13) and (3.14) the Johnston triangular plot of Fig. 3.10(a) describing the variation of cross flow through the boundary layer is obtained.

Horlock and Perkins[3.61] have extended the Johnston profile, to account for the presence of blade boundaries, by solving the secondary flow field equation using the Squire and Winter simplification and assuming the simple form for components v and w given by Mellor and Wood[3.62], w varying linearly and v parabolically across the staggered spacing. Pitchwise averaging gives

$$\bar{v} = -2\varepsilon(U-u) + \frac{4\sqrt{3}\overset{*}{\delta}\varepsilon U}{s\cos\mu}\exp\left(\frac{-2\sqrt{3}z}{s\cos\mu}\right) \quad (3.15)$$

As the spacing approaches infinity Johnston's result is obtained.

Solutions are needed for at least four variables, δ', Π, c_f and ε_w. This involves the solution of four simultaneous (differential) equations. The most promising combination appears to be:-

(i) The axial momentum equation

$$\frac{d}{dx}(\bar{c}_x^2\theta_{xx}) + \overset{*}{\delta}_x\bar{c}_x\frac{d\bar{c}_x}{dx} = Dx + \frac{\tau_{wx}}{\rho} \quad (3.16)$$

(ii) The tangential momentum equation

$$\frac{d}{dx}(\bar{c}_y^2\theta_{yx}) + \overset{*}{\delta}_x\bar{c}_x\frac{d\bar{c}_y}{dx} = Dy + \frac{\tau_{wy}}{\rho} \quad (3.17)$$

Dx and Dy are the axial and tangential force deficits defined in ref. 3.61.

(iii) The entrainment equation (Head[3.63])

$$\frac{\partial}{\partial s}(\delta' - \delta_s^*) + \frac{\partial}{\partial n}(\delta_n^*) = E - (\delta' - \delta_s^*)\cdot\left(\frac{1}{U}\frac{\partial U}{\partial s} - K_1\right) \qquad (3.18)$$

(iv) The shear stress law in differential form

$$\frac{1}{\delta'}\cdot\frac{d\delta'}{dx} + 2\frac{d\Pi}{dx} + \left\{\frac{1}{w} + \frac{0.4}{w2}\right\}\frac{dw}{dx} = -\frac{1}{U}\cdot\frac{dU}{dx}\,. \qquad (3.19)$$

With the Mager profile four equations of the form

$$a_i\frac{d\delta'}{dx} + b_i\frac{d\Pi}{dx} + c_i\frac{dc_f}{dx} + d_i\frac{d\varepsilon_w}{dx} = f_i$$

$$i = 1,2,3,4$$

are obtained. Solution of these simultaneous equations is numerical, using the Runge-Kutta method.

With the dependent variables determined as a function of x all other boundary layer parameters may be obtained.

If the Johnston or Horlock cross-flow profiles are used an additional variable is introduced, and ref. 3.61 provides the associated additional equation for apex height. The greatest difficulties in deployment of the momentum equations arise in the specification of force deficit terms Dx, Dy in the axial and pitchwise directions.

Horlock and Perkins[3.61] argued, from secondary flow theory, that neglect of the Dx term could be justified. The same analysis also gives approximate expressions for Dy which tends to zero for widely-spaced blades.

The principal departure of the Horlock theory from that of Mellor and Wood[3.62] is precisely in the treatment of the blade force terms. Both theories agree on the need to account for such terms but the components of Horlock's Dx cancel each other. Mellor and Wood neglect Horlock's third component or streamwise force deficit but add components to represent skin friction. Blade forces are treated by postulating normality of forces to the blade surface together with the assumption that the discharge flow is collateral.

The experimental evidence of Lewis and Yeung[3.64] indicates that as tip clearance is increased the ratio of lift on the end wall to lift at the blade tip falls off as $e^{-14\,t_c/c}$. The author's results[3.65] indicate that for a fairly tight clearance of 1.67% chord the pressure distribution is substantially retained between tip and end wall, regardless of incidence. These results were obtained by a combination of on-rotor measurement and transducer measurements at the compressor casing.

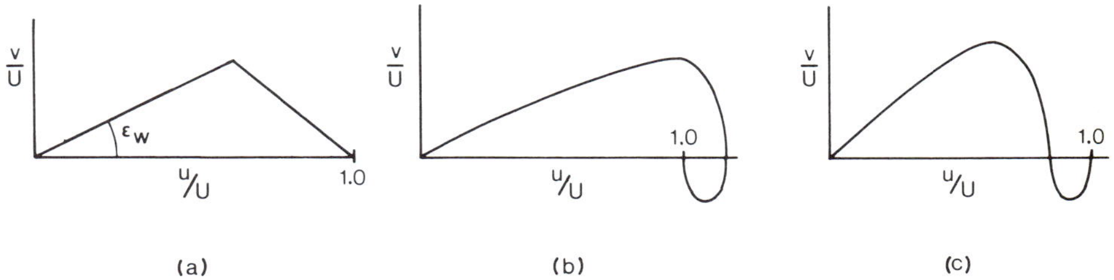

Fig. 3.10. The variation of crossflow is illustrated by a Johnston plot in (a) and by crossover plots in (b) and (c).

It therefore appears that if the tip clearance is reasonably tight (less than 2% chord) the blade forces are substantially retained to the casing. As tip clearance is increased the end wall lift falls off in relation to the tip lift as $\exp(-14\,t_c/c)$.

Although the treatment of force deficits is one area of controversy the question of a suitable cross-flow profile is also unresolved. Horlock's profile would seem to be adequate for some closely-spaced cascades but Langston's measurements [3.59] in a turbine cascade passage have clearly shown that as a strong passage vortex is traversed a corresponding crossover behaviour is present in the wall boundary layer. This is of the type indicated in Figs. 3.10(b) and 3.10(c) and is represented by

$$v/u = \tan\left[(\varepsilon_W - az)\exp(-\gamma^* z)\right] \quad . \qquad (3.20)$$

If the exit flow from a blade row is collateral then consideration of the velocity triangles for wall boundary layer flows entering a downstream blade row having relative motion indicates a marked departure from collateral. The restoration of collaterality involves an additional force deficit term. This is a feature of skewed boundary layers, which have been investigated in compressor cascades by Soderberg[3.66] and in turbine cascades by Carrick[3.67].

Railly and Sharma [3.68] have subsequently produced a theoretical treatment in which "apparent stresses" are introduced, permitting departure from the collateral flow condition. A generalised theory by Hirsch has given accurate velocity profile predictions for single stage compressors[3.69].

3.5.2 Multi-stage machines and tip clearance. Methods such as those of Horlock and Mellor [3.62] which include a representation of crossflow profiles should be capable of representing the end-wall boundary layers of cascades and single-stage machines. However the physics of annulus wall flows in multi-stage machines, with changing reference frames, tip clearance, vorticity transport and other phenomena, is exceedingly complex. One cannot rely on estimates of force deficits in such a situation and the prospect of applying either the full integral equation - cross-flow approach or the use of differential methods to turbomachines having many interacting rows of blades is a daunting one. Progress in stage-matching and flow-setting to date has been mainly by the use of empirical blockage and work-done factors. Several notable successes have been scored in the application of simplified theoretical approaches.

The existence of a non-zero defect force is a necessary condition for the establishment of equilibrium or repeating stages in a multi-stage machine. In 1962 Railly and Howard [3.70] used both momentum equations, lacking a realistic model of blade force variation, but omitted all force deficit terms. The resulting agreement with experiment was not good. Stratford's [3.71] response was to simplify the problem by omitting not only the force deficit terms but also any consideration of cross-flow, therefore dispensing with the momentum equation in the normal direction. This approach was also adopted by Jansen [3.72]. The results certainly give a first order indication of the rate of growth of axial momentum thickness along the walls. Thus, taking account of pressure gradients and some blade forces, an assessment of annulus blockage may be made. In Stratford's approach the body force term is identical with the blade force term, static pressure being assumed constant radially. However, Gregory-Smith and Marsh [3.73] have usefully distinguished between the two forces. If the blade force is assumed constant then secondary flow effects will cause a change in magnitude in the body force term.

Horlock and Perkins[3.61] and De Ruyck and Hirsch [3.69] have analysed the boundary layer displacement thickness in gas turbine compressors. The incorporation of such theories into a throughflow calculation procedure resulted in good predictions of performance for compressors having up to twelve stages. The predominant effect of tip clearance on displacement thickness seems to have been perceived independently by Mellor and Strong[3.74], Smith[3.75] and Tayler[3.76].

Smith's results for casing boundary layers in a four-stage compressor, represented by the curves of Fig. 3.11, demonstrate the dependence of displacement thickness on loading and tip clearance. The rate of increase of displacement thickness is considered to be infinite as stall is approached. The trend with tip clearance, obtained from a wide variety of differing tests, is

$$\delta^* - \delta^*_{t_c=0} = (\Psi'/\Psi'_{max})\, t_c \;. \tag{3.21}$$

More recently investigators have suggested an even stronger dependence of displacement thickness with, instead of a maximum coefficient of unity on the right hand side of equation (3.21) a multiplier as high as 1.5.

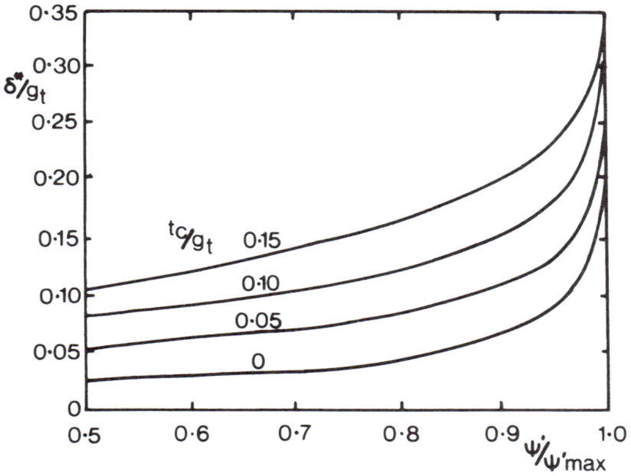

Fig. 3.11. Trends of casing boundary layer displacement thickness variation over the characteristic as reported by Smith.[3.75]

On the basis of single and multi-stage compressor measurements Tayler advocated a multiplier of unity, independent of position on the characteristic. Tayler found that to reduce the clearance below 1% of blade height did not further reduce the wall boundary layer thickness. His main conclusion though, was, "at higher tip clearances, the difference between displacement thickness and tip clearance is approximately equal to the displacement thickness measured at the minimum clearance". This is identical to the relationship of equation (3.21) at the peak loading condition.

There exists a mild controversy as to whether the efficiency of an axial flow compressor reaches an optimum value as tip clearance is reduced, and then deteriorates with further reduction, or whether the improvement continues all the way to the wall.

Academic investigators have argued, and their measurements in stationary[3.77] and rotating[3.78] cascades have shown, that the opposition of secondary vorticity and tip leakage can give rise to an optimum value of tip clearance.

Possibly because multistage compressor tip clearances are usually larger, of the order of 2% of blade height, industry has detected little evidence of this optimum and has adopted the premise that "the smaller the clearance, the better the performance".

Results from a number of independent companies confirm those of tests on 8-stage compressor 'Alice'[3.79] which gave a 4% increase in efficiency as the clearance was decreased from 3% to 1% of blade height. Increases in surge margin are even more dramatic, giving around 20% improvement for the same decrease in tip clearance.

The simplest and most reliable correlation of turbine cascade tip clearance data appears to be that given be Ainley[3.80]. Ainley's expression for tip clearance loss is

$$\bar{Y}_C = C_{DC}(c/s)(\cos^2\alpha_2 / \cos^3\alpha_m) \qquad (3.22)$$

where

$$C_{DC} = (C_L^2/2)(t_c/s)(c/h)$$

and $\bar{Y}_C$ is defined as stagnation pressure loss divided by outlet dynamic head.

In reviewing turbine tip clearance losses Horlock [1.3] warns that loss coefficients for clearance flows and secondary flows "cannot really be expressed separately".

Recent academic work [3.81,3.82] generated enhanced understanding of flow mechanisms in the clearance region of axial compressor rotors. It appears that the tip leakage flow originates near the quarter chord location and peaks around mid-chord. The low energy fluid tends to migrate from the suction surface and accumulate between mid-passage and the pressure surface, coming closest to the pressure surface as stall is approached. In detailed investigations on compressor cascades Majka[3.46] observed the existence of rope-like separation bubbles having the same rotational sense as the tip leakage vortex and occupying 20% of the tip clearance width over the latter half of the blade chord.

3.6 Incorporation into Axisymmetric Scheme and Design Process

The objective of investigations in the cascade plane is usually their incorporation into the design of axial flow machines such as turbines or compressors. The design of such machines involves consideration of a complex three-dimensional flow field involving alternately moving and stationary reference frames.

No complete solution yet exists for the viscous, unsteady, three-dimensional and usually compressible flow problem thus posed. The problem was classically described by Wu [1.9] as depicted in Fig. 1.3. The three-dimensional flow field is represented by two quasi two-dimensional stream surfaces. One, labelled S1, is the cascade or blade-to-blade plane. The second, labelled S2, is the meridional plane.

The equations of motion are satisfied on two intersecting families of stream surfaces. Ideally the complete three-dimensional flow field would be obtained by an iteration process between solutions for the flow on the two sets of surfaces. Mathematically, Wu's analysis is straightforward but the difficulty lies in obtaining numerical solutions for the stream function. The programming exercise is a substantial one calling for equally substantial computer resources. For workers less well endowed some alternative, if less complete, options are available.

3.6.1 Simple radial equilibrium. Many early designs were achieved using simple radial equilibrium theory. Strictly only applicable to an axial flow machine having cylindrical hub and casing, the theory assumes that any radial movement of streamline takes place within the blade rows and that fluid leaving the trailing edge of the blade row has no radial velocity component.

With these assumptions the radial momentum equation is simply

$$\frac{1}{\rho}\frac{dp}{dr} = \frac{c_\theta^2}{r} \quad . \qquad (3.23)$$

If the stagnation enthalpy and entropy are introduced and maintained constant the simple radial equilibrium equation is

$$\frac{d}{dr}(c_x^2) + \frac{1}{r^2}\frac{d}{dr}(rc_\theta)^2 = 0 \ . \qquad (3.24)$$

The simple radial equilibrium equation may be used judiciously for multi-stage machines and may be used as a first iteration in the more complete solutions to be described.

3.6.2 Free-vortex design. If Newton's second law of motion is expressed in moment form with zero applied torque, the tangential velocity component will vary inversely with radius. Under these circumstances, simple radial equilibrium applies and equation (3.24) reduces to

$$dc_x/dr = 0 \ . \qquad (3.25)$$

This equation represents a free-vortex flow distribution. The designers of early axial compressors used this distribution, which resulted in highly twisted blades and high tip Mach Numbers. Axial pump designers still commonly use free-vortex designs but the high tip relative velocities often result in premature cavitation. Designers of axial turbines use the free vortex approach extensively. Refs. 3.83 and 3.84 are two reports describing the design procedures.

3.6.3 Actuator disc approach. If each blade row is represented by a planar discontinuity across which there is a change in tangential velocity the well-known actuator disc model results. For an isolated blade row simple radial equilibrium exists far upstream and downstream but the radial movement of the streamlines is not confined to the axial extent of the blade passages. For this reason actuator disc analysis is a significant improvement on simple radial equilibrium theory.

The main value of actuator disc analysis is in generating physical insight into problems such as those involving the interaction of blade rows with other blade rows, with regions of distorted flow, or with rotating stall cells.

The design of modern turbomachines, however, requires a computer-based conceptual and mathematical framework which is more complete than that given by actuator disc analysis. The radial equilibrium equation must be solved leading to a determination of the flow field in the S2 surface. The two principal approaches to the problem are now outlined.

3.6.4 Streamline curvature. This method, based on the work of Smith[1.6], involves the iterative numerical solution of the equations of motion continuity, energy and state for an axially symmetric flow through a turbomachine having a varying hub and tip radii. The equation for meridional curvature contains terms involving both the slope and curvature of the meridional streamlines and these quantities are estimated by using a smooth curve, such as a spline fit, passing through points of equal stream function on neighbouring calculating planes or stations. The spline fit establishes piece-wise cubics with first and second derivatives continuous across the location points and is a reasonable approximation to the shape of a streamline. The shape of the initial streamline is determined from an initially assumed mass flow distribution with radius. This distribution is re-assessed after each major iteration.

The flowpath is divided into stations chosen to correspond with axial locations where information is available or desired. A simplification of the method is to place the calculating planes only in the duct regions, there being no calculation of the flow within the blade rows. This duct flow model treats the blade rows as cascades which produce a defined change inflow direction and loss of stagnation pressure. The blade row data could be modified between iterations to allow for the effects of varying axial velocity ratio.

Repositioning of streamlines between major iterations conventionally proceeds from the hub, with suitable interpolation and continuity check procedures for these minor iterations. The major iteration procedure continues until convergence is obtained within a specified tolerance.

Placing of calculation stations outside blade rows is inadequate when sonic conditions are reached. For transonic machines the axial variations of turning and blockage are crucial and as many as ten calculation stations may be used within a blade row.

3.6.5 Matrix through-flow. For fifteen years Wu's pioneering work in describing a solution to the flow equations was not implemented due to inadequacy of available computing power. However in 1966 Marsh[1.10] described a program using this approach. The contribution made by Marsh was in his selection of a very appropriate irregular finite difference grid.

In the matrix through-flow method the grid is formed and the non-linear stream function expression is expressed as a quasi-linear equation. The finite difference equation is then formed and solved for stream function at all mesh points. This new stream function distribution is then used to form a new set of equations and the process is repeated until a convergence criterion is satisfied.

This method is sufficiently general that calculating planes can be placed anywhere within the blade rows or such regions. It has been found that the accuracy of prediction is improved by having planes within the blade rows.

3.7 Annular Cascades

An annular cascade consists of a row of blades mounted between two co-axial surfaces of revolution. In such an arrangement an annular cascade rig would in no way differ from the stator row of an actual machine.

If the blade row is permitted to rotate about the axis it may then correspond to the rotor row of an actual machine in most essential features and is termed a rotating cascade.

3.7.1 Theoretical considerations. According to potential flow theory the essential difference between planar and annular cascades is that of the 'unrolling' effect considered by Ginsburg[3.85] and Rossow[1.8]. In the simplified model used by these authors the blades were represented by radial line vortices between two concentric cylinders. The radial flows in the blade regions were determined and the modification to linear cascade flows which would be needed to cancel these radial flows was determined.

Obviously the corrections thus derived are strongly dependent on the hub/tip ratio of the annular cascade and the blade space/chord ratio. For most practical cases the corrections are small and are neglected.

Radial variations in velocity vector and hence in blade profile imply that, for machines of low hub/tip ratio, consideration of flows on a geometrical mean radius plane may not be representative of the blade row as a whole. This is not a problem if consideration of annular cascades is incorporated into the type of axisymmetric flow procedure described in the previous section.

In the general case the blade span will not be perpendicular to the hub and casing walls. The resulting sweep and dihedral effects can be quite significant. A simplified theory for these has been given by Smith and Yeh[1.7].

Progress is being made in calculating the inviscid flows in fully three-dimensional flow passages. This has even been extended to transonic annular cascades of turbine nozzle blading. Using a time marching method Denton[3.86] has economically obtained complete three-dimensional solutions.

3.7.2 Annular cascade rigs. Whether of the accelerating or decelerating type an annular cascade will generally produce a change in the swirl of the fluid. If the blading is designed for a non-axial inlet flow it will be necessary to induce an incoming swirl generally by means of a cascade of guide vanes upstream of the test cascade. If the test cascade is one of turbine nozzle vanes the production of swirl implies a change in the radial distribution of downstream velocity which should be corrected by a suitable means. Work of Deych and Troyanovskiy[3.87] leads to the conclusion that it is essential to correct the downstream profile by use of either a second set of stationary vanes or a rotor.

Another feature which is often not replicated in annular cascade testing is the level and nature of free-stream turbulence entering a blade row. Deverson et al [3.88] found that although this had little effect for conditions away from the stall region, above a certain incidence the effects on stalling were quite dramatic.

The tunnel used by Deverson et al[3.88] is an excellent example of a versatile low speed rig which might be used to investigate the performance of either stationary or rotating annular cascades. Using this tunnel, which has a downstream slave rotor on a separate driveshaft, it was possible to establish that under the test conditions and for unstalled operation there was no perceptible difference between the performance of uncambered blading in a stationary annular cascade and one synchronised with a solid body rotational flow. Care should be taken in generalising the conclusion reached by Deverson et al [3.88] since trailing edge separation was not a feature of the flow. It is well established that strong radial migration of low energy fluid under the influence of centrifugal forces occurs in the trailing edge region of most rotor blading.

In an interesting experimental investigation of supersonic compressor blading in linear, annular and rotating cascades, Paulon et al [3.89] have found some difficulty in obtaining identical performance. The linear cascade flow suffered from separations, probably as a result of the lack of boundary layer control in the working section. The annular cascade gave good results which agreed surprisingly well with theoretical pressure distributions. Measurements from a rotating cascade showed discrepancies, possibly due to the large diameter of the pressure transducers used and also the wall boundary layer.

In summary, therefore, although the annular cascade rig is more expensive to construct and run and somewhat less versatile, it has a valuable niche in providing results of arguably greater consistency and validity than the linear cascade. At the same time it is less expensive than a rotating cascade and may give better results. In essence, in a rotating cascade almost all of the difficulties of the actual machine are present.

There is, therefore, a wide range of choice, from the basic linear cascade to the machine itself. Where cascade testing ends machine testing begins. Each device has an important role in the understanding of turbomachinery flow.

REFERENCES

3.1 SCHULZE, W.M., ERWIN, J.R. and ASHBY, G.C. NACA 65-series compressor rotor performance with varying annulus-area ratio, solidity, blade angle and Reynolds' Number and comparison with cascade results. NACA TN 4130 (1957)

3.2 POLLARD, D. and HORLOCK, J.H. A theoretical investigation of the effect of change in axial velocity on the potential flow through a cascade of aerofoils. ARC CP 619 (1963)

3.3 STARKEN, H., BREUGELMANS, F.A.E. and SCHIMMING, P. Investigation of the axial velocity density ratio in a high turning cascade. ASME Paper 75-GT-25 (1975)

3.4 OHUONU, E.H. and SHAW, R. Effect of axial velocity variation on deviation for compressor cascades. I.Mech.E. Conf. Publ. 3, pg. 110 (1973)

3.5 MASEK, Z. and NORBURY, J.F. Low-speed performance of a compressor cascade designed for prescribed velocity distribution and tested with variable axial velocity ratio. I.Mech.E. Conf. Publ. 3, pg. 224 (1973)

3.6 STARK, U. and HOHEISEL, H. The combined effect of axial velocity density ratio and aspect ratio on compressor cascade performance. Trans. ASME, J. Eng. Power 103, 1 pg. 247 (1981)

3.7 HEILMANN, W. Experimental investigations and boundary-layer theoretical calculations on decelerating cascades at compressible flow, especially when changing the ratio of axial mass-flow density and the upstream intensity of turbulence. (In German). DLR Rep. 67-88 (1967)

3.8 ILYAS, M. and NORBURY, J.F. Effect of axial velocity variation on the subsonic flow through a compressor cascade. I.Mech.E. Conf. Publ. 3, pg.276 (1973)

3.9 DUNHAM, J. The effect of axial-velocity-ratio on turbomachine blade boundary layers. ARC 33,229 (1970)

3.10 FLIGG, J.A. Tests of a low-speed three-stage axial flow compressor at aspect ratios of one, two and four. AIAA Paper 66-613 (1966)

3.11 FAHMI, G.J. The effect of blade aspect ratio on the performance of axial flow compressors. Ph.D thesis, Liverpool Univ. (1967)

3.12 WISLICENUS, G.F. Fluid Mechanics of Turbomachinery. McGraw-Hill, N.Y.(1947)

3.13 WIESNER, F.J. A review of slip factors for centrifugal impellers. Trans. ASME, J.Eng. Power, 89 558 (1967)

3.14 YEDIDIAH, SH. New look at the slip in a centrifugal impeller. Proc.I.Mech.E. 188 47 (1974)

3.15 STANITZ, J.D. Design of two-dimensional channels with prescribed velocity distribution along the channel walls. I-Relaxation Solutions. NACA TN 2593 II-Green's Function. NACA TN 2595 (1952)

3.16 EMERSON, D.H. and HORLOCK, J.H. The design of subsonic diffusers for centrifugal compressors using potential flow theory. ASME Paper 66-WA/GT-9 (1966)

3.17 PAMPREEN, R.C. The use of cascade technology in centrifugal compressors vaned diffuser design. ASME Paper 72-GT-39 (1972)

3.18 KENNY, D.P. A novel low cost diffuser for high-performance centrifugal compressors. Trans. ASME, J. Eng. Power (1969)

3.19 WELLIVER, A.D. and ACURIO, J. Element design and development of small centrifugal compressors. Vol. I, AD-384 923 and Vol. II AD-384 924, AVLABS 67-30 (1967)

3.20 SMITH, J.A. Notes on some experimental researches on internal flow in centrifugal pumps and allied machines. Engineering 74 pg. 763 (1902)

3.21 CARRARD, A. On calculations for centrifugal wheels. Tech. Mod. T.XV No.3 (1923) Trans. by J. Moore in CUED/A - Turbo/TR 73 (1975)

3.22 HOWARD, J.H.G. and KITTMER, C.W. Measured passage velocities in a radial impeller with shrouded and unshrouded configurations. Also discussion by Runstadler, P.W.Jr. Trans. ASME, J.Eng. Power pg. 207 (1975)

3.23 DEAN, R.C.Jr. The fluid dynamic design of advanced centrifugal compressors. Creare Inc. TN-185 (1974)

3.24 POLLARD, D. The extension of Schlichting's analysis to mixed flow cascades. I.Mech.E., Thermo. and Fluid Mech. Conv. (1966)

3.25 SCHLICHTING, H. Berechnung der reibungslosen inkompressiblen strömung für ein vorgegebenes ebenes schaufelgitter. VDI Forschungsheft 447 (1955)

3.26 LEWIS, R.I., FISHER, E.H. and SAVIOLAKIS, A. Analysis of mixed-flow rotor cascades. ARC R&M 3703 (1971)

3.27 MARTENSEN, E. Calculation of pressure distribution over profiles in cascade in two-dimensional potential flow, by means of a Fredholm integral equation. Arch. rat. mech. anal., vol. 3 no. 3 (1959)

3.28 ECKARDT, D. Detailed flow investigation within a high-speed centrifugal compressor impeller. Trans. ASME, J.Fluids Eng., 98, pg.390 (1976)

3.29 GOULAS, A. and BAKER, R.C. Flow in centrifugal compressor impellers: A hub-to-shroud solution. J. Mech. Eng. Sci., 22, 1 (1980)

3.30 HIRSCH, CH. and WARZEE, G. Quasi 3-D finite element computation of flows in centrifugal compressors. ASME Symp. Perf. Pred. Centrif. Pumps and Comp. pg. 69 (1980)

3.31 PRINCE, T.C. and BRYANS, A.C. Three dimensional inviscid computation of an impeller flow. ASME Paper 83-GT-210 (1983)

3.32 ADLER, D. and KRIMERMAN, Y. Comparison between the calculated subsonic inviscid three-dimensional flow in a centrifugal impeller and measurements. ASME Symp. Perf. Pred. Centrif. Pumps and Comp. pg. 19 (1980)

3.33 SCHUSTER, P. and SCHMIDT-EISENLOHR, U. Flow field analysis of radial and backswept centrifugal compressor impellers. Part 2: Comparison of potential flow calculations and measurements. ASME Symp. Perf. Centrif. Pumps and Comp. pg. 87 (1980)

3.34 STURGE, D.P. and CUMPSTY, N.A. A two dimensional method for calculating separated flow in a centrifugal impeller. Trans. ASME, J. Fluids Eng., 97 pg. 581 (1975)

3.35 HAJEK, T. and GOPALAKRISHNAN, S. Prediction of flow in centrifugal impellers. ASME Symp. Perf. Pred. Centrif. Pumps and Comp. pg. 27 (1980)

3.36 MIZUKI, S., ARIGA, I. and WATANABE, I. A study on the flow mechanism within centrifugal impeller channels. ASME Paper 75-GT-14 (1975)

3.37 MOORE, J. and MOORE, J.G. Calculations of three-dimensional, viscous flow and wake development in a centrifugal impeller. Trans. ASME, J. Eng. Power, 103 pg. 376 (1981)

3.38 MALIN, M.R., ROSTEN, H.I. and TATCHELL, D.G. Three-dimensional computations of flows in centrifugal pumps and compressors. ASME Symp. Perf. Pred. Centrif. Pumps and Comp. pg. 33 (1980)

3.39 DODGE, P.R. Numerical methods for 2D and 3D viscous flows. AIAA J, 15 (1977)

3.40 FRASER, D.A., HOWARD, J.H.G. and LENNOX, W.C. A three-dimensional turbulent flow analysis method for the rotating channels of a centrifugal turbomachine. ASME Paper 82-GT-221 (1982)

3.41 SWANSON, B.W. Solutions of the three-dimensional Navier-Stokes equations in non-orthogonal coordinates to calculate the flow in a log spiral impeller. ASME Paper 82-GT-268 (1982)

3.42 GOLDSTEIN, S. Modern Developments in Fluid Dynamics. Vol. II. Oxford Univ. Press (1938)

3.43 SQUIRE, H.B. and WINTER, K.G. The secondary flow in a cascade of aerofoils in a non-uniform stream. J. Aero. Sci. 18 pg. 271 (1951)

3.44 HAWTHORNE, W.A. Secondary circulation in fluid flow. Proc.R.Soc. A205 pg. 374 (1951)

3.45 LANGSTON, L.S., NICE, M.L. and HOOPER, R.M. Three-dimensional flow within a turbine cascade passage. Trans. ASME, J. Eng. Power, 99, 1. pg.21 (1977)

3.46 MAJKA, K. An experimental study of end-wall flow in a compressor cascade. Ph.D. thesis, I.I.Sc., Bangalore (1982)

3.47 CAME, P.M. and MARSH, H. Secondary flow in cascades: Two simple derivations for the components of vorticity. J.Mech.Eng.Sci., 16,6 (1974)

3.48 GLYNN, D., SPURR, A. and MARSH, H. Secondary flow in cascades. AGARD CP-214 (1977)

3.49 GREGORY-SMITH, D.G. Secondary flows and losses in axial flow turbines. Trans. ASME, J. Eng. Power, 104, 4 pg. 819 (1982)

3.50 ADKINS, G.G. and SMITH, L.H. Spanwise mixing in axial flow turbomachines. Trans. ASME, J. Eng. Power, 104, 1 pg. 97 (1982)

3.51 JOHNSTON, J.P. On the three-dimensional turbulent boundary layer generated by secondary flow. Trans. ASME, J. Basic Eng., 81,3 (1960)

3.52 LOUIS, J.F. Secondary flow and losses in a compressor cascade. ARC R&M 3136 (1958)

3.53 DUNHAM, J. A review of cascade data and secondary losses in turbines. J. Mech.Eng.Sci. 12,1. pg.48 (1970)

3.54 SALVAGE, J.W. A review of the current concept of cascade secondary flow effects. VKI TN95 (1974)

3.55 CHAUVIN, J. Von Karman Institute Lecture Series. (January 1975)
3.56 GRIEPENTROG, H.F.L. Performance prediction for high turning low aspect ratio stator cascades in the transonic regime. Trans. ASME, J. Eng. Power, pg. 390 (1970)
3.57 DUNHAM, J. and CAME, P.M. Improvements to the Ainley-Mathieson method of turbine performance. Trans. ASME, J. Eng. Power 92, pg. 252 (1970)
3.58 PAPAILIOU, K., FLOT, R. and MATHIEU, J. Secondary flows in compressor bladings. Trans. ASME, J. Eng. Power, pg. 211 (April 1977)
3.59 LANGSTON, L.S. Crossflows in a turbine cascade passage. Trans. ASME, J. Eng. Power 102, 4 pg. 866 (1980)
3.60 MAGER, A. Generalization of boundary-layer momentum integral equations to three-dimensional flow including those of rotating systems. NACA Rep. 1067 (1952)
3.61 HORLOCK, J.H. and PERKINS, H.J. Annulus wall boundary layers in turbomachines. AGARDograph-185 (1974)
3.62 MELLOR, G.L. and WOOD, G.M. An axial-compressor end-wall boundary-layer theory. Trans. ASME, J. Basic Eng., 93 pg. 300 (1971)
3.63 HEAD, M.R. Entrainment in the turbulent boundary layer. ARC R&M 3152 (1958)
3.64 LEWIS, R.I. and YEUNG, E.H.C. Vortex shedding mechanisms in relation to tip clearance flows and losses in axial fans. ARC R&M 3829 (1977)
3.65 GOSTELOW, J.P. A new approach to the experimental study of turbomachinery flow phenomena. Trans. ASME, J. Eng. Power, pg. 97 (Jan. 1977)
3.66 SODERBERG. C.R. Unpublished notes, Gas Turbine Laboratory, MIT (1949)
3.67 CARRICK, H.B. Preliminary experimental results on the effect of skew in the boundary layer on the secondary flow in an impulse turbine cascade. ARC 35,915 (1975)
3.68 RAILLY, J.W. and SHARMA, P.B. Treatment of the annulus wall boundary layer using a secondary flow hypothesis. Trans. ASME, J. Eng. Power 99, 1 pg. 29 (1977)
3.69 DE RUYCK, J. and HIRSCH, C. Investigations of an axial compressor end-wall boundary layer prediction method. Trans. ASME, J. Eng. Power 103,1 pg. 20 (1981)
3.70 RAILLY, J.W. and HOWARD, J.H.G. Velocity profile development in axial flow compressors. J. Mech. Eng. Sci. 4 pg. 166 (1962)
3.71 STRATFORD, B.S. The use of boundary layer techniques to calculate the blockage from the annulus boundary layers in a compressor. ASME Paper 67-WA/GT-7 (1967)
3.72 JANSEN, W. The application of end-wall boundary-layer effects in the performance analysis of axial compressors. ASME Paper 67-WA/GT-11 (1967)
3.73 GREGORY-SMITH, D.G. and MARSH, H. Annulus wall boundary layers in turbomachines. Camb. Univ. Int. Rep. UCT/3 (1967)
3.74 MELLOR, G.L. and STRONG, R.E. End-wall effects in axial compressors. ASME Paper 67-FE-16 (1967)
3.75 SMITH, L.H.Jr. Casing boundary layers in multistage compressors. In Dzung, L.S. (ed) 'Flow Research on Blading' Elsevier (1970)
3.76 TAYLER, C.A.M. The prediction of wall boundary layers in multistage axial compressors. M.Sc. dissertation, Bristol Univ. (1968)
3.77 LAKSHMINARAYANA, B. and HORLOCK, J.H. Leakage and secondary flow in compressor cascades. ARC R&M 3483 (1967)
3.78 HUBERT, G. Investigations of three dimensional flow through compressor cascades and stator. Zeit. Flug., 8 pg. 226 (1960)
3.79 JEFFERSON, B.A. and TURNER, R.C. Some shrouding and tip clearance effects in axial flow compressors. Int. Shipbuilding Prog. pg. 78 (Feb. 1958)
3.80 AINLEY, D.G. and MATHIESON, G.C.R. A method of performance estimation for axial-flow turbines. ARC R&M 2974 (1951)
3.81 LAKSHMINARAYANA, B., PONAGARE, M. and DAVINO, R. Three-dimensional flow field in the tip region of a compressor rotor passage - Part 1: Mean velocity profiles and annulus wall boundary layer. Trans. ASME, J. Eng. Power 104, 4 pg. 772 (1982)

3.82 HUNTER, I.H. and CUMPSTY, N.A. Casing wall boundary-layer development through an isolated compressor rotor. Trans. ASME, J. Eng. Power 104, 4 pg. 805 (1982)
3.83 TAYLOR, E.S. Comparison of free vortex and constant nozzle angle turbines. MIT GTL (unpublished) (1957)
3.84 HORLOCK, J.H. and DIXON, S.L. The off-design performance of a free vortex turbine stage. Liverpool Univ. Rep. ULME B12 (1965)
3.85 GINSBURG, T. Untersuchungen uber die dreidimensionale potentialstromung durch axiale schaufelgitter. Mitt. Inst. Aero. (Zurich) 22 (1956)
3.86 DENTON, J.D. A time marching method for two- and three- dimensional blade to blade flows. ARC R&M 3775 (Oct. 1974)
3.87 DEYCH, M.E. and TROYANOVSKIY, B.M. Investigation and calculation of axial-turbine stages. USAF Trans. FTD-MT-65-409 (1965)
3.88 DEVERSON, E.C., MARSH, H. and OXFORD, J.T.B. The performance of an aerofoil in stationary and rotating cascades. Cambridge Univ. Int. Rep. (1967)
3.89 PAULON, J., REBOUX, J. and SOVRANO, R. Comparison of test results obtained on plane and annular, fixed or rotating supersonic blade cascades. Trans. ASME, J. Eng. Power 97, 2 pg. 245 (1975)

CHAPTER 4

High Speed Cascade Testing

The development of materials technology and stressing techniques gives the turbomachinery designer the option of high rotational speeds. The use of these affords the possiblility of adding or extracting 200 MW or more in a single stage of blading and this places a correspondingly high premium on even small efficiency improvements[4.1].

In compressor or fan design, high by-pass ratio engines for airliners have transonic fans with a pressure ratio of around 1.6. This is the key feature of these engines and therefore of the economics of long-distance air transportation. The first such engine was the JT9D and ref. 4.2 describes the fundamental role that cascade testing played in its development. Research in supersonic compressors is aimed at producing efficient axial stages having pressure ratios up to six[4.3]. This would open up new applications in transportation and industry.

In facilitating such developments the high speed cascade tunnel has followed an evolutionary course from the early steam nozzle tests. Such tests were conducted by most turbine manufacturers but the most consistent effort was that of the British Steam Nozzle Research Committee[2.12].

The high speed cascade tunnel was developed rapidly as a tool for generating the essential information on compressor and turbine blades for design of the early jet engines. The steam turbine nozzle testers had been crude affairs whereas the new cascade tunnels had all the sophistication of the contemporary wind tunnel.

Existing high speed cascade tunnels in the western world are listed in Table 1. There are almost certainly an equivalent number in Eastern Europe and the USSR. Information on Soviet bloc tunnels is less accessible but valuable sources of cascade data are available in such works as[3.87] and[4.4].

All the complexities of low speed testing apply to high speed testing and many more. Because of this and the considerable compressed air requirements the cost can be high. However since the performance of transonic blading, having both subsonic and supersonic flow regions, has far exceeded early expectations, the economic benefit of high speed cascade testing has been substantial.

The uses of high speed cascade testing are various:

(i) Generation of systematic design information is best achieved in this way.

The correlations [2.15, 3.80], which have served designers well for so many years were largely based on the earliest high speed cascade tests. Subsequent proprietory and published testing has added to that stock of data.

(ii) Lower costs, quicker results and simpler configuration changes are available from cascade tests as compared with full-scale rig tests.

(iii) More detailed flow measurements and a clearer interpretation of flow patterns are available from cascade testing. This enables any local source of loss or flow separation to be rectified.

(iv) Flow computation methods used in blade design cannot be considered accurate or reliable until adequate comparisons with experimental results are available. These are most readily obtained from cascade testing.

(v) Flow prediction is presently limited by physical understanding of such questions as shock-boundary layer interaction and base pressure modelling. This is most conveniently attained in a cascade tunnel.

(vi) There is an educational value to observation of cascade performance, especially if accompanied by suitable flow visualisation. The junior engineer more quickly understands the characteristic behaviour of blading by such observations than by overall machine testing.

Although the latter objective is possibly the least easy to quantify, it is arguably the most important. Any company or educational establishment training engineers for industries such as aerospace or power generation should not neglect the high speed cascade tunnel for describing the aerodynamic basis of the end products.

4.1 Subsonic and Transonic Tunnels

Because their power requirements are modest subsonic cascade tunnels are usually continuous-running rather than of the intermittent type.

The first British high speed cascade tunnel was the ingenious No. 2 tunnel of the National Gas Turbine Establishment. During World War II C.A. Parsons and Co. took over 100,000 readings on that tunnel [4.5]. Its features, for example, included incidence variation by drum rotation, wall boundary layer bleed ahead of top and bottom blades and remotely actuated traverse gear. The tunnel was fed from a central air supply and had a dump discharge into the laboratory. This was, and still is, the most common and generally useful arrangement.

Some workers at that time argued that a better arrangement consisted of an intake to the cascade direct from the laboratory atmosphere with the suction of the main air flow taking place downstream of the cascade. Todd [4.6], for example, advocated the use of downstream suction pumps for the following reasons:
(i) Whereas turbulence can be applied at will by the inclusion of wire mesh grids in the stream, its removal is not so simple.
(ii) For turbine cascades the required power input is less than with a tunnel operating from an upstream compressor.
(iii) Oil from an upstream compressor may contaminate the blades and any transparent side-walls.
Such a tunnel, in continuing industrial use for varying incidence tests on moving blade profiles, is shown in Fig. 4.1.

Nevertheless sucked tunnels do have disadvantages compared with blown tunnels in that downstream traverse gear is less accessible and condensation effects through the blading can be less easily controlled.

Table 4.1 High Speed Linear Cascade Tunnels

ORGANISATION	CITY	CONTACT OR REFERENCE	SIZE mm	MAX M	FEATURES (see over)
U.K.					
Amal. Power Eng.	Bedford	Bird	12 x 400	2.1	-
C.A. Parsons	Newcastle	Smith	76 x 250	1.4	abc
CEGB	Marchwood	Curtis	150 x 237	1.8	ab
CERL	Leatherhead	Walters	150 x 300	2.0	abe
Cranfield Inst.Tech.	Cranfield	Fletcher	Various	1.0	b
G.E.C. Turbine Gen.	Manchester	Cox	150 x 880	1.9	be
G.E.C. Turbine Gen.	Manchester	Cox	50 x 270	1.9	bc
G.E.C. Turbine Gen.	Manchester	Cox	100 x 300	1.6	-
G.E.C. Turbine Gen.	Rugby	Cox	125 x 600	1.8	ab
Liverpool University	Liverpool	Norbury	50 x 215	1.4	c
NGTE No. 2	Farnborough	Dunham	Various	1.0	b
NGTE No. 3	Farnborough	Dunham	50 x 200	1.0	b
NGTE No. 4 Hi.Temp.	Farnborough	Dunham	112 x 300	1.0	b
Pametrada	Newcastle	Colclough	15 x 24	1.5	b
Rolls Royce	Derby	-	Various	0.9	bc
Steam Nozzles R.C.	Various	Guy	Various	1.3	e
Univ. of Oxford, 1	Oxford	Schultz	74 x 220	1.5	af
Univ. of Oxford, 2	Oxford	Schultz	300 x 500	1.45	abcd
Univ. of Sussex	Brighton	Bayley	85 x 140	0.8	-
Whittle Lab.	Cambridge	Camus	100 x 150	1.4	ab
U.S.A.					
Detroit Diesel-Allison	Indianapolis	Riffel	74 x 105	1.5	b
G.E.	Cincinnati	Smith	100 x 300	1.8	abc
G.E.	Schenectady	Maddaus	150 x 500	1.8	ab
G.E.	Schenectady	Mani	50 x 500	2.0	ab
MIT 260°C	Cambridge	Sotak	51 x 280	2.0	bd
MIT 540°C	Cambridge	Sotak	51 x 280	2.0	bd
NASA, Langley	Langley, Va	-	56 x 56	1.8	a
NASA, Lewis	Cleveland	-	300 x 900	1.0	abc
Pratt & Whitney	Hartford	Stubner	75 x 125	1.6	-
UTRL	Hartford	Carta	100 x 175	2.2	ab
Univ. of Notre Dame	South Bend	Roberts	38 x 108	1.6	-
Univ. of Cincinnati	Cincinnati	Tabakoff	36 x 100	1.0	d
Westinghouse	Philadelphia	Desai	75 x 350	1.4	b
Westinghouse	Pittsburgh	JaiKrishnan	125 x 450	1.7	abe
JAPAN					
Hitachi	Ibaraki	Uenishi	120 x 450	0.75	b
Hitachi	Ibaraki	Uenishi	120 x 450	1.35	b
Hitachi	Ibaraki	Uenishi	120 x 450	2.0	abe
Kyushu Univ.	Kyushu	Inoue	50 x 160	1.3	a
Mitsubishi	Nagasaki	Kuramoto	90 x 270	2.0	bf
Mitsubishi	Nagasaki	Yano	90 x 350	1.2	b
NAL, Tokyo	Tokyo	Minoda	90 x 400	1.1	b
Toshiba	Yokohama	Ikeda	90 x 130	1.8	abe
Toshiba	Yokohama	Araki	45 x 168	1.8	bd

Table 4.1 High Speed Linear Cascade Tunnels (cont.)

ORGANISATION	CITY	CONTACT OR REFERENCE	SIZE mm	MAX M	FEATURES
WEST GERMANY					
DFVLR, Göttingen	Göttingen	Lehthaus	130 x 380	1.8	d
DFVLR, Porz Wahn, 1	Köln	Starken	150 x 250	2.5	abc
DFVLR, Porz Wahn, 2	Köln	Starken	175 x 500	1.4	abc
Hoch.der Bundeswehr	München	Fottner	300 x 500	1.05	abc
T.H. Aachen, 1	Aachen	Gallus	50 x 70	2.0	bc
T.H. Aachen, 2	Aachen	Gallus	80 x 80	2.0	ab
T.H. Hannover	Hannover	Bammert	180 x 600	0.7	b
BELGIUM					
Ecole Roy. Mil.	Brussels	Decuypere	50 x 200	2.5	abe
VKIFD C2	Brussels	Sieverding	50 x 200	2.5	bd
VKIFD C3	Brussels	Sieverding	100 x 200	2.5	bd
VKIFD CT2	Brussels	Ligrani	100 x 250	2.5	f
FRANCE					
IMFM	Marseille	Chauvin	40 x 160	2.05	abc
ONERA	Paris	Fabri	247 x 400	1.85	bc
ONERA	Paris	Fabri	120 x 600	1.45	abd
INDIA					
BHEL	Hyderabad	Vishnu	90 x 100	2.0	be
IIT, Bombay	Bombay	Sane	80 x 150	1.5	b
NAL	Bangalore	Swamy	150 x 500	1.5	bc
AUSTRALIA					
NSWIT	Sydney	Revel	12 x 50	1.4	d
CANADA					
McMaster Univ.	Hamilton	Wade	62 x 300	1.2	bd
SWEDEN					
Stal-Laval Turbin	Finspong	Widell	100 x 400	1.4	b
SWITZERLAND					
E.P.F. Lausanne	Lausanne	Suter	45 x 100	2.5	-

KEY TO FEATURES

a Independent Reynolds' No. & Mach No. variation.
b Variable incidence mechanism.
c Sidewell suction.
d Blowdown or intermittent tunnel.
e Wet steam cascade.
f Isentropic light piston or shock tunnel.

Fig. 4.1. Subsonic turbine rotor cascade tunnel of GEC Turbine Generators Limited.

One interesting hybrid arrangement is the new Oxford University blowdown tunnel[4.7]. Figures 4.2 and 4.3 show the cascade mounted within the settling chamber and variable incidence flaps attached. The discharge jet is retrieved by a catcher and may be boosted by two ejector stages.

The most usual resolution of the debate was to be in the form of the closed circuit variable density tunnel. Since Mach Number and Reynolds' Number are each proportional to velocity, in the earlier tunnels it proved difficult to separate their effects on performance. The solution is to systematically vary the ambient pressure level, and hence the density, within the tunnel. Tunnels at Rugby[4.8] and Cambridge[4.9] (UK) and Evendale, Ohio[4.10] (USA) use this approach.

Condensation problems can severely affect cascade operation. A visible fog of droplets often emerges from a cascade and condensation shock waves[4.11] may give spurious results. For repeatable work to be possible these effects must be eliminated and this is most readily achieved in open jet discharge tunnels or in variable density tunnels. Dry and wet-bulb air temperatures are available as variables.

Many cascade tunnels employ a heat exchanger downstream of the compressor to cool the air to laboratory temperature. The tendency for air to condense is assisted by aftercooling and therefore control over the degree of aftercooling is essential. Only a limited understanding of the condensation problem can be obtained from a study of state points since the relative humidity of the air at compressor inlet is a further variable and also condensation may be induced in condensation shocks. Smolderen[4.12] has produced an empirical correlation facilitating prediction of whether condensation shocks are likely to occur.

Drying of the air by passing it through a bed of activated alumina or silica gel dessicant is another possibility. If all the air were to be passed through a dessicant, bed size and power losses would be prohibitive. In a closed circuit tunnel the air is being continually recirculated and it is usually adequate for 10% or less of the air

Fig. 4.2. Oxford University blowdown tunnel showing boost ejectors.

Fig. 4.3. Cascade and inlet section within settling chamber.

to pass through the dryer at any time. The dryer can be mounted in parallel with the test section thus making virtually the overall compressor pressure ratio available for the drying process. This arrangement is succesfully used in the Cambridge and Marchwood tunnels. Two alternate dryer vessels are provided, one of which can have its contents reactivated by heating whilst the other is in use.

Special features desirable in the test section are numerous. The first essential aims of all cascade testing, and ones which were not always met in early tunnels, are good periodicity in a pitchwise direction and good uniformity of the flow in a spanwise direction.

At subsonic speeds good periodicity dictates the use of a sufficient number of blades in the cascade and a means of coping with the boundary layers from the top and bottom upstream walls, usually by auxiliary compressor suction through appropriate slots. This "end passage boundary layer" problem is especially critical for compressor cascades. The minimum number of blades which will give good periodicity of the flow (as traversed downstream from one wake to another) has generally been put at around seven. However this is very dependent on the care which can be taken in making adjustments to the tunnel. For routine testing, which may be hurried in nature, a higher number of blades is desirable. For basic research, in which more care can be taken in setting slot suction and other variables, and for cascades having low loading, good periodicity might be obtained with a smaller number of blades.

Good spanwise uniformity is as important to the quality of the results and can be more difficult to attain. For turbine blades, having a generally favourable pressure gradient, the blade aspect ratio can usually be chosen sufficiently high that a substantial region of two-dimensional flow is available over most of the span. The desirable aspect ratio is higher for impulse rotor blades than for nozzles which have considerable acceleration. Impulse blades are particularly prone to secondary flow effects[4.13]. A value of two may be adequate for many nozzle tests. In strongly accelerating nozzles very low aspect ratios below unity have been used successfully [4.14, 4.15].

Techniques for obtaining good two-dimensionality in compressor cascades often include the use of porous side-walls. Even with the best side-wall suction arrangements it would be prudent not to use aspect ratios below two [3.8]. With low aspect ratios even if desired axial velocity ratios are obtained the suction surface boundary layer may still be strongly three-dimensional.

A problem exists if it is desired to use schlieren visualisation as well as side-wall suction. No porous material has yet been found which has the desired optical properties. At DVFLR [3.7] however, perforated transparent walls have been used giving simultaneous schlieren visualisation and side-wall suction. The quality of both is somewhat compromised.

Many subsonic cascade tunnels are workhorses, permitting large quantities of reliable data on blade performance to be obtained. Most have automatic traversing and data acquisition.

For the upstream flow a familiar feature is the traversing of a yaw/pitot probe in a proprietory actuator. These work well, enabling rapid nulling of the yawmeter, and can be used in conjunction with manometers or pressure transducers for faster response time and automatic data acquisition. Downstream of the cascade the traverse gear is more usually tailor-made for a particular cascade tunnel. However, the principle is similar and, although manometers may be used, nulling is most quickly and accurately attained by applying the limb pressure of the yaw tube to a differential pressure transducer and either a galvanometer or an automatic nulling device. Static pressures from blades and side-wall taps may be monitored either on a manometer bank or by scanning valve.

As the sonic velocity at any point on the blade surface is exceeded shock waves appear. These grow and move; whilst greatly increasing the compression or expansion properties of the blading, their interaction with boundary layers can cause serious local separation and substantial degradation of performance.

The first essential feature, then, of any cascade tunnel which will operate with supersonic local Mach Numbers, is for adequate visualisation of expansions and shock waves. Visualisation of shock waves is attained most simply by the shadowgraph method but to obtain any resolution of density gradients, and of the effects of shock-wave interactions with boundary layers, use of the schlieren method is preferable.

A blade row is transonic if either the inlet or outlet flow field is subsonic but there are supersonic regions within the flow. Many turbine cascades operate with a subsonic inlet and a supersonic discharge. Conversely some compressor or fan cascades operate with a supersonic inlet flow and a subsonic discharge. As a result of this fundamental difference transonic tunnels for turbine blading are different in several features from those for compressor blading. A general purpose cascade tunnel, such as that of G.E. Evendale[4.10] can accommodate both types of blading, but only at a considerable cost in complexity.

For both turbine and compressor cascade testing the techniques of conventional transonic wind tunnels have been adopted successfully. These include slotted and perforated walls either upstream or downstream of the cascade.

4.2 Testing of High Speed Compressor Cascades

The inlet Mach Number relative to the rotor blading of contemporary axial flow compressors can be as high as 1.8. Accordingly there exists a need for cascade data from blading for a wide range of transonic and supersonic inlet conditions. Provision of a supersonic inlet flow for fan and compressor cascade testing is by means of a correctly-designed converging-diverging nozzle ahead of the working section. This may be of fixed nozzle block form for a given supersonic Mach Number, or a flexible form allowing continuous adjustment.

If perforated and porous upstream walls, having suitable bleed arrangements, are used with compressor cascades the bleed flow itself may be used to give some degree of expansion and fine control over Mach Number. In the author's experience no upstream converging-diverging nozzle is needed for Mach Numbers up to around 1.15, the required expansion being provided by bleed flow.

The first essential test of the wind tunnel working section is that it should behave as a converging-diverging nozzle. There should be a shock system at the nozzle discharge and as the downstream flow is throttled static pressures in the working section should not be dependent upon changes in back pressure. A wide range of back-pressure variation should exist for which no change is detected in working section Mach Numbers. Supersonic tunnel operation should be limited to this range. If the range is inadequate then work is needed on the tunnel flow path to improve its operation as a converging-diverging nozzle.

With this basic condition met it is possible to consider the insertion of a compressor cascade and its effects on the flow.

The discussion of this section will be limited to cascades having supersonic relative inlet flows but subsonic axial velocities. This is the general case which attains in axial compressors. Some research has been performed on compressors having supersonic axial velocities but such work is still in its infancy[4.16].

Complications arise with a supersonic inlet Mach Number. At the cascade inlet these are mainly associated with the shock waves emanating from the blade leading edges. These shocks are reflected from the upper wall of the working section and can thus cause interference with the upstream flow field. For this reason it is necessary to use slotted or porous-wall techniques in the working section.

A more substantial peculiarity is that of the 'unique incidence' condition [4.17]. The decrease of operational incidence range with increasing subsonic Mach Number is shown in Fig. 4.4. Once the cascade is 'started' and is operating supersonically there is only one incidence possible for a periodic flow pattern. This is true for a rotating compressor or for a linear cascade. However ref. 4.18 demonstrates that the change-over from subsonic to supersonic operation is a continuous one with no performance discontinuities.

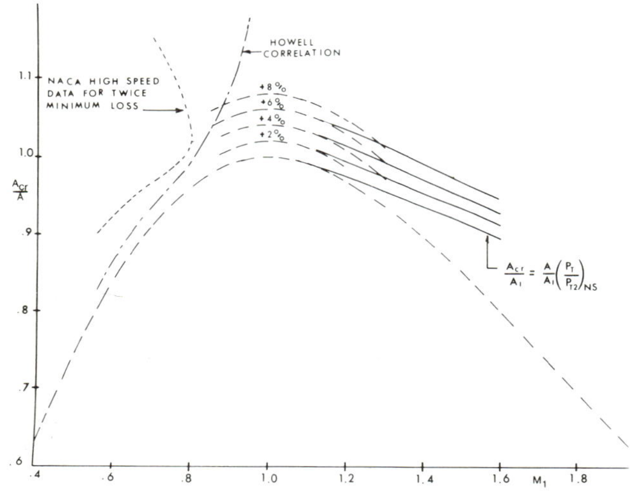

Fig. 4.4. Throat area of compressor cascades over a range of inlet Mach Number.

In testing cascades with a supersonic inlet flow only a semi-infinite cascade is represented. Essentially the infinite series of shocks which would lie upstream of a compressor blade is truncated at the shock from the first blade of the cascade, i.e., the one furthest upstream in the tunnel. This first blade in fact has a key role in establishing the supersonic inlet flow field. An additional degree of inlet Mach Number variation may be obtained merely by varying the stagger angle of that blade. Thus varying the cascade drum angle achieves not a variation of incidence angle, which is set by the unique incidence condition, but rather a variation in inlet Mach Number.

For supersonic inlet flow with subsonic axial velocity the flow will adjust itself to adopt the unique incidence even when the blades are set at quite different angles. In this event a static pressure gradient will exist along the leading edge plane of the cascade. For periodicity of inlet flow to be retained the cascade should therefore be capable of inlet angle variation. Once the axial velocity becomes sonic it is again possible to set the cascade for a range of incidence variation [4.16].

The phenomena of choking and starting are of considerable importance for the operation of high-speed compressor blading.

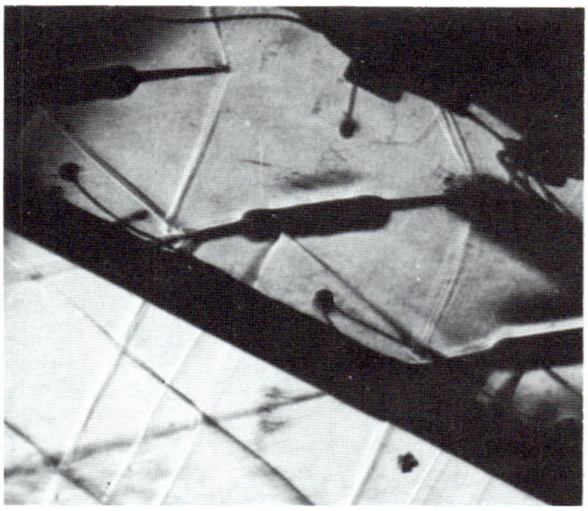

(a) $C_{p_2} = -0.119$, $\bar{\omega} = 0.125$

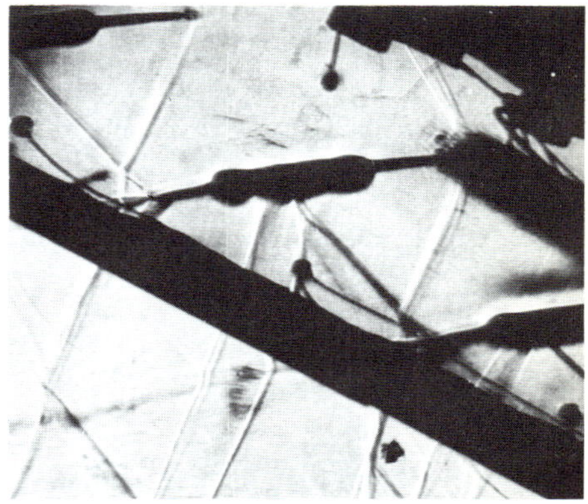

(b) $C_{p_2} = 0.000$, $\bar{\omega} = 0.096$

(c) $C_{p_2} = 0.204$, $\bar{\omega} = 0.088$

(d) $C_{p_2} = 0.333$, $\bar{\omega} = 0.130$

Fig. 4.5. Effect of increasing back pressure on transonic compressor rotor cascade at inlet Mach Number of 1.2.

Typically the run will commence with low back pressure and the flow through the cascade will be choked. Such operation for a transonic compressor rotor tip cascade is shown in the photographs of Fig. 4.5(a) and 4.5(b) in which the inlet Mach Number is 1.2 and the inlet angle is 65°. An increase in back pressure results in forward movement of the passage shock until it reaches the operating condition of Fig. 4.5(c). The flow is now started and no longer choked. The unique incidence condition sets the incoming flow. A further increase in back pressure results in the passage shock being pushed forward beyond the leading edge of the blade (Fig. 4.5(d). The flow is now unstarted and 'spills' from one passage to the next. The unique incidence condition no longer holds. In a transonic compressor operating under this condition the mass flow, which would have been constant whilst the shock remained attached to the blade, would now begin to fall off, with an associated increase in incidence.

The above description applies to correctly designed blade passages for which the throat is in the leading edge region. Blading may be encountered which is more difficult to start and for which a more detailed account of the flow processes is desirable. For such insight the reader is referred to the excellent description of starting and choking processes in cascades given by Lichtfuss and Starken[4.19].

A further complication is that if all passages in a supersonic compressor were throttled equally the upper and lower floor boundary layers would cause excessive contraction in those end passages. Upon application of back-pressure a shock would be forced through the end passages before full throttling could be accomplished in the centre passages.

Some operators of supersonic compressor cascades have used a free jet and throttle box arrangement [4.20]. In such a configuration the tunnel discharges a supersonic jet into the tank. The cascade box is mounted in the jet. In order to impose a back pressure the discharge of this box is in the form of a throttle valve which can be controlled remotely by the operator.

A different arrangement is shown in Fig. 4.6. The compressor cascade is here attached directly to the tunnel working section as for conventional subsonic testing. In order to avoid the tunnel end wall boundary layer problem the entire flow entering each end passage is separately bled off. As shown in Fig. 4.7 this gave improved back-pressuring range as compared with a throttle box arrangement.

A further problem in testing supersonic compressor cascades is that unless steps are taken to ensure that the axial density velocity ratio is correctly controlled, severe side wall boundary layer separation may result. This would give difficulties in changing back pressure over the range of operation.

The usual way of overcoming this problem is to employ converging side walls. At the very least it could be argued that the effective area contraction prevalent in the rotating high-speed compressor should be replicated. In the examples in Figs. 4.5 to 4.7 an area contraction of 14% was used. Although there are various sources of interference with the schlieren visualisation no difficulty in observation is caused by using converging side-walls.

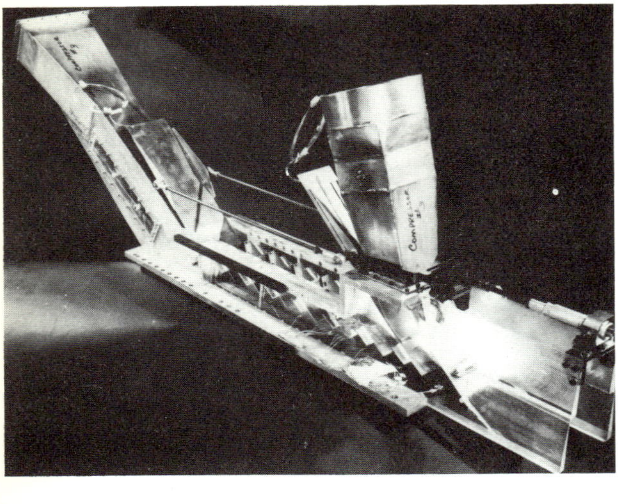

Fig. 4.6. Transonic compressor rotor cascade showing end passage bleed ducts.

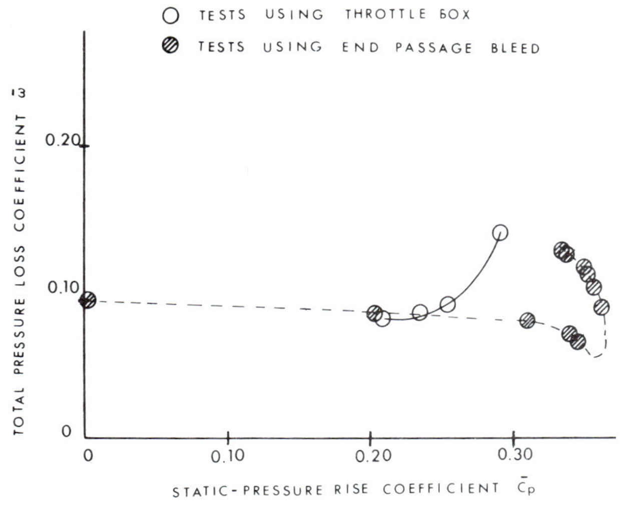

Fig. 4.7. Loss coefficient as a function of static pressure rise coefficient.

4.3 Testing of High Speed Turbine Cascades

Although for most aircraft and industrial gas turbines the discharge Mach Number of turbine blading rarely exceeds 1.2, in the last stages of large steam turbines levels as high as 1.8 may be attained [4.1]. The nozzles of small impulse turbines may be designed for substantially higher supersonic velocities.

Inlet relative velocities to turbine blading will generally be subsonic. Exceptions to this may include the rotor blading of impulse turbines and the tip sections of last stage steam turbine blading.

These latter cases, exhibiting supersonic conditions at both inlet and discharge, are particularly difficult ones. Both the complexities of supersonic inlet flows, discussed in the previous section, and those of supersonic discharge flows, to be described, will be present.

The principal difficulty in simulating supersonic discharge flows of turbine nozzle and rotor blading in a linear cascade is in attaining pitchwise periodicity of the flow field. Traverses will typically be made in a plane about a chord downstream from the trailing edge plane. The aim will be to establish the turning angle and loss coefficient of the blade row.

In subsonic turbine blading there is not usually too much difficulty in attaining periodicity. A reasonable number of blades and possible attention to boundary layer bleed in the end passages are usually quite sufficient. It is, however, quite important to ensure that the blading is dimensionally stable and that the throat width of each passage is equal.

The discharge of a supersonic jet from a cascade into ambient air is a more demanding boundary condition Expansions and shock waves emanating from the trailing edge region of the blades impinge upon a highly turbulent shear layer. These are then reflected back into the main stream flow, usually in the region of the traverse plane, causing a strong and spurious flow disturbance.

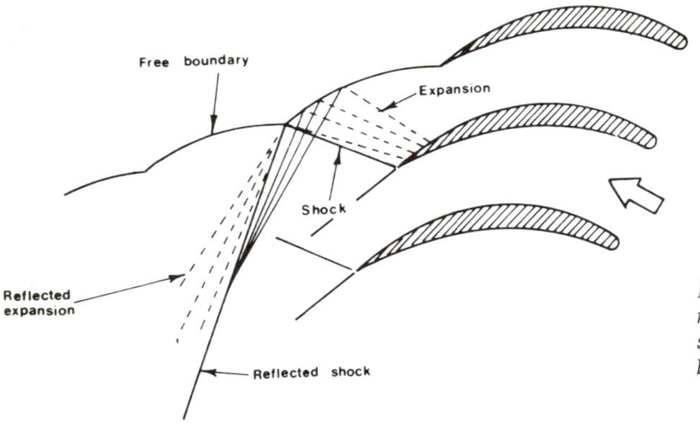

Fig. 4.8. Schematic representation of reflected shock pattern with free boundary.

If a turbine blade has a convex region downstream of the throat expansion waves will be generated from the blade surface and impinge upon the free boundary of the downstream supersonic flow. These will be reflected as compression waves which usually run together to form a reflected shock. A shock wave will also be generated from the trailing edge region. Its impingement on the free boundary results in the reflection of a fan of expansion waves.

The situation is depicted schematically in Fig. 4.8. Depending upon blade shape, flow angle and Mach Number, the reflected shocks and expansions will vary in strength and effect.

An example of this type of disruption is given in Fig. 4.9. Flow periodicity was bad in a plane one chord downstream and even worse half a chord downstream of the trailing edge plane. The cause of the trouble was the reflection of trailing edge shocks from the free shear layer as expansions, thus creating regions of excessively low static pressure. This problem would not be solved simply by increasing the number of blades. Hauser et al [4.21] described cascade tests in which a large number of blades was used. The visible problem in that case was of incident expansions being reflected from the free shear layer as shocks with equally catastrophic effect.

Fig. 4.9. Schlieren photograph of unbounded downstream flow field in a turbine nozzle cascade; dominated by reflected expansions.

An alternative configuration from that of the supersonic discharge jet emerging into ambient air, with a turbulent and uncontrollable shear layer at the interface, is for the jet to be bounded by a solid surface. A movable wall is hinged from the trailing edge of the end blade. This wall becomes a streamline. The flow direction and pressure are thereby prescribed and expansion of the flow to the corresponding angle becomes the downstream boundary condition[4.14].

Expansions generated from the convex blade surface will be reflected as expansions and the shocks generated in the trailing edge region will be reflected as shocks. This configuration is therefore usually as damaging to flow periodicity as the free jet discharge case.

Nevertheless configurations employing a movable wall as a downstream flow boundary have been employed with reasonable success[4.15, 4.22]. The schlieren photographs of Fig. 4.10 show the results of cascade tests on nozzle blades for Curtis turbines. The pressure ratios corresponding to small changes in tailboard angle are also given. The strong reflected shock structure is particularly evident in these examples.

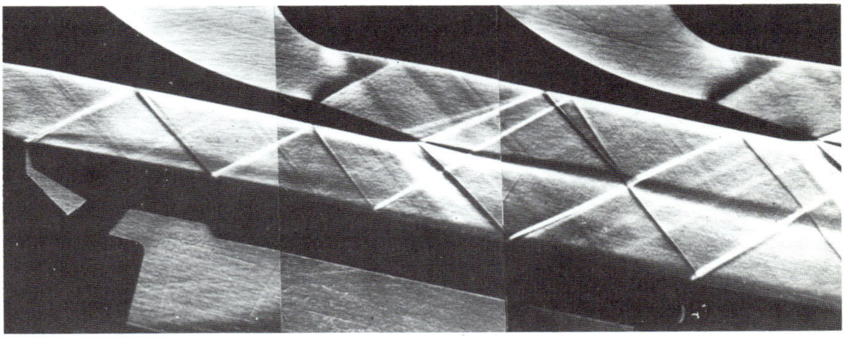

a) Tailboard angle 78°: Pressure ratio 4.25.

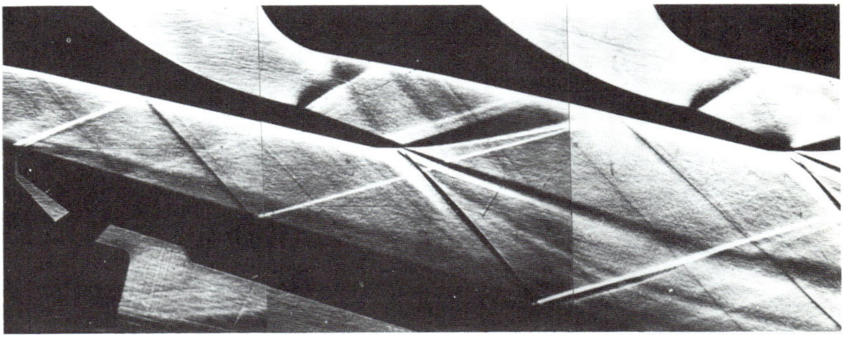

b) Tailboard angle 73°: Pressure ratio 8.

Fig. 4.10. Flow in Curtis turbine nozzles[4.22].

Since reflected waves have proved to be so troublesome, the problem is one of identifying a downstream boundary configuration which might be capable of successful and routine incorporation in cascade testing. The use of a perforated tailboard of controlled porosity[4.23] is recommended as a versatile solution to this problem.

If a shock is incident on an undistorted shear layer it will be reflected as an expansion fan (Fig. 4.11(a)). Conversely if such a shock impinges on a flat plate the reflection is in the form of a shock (Fig. 4.11(b)). It is not too difficult to devise a suitable perforated or porous surface such that the nett reflected disturbance at a suitable distance from that surface will be zero (Fig. 4.11(c)). The only reflected shocks or expansions present in the flow are local ones of a length and scale corresponding to the porosity of the surface. Mutual cancellation is effected within that length scale.

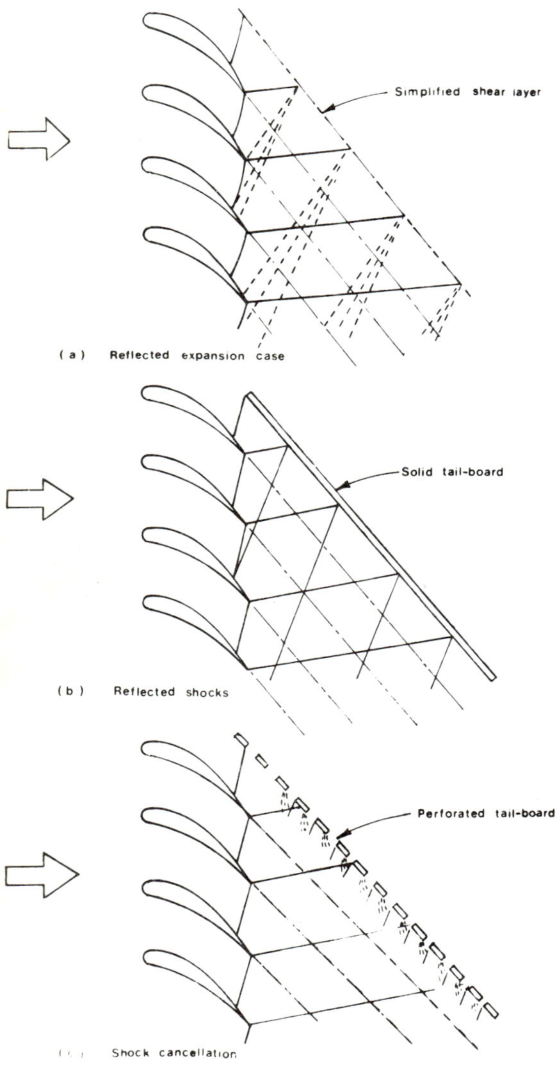

Fig. 4.11. Schematic representation of effect of differing downstream flowfield boundaries.

Such a tailboard, of adjustable porosity and inclination, was devised for the cascade of Fig. 4.9. This resulted in cancellation of the strong reflected wave field and substantial improvement in downstream periodicity.

The difficulty of condensation shocks is especially acute with supercritical turbine blading where the low discharge temperatures make condensation quite likely. If it is desired to eliminate condensation shocks the use of a closed-circuit continuous running tunnel with adequate drying facilities is recommended. Alternatively cascades may be run 'hot' at such a temperature that condensation is not a problem.

In the last stages of steam turbines condensation phenomena, including droplet production from nucleation and condensation shocks within the blading, appear to occur in operation and lead to decreased efficiency in addition to the other well-known problems such as blade erosion. Investigations on these problems have been reported in ref.[4.24] and calculations of drop size have been performed by Moore et al[4.25].

In experimental investigations of wet steam behaviour in cascades a wet steam tunnel is a useful tool. That shown in Fig. 4.12 is that of the British Central Electricity Research Laboratories. Droplet size can be varied between 0.2 and 2.0 mm and a wide

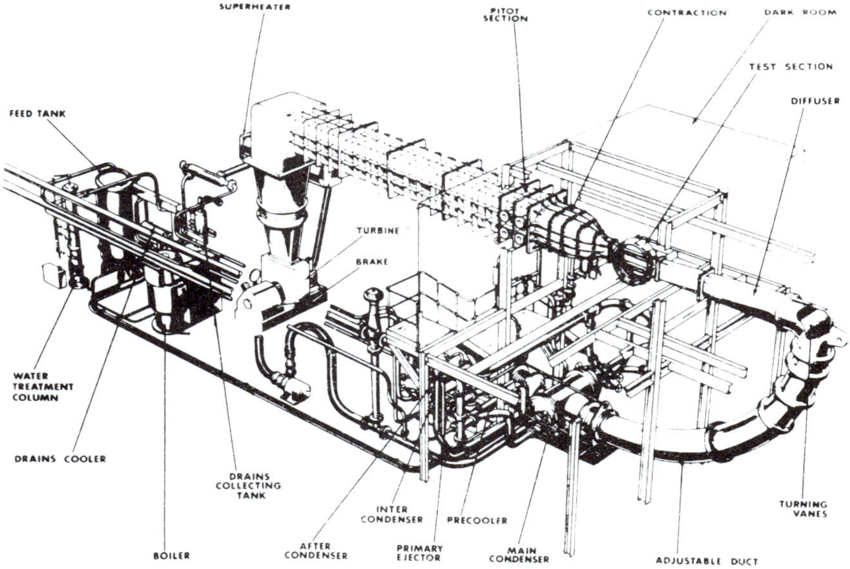

Fig. 4.12. The C.E.R.L. wet-steam tunnel.

range of supersonic Mach Numbers will be attainable at the cascade discharge. Most relevant dimension less parameters can be modelled well in this tunnel, including the specific heat ratio. At high discharge Mach Numbers it may be desirable to replicate the specific heat ratio of various steam qualities without the attendant moisture problems or the other inconveniences of working with steam. Forster[4.8] has described the advantages of testing in a Freon 12-air mixture.

4.4 Instrumentation and Observation Techniques

The measurement of time-averaged pressures and flow angles in the vicinity of subsonic cascades is straightforward. However as the loading of compressor blading increases or in the wake region of turbine blading static pressure variations become significant. For this reason it is desirable to incorporate a reliable static pressure measuring device into the probe head. Workers in many countries have produced combination probes capable of measuring total pressure, static pressure and flow angle.

Probes for supersonic flow need careful evaluation. The problem is most acute in the discharge region of turbine blading where shocks, severe expansions and wakes are present. Under these conditions even a pitot tube can be sensitive to incidence. Internally chamfering the tube inlet reduces this sensitivity.

Pitot tube measurements in supersonic flow are affected by the shock standing upstream of the probe. Correction for this shock presumes knowledge of the local static pressure. The Goodyer probe[4.26] overcomes this difficulty by isentropically decelerating the supersonic flow to a subsonic condition upstream of the measurement orifice.

More work is still needed on suitable probes for measuring static pressure in a transonic flow. Conrad probes seem to offer most promise but flow blockage from a thick probe and support assembly can be a problem.

There are a number of techniques available permitting accurate pressure measurements to be made in wet steam[4.27] . These usually involve purging the probe with dry air or inert gases before or during a reading. Direct measurement of velocities and turbulence levels in low speed flows is accomplished by hot wire anemometry. Susceptibility to damage has not prevented the use of hot wires in a wet steam cascade at Mach Numbers up to 1.3 (ref. 2.33).

New semiconductor temperature transducers are well suited to measuring static and stagnation temperatures in a high speed flow. A range of specialised techniques is available for measurements in hot combustion gases.

In ref. 4.28 the advantages of using a transient cascade facility are described. The test may be started with cold blading; by using air with a stagnation temperature of 520K a gas-to-wall temperature ratio of 1600K : 1100K is simulated. Thin film gauge quartz inserts are used to measure heat transfer rate distribution on light alloy blades. Alternatively, blades are machined from MACOR glass ceramic which also takes a deposited thin film gauge.

The most readily useful information for high speed cascades is obtained from flow observation by the Toepler-Schlieren method. The most common version is single pass schlieren. The light source, which may be a mercury arc lamp for black and white visualisation, a polychromatic light for colour observation or a spark source for instantaneous observation, is focused on to an adjustable slit. This slit is at the focus of the first parabolic mirror. A beam of parallel light is thus produced which passes through the cascade working section. The working section contains the blades, simply supported between transparent sidewalls. A second parabolic mirror then focuses an image of the slit onto a knife-edge cut-off. The cut-off is adjusted so that in the absence of air flow about half of the incident light is cut off, reducing the intensity of light projected onto a viewing screen.

Density gradients in the working section cause the light rays to bend so that part of the image of the slit is displaced either away from or towards the knife-edge. This has the effect of producing darker or lighter areas on the screen. The schlieren effect is only produced by density gradients in a direction perpendicular to the knife-edge.

Colour schlieren, although not useful for reporting in published work, may aid discrimination between shock and expansion waves. Colour schlieren is most readily obtained by replacing the knife-edge by a tri-colour filter having a narrow central strip.

Shadowgraph visualisation is similar to schlieren but without the cut-off. Only very strong density changes, such as those from shock waves, or very strong expansions are revealed.

Single pass schlieren systems suffer from two major deficiencies. Firstly, loaded blading which is structurally supported in transparent sidewalls, must impart a stress to the walls at the support point. The resulting concentration is often visible as a photoelastic stress pattern and may jeopardise flow observation. Secondly, it is often desired to measure static pressures on the sidewalls or from the blades. Tapping leads will invariably interfere with the schlieren picture and usually contrive to conceal the most important shock wave under investigation.

For these reasons some investigators have preferred to use the double-pass schlieren system in which the blades are cantilevered from a thick and optically-surfaced steel wall. All pressure tapping leads are led off behind the steel wall and do not interfere with visualisation. The beam is simply reflected back along itself by the wall. A half-silvered mirror acts as a beam splitter and the returning beam is directed to a cut-off as with conventional schlieren.

Most schlieren surveys are strictly of a qualitative nature. Interferometry is a technique which can give a quantitative result. The Mach-Zehnder interferometer operates on the principle of ray amplitude division. Parallel light from the source is divided by a beam splitter so that half of the amplitude of each ray passes through the working section whilst the remainder passes through a reference medium. Having traversed different optical paths, a phase difference exists between corresponding rays dependent on the fluid density in the working section. When the rays are recombined at the screen optical interference occurs with alternate dark and light bands relating to fluid density contours. Some excellent results have been obtained from testing on compressor cascades[4.29] and turbine cascades[4.30] but the device is expensive to instal and unduly sensitive to vibrations.

A simpler form of interferometer would probably find more widespread usage. Fortunately the polarization interferometer is simpler and less sensitive to vibration. It can be used with a schlieren cut-off[4.31] to combine the qualitative field information of the schlieren technique with some of the quantitative information of interferometry.

It might be expected that lasers would be used to improve the quality of flow measurements and that is indeed the case. Ref.4.32 describes the use of a laser as a coherent light source for interferometric and holographic measurements on a turbine cascade.

REFERENCES

4.1 COX, H.J.A., FORSTER, V.T. and HOBSON, G. Aerodynamic development of steam turbine blading. I.Mech.E. Conf. "Steam Turbines for the 1980's" Paper C230/79 (1979)

4.2 MIKOLAJCZAK, A.A., MORRIS, A.L. and JOHNSON, B.V. Comparison of performance of supersonic blading in cascade and in compressor rotors. Trans. ASME, J. Eng. Power, pg.42 (1971)

4.3 FUCHS, R. and STARKEN, H. Experimental investigations of supersonic cascades designed for high static pressure ratios. ASME Paper 77-GT-37 (1977)

4.4 DEYCH, M.E., FILIPPOV, G.A. and LAZEREV, L.YA. Atlas of axial turbine blade characteristics. Mash. Pub. Ho., Moscow (1965). CEGB(UK) Trans. 4563

4.5 HOWELL, A.R. Fluid dynamics of axial-flow compressors. Proc. I.Mech.E. 153 pg. 441 (1945)
4.6 TODD, K.W. Practical aspects of cascade wind tunnel research. Proc. I.Mech.E. 157 (1947)
4.7 BAINES, N.C., OLDFIELD, M.L.G., JONES, T.V., SCHULTZ, D.L., KING, P.I. and DANIELS, L.C. A short-duration blowdown tunnel for aerodynamic studies on gas turbine blading. ASME Paper 82-GT-312 (1982)
4.8 FORSTER, V.T., ARCHER, B.V. and UNSWORTH, R.G. Development of experimental turbine facilities for testing models in air or Freon. I.Mech.E., Conf. Pub. 3, pg. 84 (1973)
4.9 GOSTELOW, J.P. and WATSON, P.J. A closed circuit variable density air supply for turbomachinery research. ASME Paper 76-GT-62 (1976)
4.10 ERWIN, J.R. Experimental techniques - in Hawthorne, W.R. (ed) - Aerodynamics of Turbines and Compressors. Oxford Univ. Press (1964)
4.11 VEREDIKTOV, V.D. Two-phase flow in turbines and reaction nozzles. NASA TTF-613
4.12 SMOLDEREN, J.J. Condensation effects and air drying system for supersonic wind tunnels. AGARDograph-17
4.13 FORSTER, V.T. Discussion on paper by H.R.M. Craig and H.J.A. Cox. Proc. I.Mech.E. 185, pg. 407 (1971)
4.14 STRATFORD, B.S. and SANSOME, G.E. The performance of supersonic turbine nozzles. ARC R&M 3273 (1959)
4.15 ADLER, M., BOTTLE, D.W. and CARLTON, R. Investigation of supersonic flow through the tip section of low pressure blades of large steam turbines by means of cascade tests. Proc. I.Mech.E. 178, pg. 25 (1963)
4.16 BREUGELMANS, F.A.E. High speed cascade testing and its application to axial flow supersonic compressors. ASME Paper 68-GT-10 (1968)
4.17 KANTROWITZ, A. The supersonic axial flow compressor. NACA Rep. 974 (1950)
4.18 SCHREIBER, H.A. and STARKEN, H. Evaluation of blade element performance of compressor rotor blade cascades in transonic and low supersonic flow range 5th ISABE, Bangalore (1981)
4.19 LICHTFUSS, H.J. and STARKEN, H. Supersonic cascade flow. Prog. Aero. Sci., vol. 15, Pergamon, Oxford (1974)
4.20 BLANTON, J.W. Demonstration of supersonic compressor technology. 1st ISABE, Marseilles (1972)
4.21 HAUSER, C.H., PLOHR, H.W. and SONDER, G. Study of flow conditions and deflection angle at exit of two-dimensional cascade of turbine rotor blades at critical and supercritical pressure ratios. NACA R&M E9K25 (1950)
4.22 BIRD, P.J. A high-speed cascade tunnel for the development of supersonic steam turbine nozzles and blading. APEX, Tech. Journ. of Amal. Power Eng., no. 2 pg. 17 (Aug. 1969)
4.23 GOSTELOW, J.P., HOBSON, D.J. and WATSON, P.J. Preliminary evaluation of the influence of tailboard porosity on shock reflections in a supercritical nozzle cascade. Cambridge Univ. CUED/A Turbo/TR64 (1974)
4.24 IKEDA, T. and SUZUKI, A. Some findings on the flow behaviour of last-stage turbine buckets by linear cascade tests in steam. I.Mech.E. Conf. Pub. 3, pg. 46 (1973)
4.25 MOORE, M.J., WALTERS, P.T., CRANE, R.I. and DAVIDSON, B.J. Predicting the fog-drop size in wet steam turbines. I.Mech.E. Conf. Pub. 3, pg. 101 (1973)
4.26 GOODYER, M.J. A stagnation pressure probe for supersonic and subsonic flows. Aero. Quart. (May 1974)
4.27 RYLEY, D.J. Wet steam property measurements: current problems - a review. Int. J. Mech. Sci. 8,581 (1966)
4.28 SCHULTZ, D.L., JONES, T.V., OLDFIELD, M.L.G. and DANIELS, L.C. A new transient cascade facility for the measurement of heat transfer rates. AGARD Conf. Proc. 229 (1977)
4.29 CELIKOVSKY, K. Investigation of flow through supersonic axial compressor cascade. 5th ISABE, Bangalore (1981)

4.30 YANO, T. Research on steady flow performance of exhaust gas turbine. Mitsubishi Tech. Bull. 01015R (1964)

4.31 BIRD, P.J. Transonic flow measurements in a two-dimensional nozzle using a schlieren interferometer. Proc. Symp. on Meas. Tech. in Transonic and Supersonic Flow. CERL(UK) RD/L/N 166/79 (1980)

4.32 BRYANSTON-CROSS, P.J., LANG, T., OLDFIELD, M.L.G. and NORTON, R.J.G. Interferometer measurements in a turbine cascade using image-plane holography. Trans. ASME, J. Eng. Power 103, 1 pg 124 (1981)

CHAPTER 5

Theories for Incompressible Potential Flow

The systematic improvement of blade shapes has only become possible as a result of advances in calculating the flow through plane cascades. For incompressible flows the problem is posed mathematically as the solution of the Laplace equation (1.6) in a potential field for specified boundary conditions.

In cascade theory the plane field conventionally stretches to infinity in all directions and is bounded only by the blades. Although a cascade is a single infinite row of similar bodies it is not necessary to limit the discussion to any particular shape of body and thus bodies having one or more sharp corners (Hölder discontinuities) are possible. The field under consideration will be assumed uniform and steady and will be specified at upstream or downstream infinity or both. Since the cascade is an infinite row, it is possible to postulate a change in the direction of the flow field as it traverses the cascade.

The method of solution varies according to the number of discontinuities on the profile boundary. The general case is taken to be that in which there are none and hence no singularities in the flow field. Compressor and turbine blades used in turbomachines are often made in this way for stress and manufacturing reasons. If there are no singularities then the potential solution for such a cascade is indeterminate. Both upstream and downstream flow conditions have to be specified. This is not a disadvantage but, on the contrary, is implicit in any accurate simulation of the blade shape and flow field. The problem for blades with rounded trailing edges, for instance, can not be uniquely solved until the role of viscosity is introduced.

Of course, cascades of blades having an infinitely thin leading or trailing edge may also be considered. The Kutta-Joukowsky condition [5.1, 5.2] applies to such cases, as it does to most aerofoils used in aeroplane wings. The only way in which a fluid can successfully negotiate the surface of an aerofoil having a point of infinite curvature is for the two streams of fluid to either divide or rejoin at that point. The potential gradient in such an instance will remain finite. In this type of solution either the upstream or the downstream flow condition may be prescribed but not both. This will be sufficient, when used with the Kutta-Joukowsky condition, to determine the circulation around the body.

Although this latter case of pointed blades does not usually correspond to reality, it does permit rapid calculation of flows for blades of simplified form without the need to consider the role of viscosity.

5.1 Solutions to the Direct Problem for Incompressible Potential Flow

In the direct problem it is required to find a solution to the flow about a cascade of given geometry. More specifically the following information is given :-

1. Cascade geometry, specified as $y = f(x)$.
2. Inlet flow angle (conventionally at upstream infinity).
3. Discharge condition, given as either
   (i) flow angle at downstream infinity
   or (ii) the circulation around a cascade blade
   or (iii) satisfaction of the Kutta-Joukowsky condition for a blade with a pointed trailing edge.

One can expect to find the following :-

4. Velocity distribution around the cascade, normalised by a reference velocity or given as a pressure coefficient.
5. Further information depending upon the choice of discharge condition in (3) above :-

If discharge condition (i) is used then the solution gives the circulation around a blade and the position of rear stagnation point.
If (ii), downstream flow angle and stagnation point position are obtained.
If (iii), downstream flow angle and circulation are obtained.

Most of the older methods to be described take information in categories 1,2 and 3 (iii) and accordingly result in computation of 4 and 5 (iii) although later in the chapter theories which are more flexible in their scope will be considered.

5.1.1 Historical development. The pioneer of cascade theory was Sergei A. Chaplygin, whose work was performed in the year 1911[5.3]. Chaplygin was able to set out an approach to the cascade problem which showed great foresight. It is, unfortunate that there were many deviations from the path before the problem was finally solved, since Chaplygin had given a solution for flat plate cascades which he expanded upon republication in 1933 to cover aerofoils with camber and thickness. His approach forms the basis of the work of Merchant and Collar[5.4] which, in turn, was the basis of the exact solution of section 5.2.

Shortly afterwards, Joukowsky[5.5] studied the problem, synthesising the flow around a flat plate cascade from the appropriate basic flow patterns.

In 1917 Grammel[5.6] obtained a solution for a set of approximately straight blades set at any stagger angle. The blades were exactly straight only at zero and 90° stagger. McWood and Glauert[5.7] used cascade theory and experiments to obtain modified lift coefficients for use in designing airscrews.

Koenig[5.8] achieved a solution of the cascade problem in 1922 in a manner which would apply for any shape of blade. Although the mathematics is intractable, his method was useful in laying a foundation for design methods such as that of Simonov[5.9] mentioned in section 5.4. Koenig's only published result is for straight blades at 45° stagger. This result agrees with the later, exact, work of Merchant and Collar.

Early German work includes the theories of references[5.10],[5.11],[5.12] and[5.13]. In 1929 Numachi[5.14] computed blade interference using the Grammel formula. His values at zero and 90° stagger agree with the exact results of Merchant and Collar, but are discrepant at other staggers for which the Grammel blades are not straight.

In the early nineteen-thirties Weinig[5.15] discovered the transformation function from an infinite cascade of flat plates to a single circle. For the first time, it was possible to obtain a picture of the entire flow pattern through a cascade of straight-line aerofoils over the complete range of stagger and space-chord ratio. The transformation permitted the calculation of the lift coefficient of any flat plate cascade and showed that the lift coefficient is given by

$$C_{L\infty} = 2\pi k \cdot \sin\alpha$$

where k is a function of stagger and space-chord ratio and α is the angle of attack with respect to the equivalent free stream velocity vector. The theory further showed that, for every cascade of any profile whatsoever, there exists an equivalent flat plate cascade producing the same turning angle.

Betz[5.16] and von Karman and Burgers[5.17] have also given expositions of a similar type of theory. In 1931 Betz showed that if the space-chord ratio was large the flow around any of the aerofoils could be approximated by replacing the other aerofoils by vortex filaments and passing from an isolated aerofoil with known properties, by a method of successive approximation, to an equivalent aerofoil in cascade. This method could be applied to cascades of slightly curved aerofoils, but needed considerable labour.

The last works to be published before the Second World War conveniently close the epoch of 'early work'. Pistolesi[5.18] gave, in 1937, a valid method for blades normal to the axis, having a slight camber. The method is an extension of the Birnbaum-Glauert isolated aerofoil theory. He applied the same method to arbitrarily staggered profiles but the mathematics became very complicated. He obtained Grammel's formula, but for blades of indeterminate thickness, and was thus prompted to ask whether the curved blades obtained by Grammel should not be reflected about their axis. Pistolesi's work was extended by Lieblein[5.19] to include the effect of finite thickness distribution. Lieblein's approximations became increasingly inexact as thickness and camber increased.

A rapid approximate theory due to Erwin and Yacobi[5.20] was also based upon the known flow around an isolated aerofoil. The principle of superposition was used, the velocity on the cascade blade surface being composed of the average passage velocity, with perturbations due to thickness, camber and incidence. The method was therefore only intended for use at around design incidence, where perturbations on the average velocity are not excessively large. The average velocity was determined from the passage area change and the perturbation velocities related to corresponding perturbations on the identical isolated aerofoil.

Wu and Brown[5.21] gave a simple solution which can be used for either the direct or the inverse problem and which is best suited to closely-spaced turbine blading. As in the Erwin and Yacobi approach the velocities on the mean stream line must be determined. In the Wu and Brown method the properties away from the mean stream-line may be obtained from a Taylor series expansion. The first and second gradients of the velocities in the y direction, which are needed for the Taylor series, are conveniently obtained from the equations of continuity and irrotationality.

Other authors to have published approximate or simplified theories include Fanti, Kemp and Nilson[5.22], Ferrari[5.23], Hlavka[5.24], Hudimoto, Kamimoto and Hirose[5.25], Krüger[5.26] Murai[5.27] and Tanabe and Horlock[5.28]. The author's suggestions for really rapid methods would be those of references 5.20 and 5.21 for compressor and turbine cascades respectively.

During the twenty years after the Second World War the problem tended to be attacked in two distinct ways. These were the conformal transformation and singularity methods to be described in the following sections. It was not until the divergence between transformation and singularity methods was resolved that the problem was finally solved in the form of mathematically rigorous developments of the singularity approach such as that of Martensen[3.27] described in section 5.1.4.

Substantial progress has been made in recent years in the application of numerical methods, such as the finite element and finite difference techniques, to cascade flows. Examples of the application of finite element techniques are to be found in Baskharone and Hamed[5.29], Habashi et al[5.30], and Whitehead and Grant[5.31]. Finite difference solutions have been advanced by Kavanagh and Serovy[5.32], Katsanis[5.33], Daiguji[5.34], and others. Most of these numerical methods are readily applicable to compressible flows.

5.1.2 Conformal transformation methods. In 1932 Theodorsen[5.35] had produced a solution to the direct problem for isolated aerofoils of arbitrary shape. A Joukowsky transformation was used to map the aerofoil onto a near-circle; this in turn was mapped onto an exact circle using an integral equation which was solved iteratively. This general approach was to find subsequent application for cascades, notably in the work of Howell[5.36] and Garrick[5.37].

Howell firstly used the transformation $\zeta = \tanh z$ to collapse the cascade into an isolated S-shaped contour. Two or more Joukowsky transformations $\zeta' + a^2/\zeta' = \zeta$ were then used, with appropriate changes of axes between, to transform the S-shaped contour into an irregular circle. When the best possible irregular circle had been obtained it was then transformed into an exact circle using the Theodorsen transformation

$$\sum_1^n (C_r \zeta^{-r}) = \ln(z - \zeta) \tag{5.1}$$

where C_r is a complex constant.

The calculation is performed at every point for which the aerofoil is defined and the choice of axes and value of a in the Joukowsky transformation are best obtained graphically. The first $\zeta = \tanh z$ transformation is difficult to apply at low values of space/chord ratio since the ends of the S-shaped profile become very distorted, tending to curl around at the infinity points.

Howell's method was used by Carter and Hughes[5.38], simplified by Vazsonyi[5.39], and programmed by Pollard and Wordsworth[5.40].

Garrick also gave a solution based upon the Theodorsen transformation. His initial transformation was

$$z = \frac{s}{2\pi}\left[e^{-i\chi}\ln\left(\frac{e^b + \zeta_1}{e^b - \zeta_1}\right) + e^{i\chi}\ln\left(\frac{\zeta_1 + e^{-b}}{\zeta_1 - e^{-b}}\right)\right]. \tag{5.2}$$

The parameter b depends strongly upon the space/chord ratio.

Equation (5.2) represents the transformation from a circle to a cascade of straight lines similar to that used by Weinig[5.15]. Garrick used it in the reverse direction to pass from a cascade of aerofoils to a contour which resembles a circle. Having attained such a contour the procedure essentially follows that of Howell with the use of a Theodorsen transformation to attain an exact circle.

Hall and Thwaites[5.41] were able to improve Garrick's method in two ways. Firstly they examined the relationship between the arbitrary aerofoil cascade and the straight line cascade. It was found that by correct positioning of the aerofoil cascade with respect to the straight line cascade it was possible to transform the aerofoil cascade into a good near-circle. Secondly, Thwaites[5.42] avoided the explicit determination of the Laurent series coefficients in the Theodorsen transformation by using the formulae of Watson[5.43].

In 1944 Mutterperl[5.44] mapped the arbitrary cascade directly to a straight line cascade which was then transformed into a circle by an equation similar to that used by Murai (1955) [5.27]. Mutterperl's method is essentially a simplification of Garrick's method but is really only suited to aerofoils of low camber and thickness, due to difficulties in harmonic analysis at the leading edge.

In an interesting paper Traupel[5.45] mapped the cascade flow field onto the interior of a closed curve using the transformation

$$\zeta' = (e^z + 1)/(e^z - 1) \quad . \tag{5.3}$$

The cascade plane infinity points are brought into the finite plane. At least four transformations are required, resulting in contours of the type depicted in Fig. 5.1.

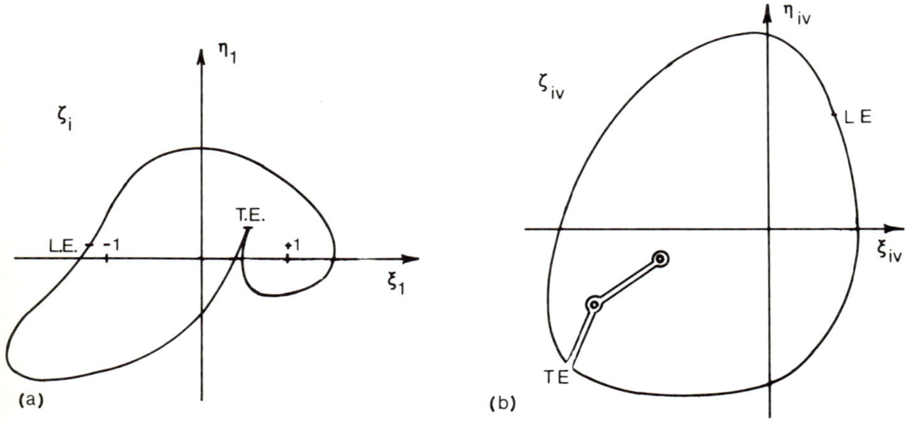

Fig. 5.1. Stages in Traupel's transformation.

With the type of curve shown in Fig. 5.1(b) Traupel was able to use Green's second formula to give an integral equation which was solved by an iterative procedure. Traupel actually recommends graphical means for the first iteration but in a well-argued appendix is able to demonstrate convergence.

5.1.3 Singularity methods. The flow field resulting from an infinite row of singularities forms the basis of many cascade theories. Its derivation is outlined below.

The velocity induced at a field point z by a vortex of strength Γ located at l is given by

$$c(z) = c_x(z) - i c_y(z) = \frac{\Gamma}{2\pi i(l - z)} \quad . \tag{5.4}$$

Alternatively this may be expressed as the velocity induced by a vortex element $\gamma(\lambda')d\lambda'$ where λ' is a parameter representing surface P_o

$$dc(z) = \frac{\gamma(\lambda')d\lambda'}{2\pi i(l - z)} \quad .$$

It is possible to construct a closed profile from a distribution of vortex elements. Accordingly the flow field around an aerofoil P_o is obtained by integration

$$c(z) = c_x(z) - ic_y(z) = c_o(z) + \frac{1}{2\pi i}\oint_{P_o} \frac{\gamma(\lambda')\,d\lambda'}{(l - z)} \qquad (5.5)$$

where $c_o(z)$ is the flow field at infinity.

A condition of the above is that the vortex density function $\gamma(\lambda')$ is real and such that the contour P_o becomes a streamline. It may be shown that $\gamma(\lambda')$ represents the velocity of the flow field at contour point l .

An infinite row of vortices may be constructed by placing vortices of strength Γ a distance s apart along the y axis. The combined effect of a row of equidistant vortices at points $z = l \pm kis$, where k is an integer, is obtained by summation :-

$$c(z) = c_x(z) - ic_y(z) = \frac{\Gamma}{2\pi i}\sum_{k=-\infty}^{\infty} \frac{1}{l - z - kis} \, .$$

This summation is a standard one which gives

$$\sum_{k=-\infty}^{\infty} \frac{1}{l - z - kis} = \frac{\pi}{s}\coth\frac{\pi}{s}(l - z) \, .$$

Just as the flow around the aerofoil (eqn. 5.5) was obtained by contour integration, so for the cascade,

$$c(z) = c_x(z) - ic_y(z) = c_o(z) + \frac{1}{2is}\oint_{P_o} \gamma(\lambda')\coth\frac{\pi}{s}(l - z)\,d\lambda' \, . \qquad (5.6)$$

Whilst the above expression was obtained for distributions of vorticity a similar result is also obtained by considering source distributions $q(\lambda')$.

$$c(z) = c_x(z) - ic_y(z) = c_o(z) + \frac{1}{2is}\oint_{P_o} q(\lambda')\coth\frac{\pi}{s}(l - z)\,d\lambda' \, . \qquad (5.7)$$

Equations (5.6) and (5.7) are the fundamental equations of potential flow theory for cascades and many of the solution methods represent differing approaches to the solution of these for the boundary conditions placed upon the singularity distribution.

Singularity methods had been developed by Birnbaum[5.46], Glauert[5.47], and Allen[5.48], for aerofoils of small thickness and camber with cusped trailing edges. Katzoff et al[5.49] used the Theodorsen transformation to determine the pressure distributions around an isolated aerofoil. They then corrected that pressure distribution for the interference effects due to the other aerofoils in the cascade, which have been computed using suitable distributions of sources and vortices. This iterative procedure used standard charts for the interference effects and became obsolete when computers were introduced.

Schlichting's method [3.25], based on equations (5.6) and (5.7) was a cascade approach, rather than a modified aerofoil theory, and was ideally suited to computers. The method was programmed by Pollard and Wordsworth [5.40] and by Chauvin and Breugelmans [5.50]. Modifications have been made by Richter [5.51] and by Lewis and Pennington [5.52] to overcome some of its basic weaknesses.

In the Schlichting method each blade is represented approximately by a distribution of sources, sinks and vortices placed on the chord line or the camber line. The relationship between these distributions $q(x), \gamma(x)$ and the cascade geometry is derived from the kinematic flow conditions. These are illustrated in Fig. 5.2.

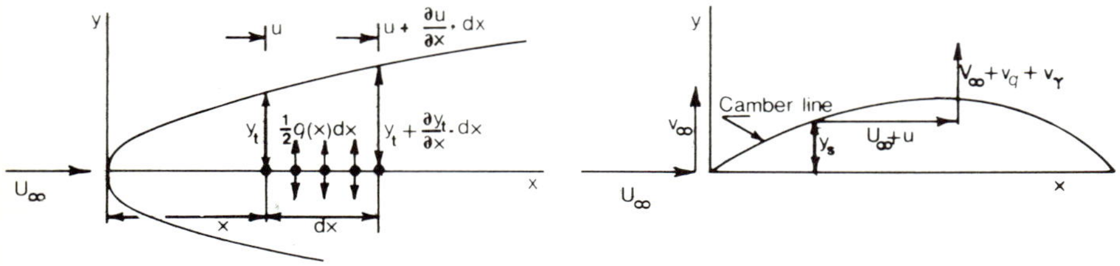

Fig. 5.2. Kinematic flow conditions for Schlichting's method.

Firstly for a thick blade, of zero camber, where $y_t = \frac{1}{2}(y_u - y_l)$, application of the continuity equation to Fig. 5.2(a) gives

$$\frac{dy_t}{dx} = \frac{1}{2}\frac{q(x)}{U_\infty + u}\,. \qquad (5.8)$$

Now for the camber line only, $y_s = \frac{1}{2}(y_u + y_l)$. For tangency in Fig. 5.2(b)

$$\frac{dy_s}{dx} = \frac{V_\infty + v}{U_\infty + u}\,. \qquad (5.9)$$

In the above equations both u and v have source, sink and vortex components.

It is convenient to put

$$x/c = \tfrac{1}{2}(1 - \cos\varphi)\,. \qquad (5.10)$$

Following Glauert [5.47] the vortex and source distributions are expanded in series form

$$\gamma(x) = 2U_\infty\left[A_0 \cot\varphi/2 + A_1 \sin\varphi + A_2 \sin 2\varphi + \ldots\right] \qquad (5.11)$$

$$q(x) = 2U_\infty\left[B_0(\cot\varphi/2 - 2\sin\varphi) + B_2 \sin 2\varphi + \ldots\right] \qquad (5.12)$$

where A_0, A_1, A_2, B_0, B_2 are Fourier coefficients which have to be calculated by fulfilling the kinematic flow conditions. It is interesting to note that the first term in $\gamma(x)$ represents a flat plate with incidence and the first term in $q(x)$ gives a symmetrical Joukowsky profile at zero incidence.

The induced velocities are evaluated using equations (5.6) and (5.7) for the vortex and source distributions respectively.

It is now possible to introduce the vortex and source distributions into the kernels resulting in a large number of integrals depending upon φ, s/c and χ. The results of these integrals can be tabulated and used as constant parameters. For a given fluid inlet angle the only variable is then the blade shape, which is specified by y_s' and y_t' at each given point. For each point on the profile a pair of simultaneous equations is produced. The solution of these to give velocities around the profile is straightforward; the only slight difficulty is that since the singularities were distributed along the chord line, the induced velocities are calculated along the chord line. For the velocity on the profile, a simple correction given by Riegels[5.53] is used.

The main assumptions and approximations of the Schlichting method are as follows:-

(i) a distribution of singularities is used to match the blade profile at a finite number of points

(ii) this number of matching points is restricted by the stability of the Fourier series which is used to represent the singularity distribution

(iii) separate distribution of blade camber line and thickness distribution

(iv) distribution of singularities along the chord line

(v) the blade shape is introduced, not in the form of (x,y) coordinates, but in the form (x, dy/dx) and since the gradients of an arbitrary profile are difficult to measure, errors in profile specification are unavoidable

(vi) the continuity equation which describes the thickness distribution contains a term in du/dx which is usually neglected. This leads to errors near the leading edge.

Each of the singularity methods was complicated and tedious to apply. There was thus an incentive, especially when using the methods for design purposes to look for simplifications which would expedite the calculations.

In deriving equation (5.8) a term in du/dx was neglected which was admitted to be not insignificant, especially near the leading edge. Mellor[5.54] took this simplification a stage further, neglecting the term $u \cdot dy_t/dx$ thus leaving equation (5.8) in the form

$$\frac{dy_t}{dx} = \frac{1}{2}\frac{q(x)}{U_\infty} \quad . \tag{5.13}$$

This represented a remarkable simplification of Schlichting's analysis since, dy_t/dx being known, q is solved directly and the profile contour is now solved in integral form. However, in the written discussion of Mellor's paper, Scholz showed that the neglect of u was not really justifiable and that u must be treated as a function of y. The Taylor series expansion in Scholz's own paper[5.55] shows this point quite clearly. Simplified theories in refs 5.54, 5.56 and 5.57 neglect the value of u as compared with U.

5.1.4 The Martensen method. The starting point for the Martensen method is equation (5.6) giving a representation of the flow field by an appropriate distribution of vorticity,

$$c(z) = c_x(z) - ic_y(z) = c_o(z) + \frac{1}{2is}\oint_{P_0} \gamma(\lambda')\coth\frac{\pi}{s}(\,l - z)\,d\lambda' \tag{5.6}$$

with $z = f(\lambda)$ and $l = f(\lambda')$. λ' gives a convenient representation of the profile analogous to equation (5.10) of the Schlichting theory.

It can be shown from Stokes' theorem that the function $\gamma(\lambda)$ is given by

$$\gamma(\lambda) = 2\frac{\partial c}{\partial z} = 2cf'(\lambda) \ . \tag{5.14}$$

The integration extends over the contour P_o (Fig. 5.3) the integrand having a period is. It is now necessary to choose a real vorticity distribution in such a way that the profiles become streamlines and the vorticity distribution is identical to the surface velocity distribution. This is achieved by forcing disappearance of any velocity within the profile. To do this one can postulate that either the normal velocity or the internal tangential velocity should tend to zero as the surface is approached.

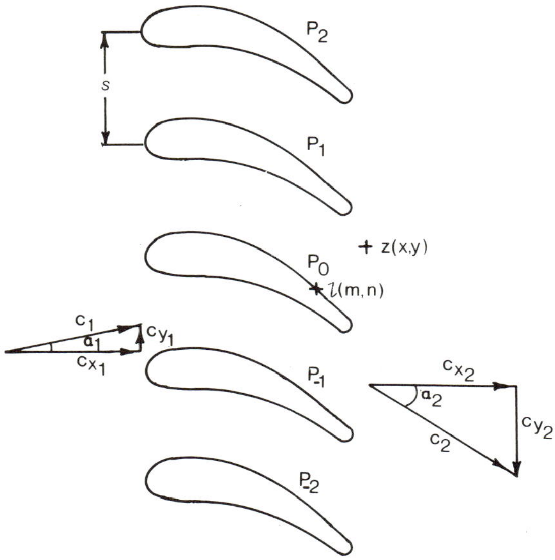

Fig. 5.3. Notation for the Martensen method.

If the former condition is chosen then

$$2\,\mathrm{Im}[c_0 f'(\lambda)] = \oint_{P_0} \mathrm{Im}[K(\lambda,\lambda')]\gamma(\lambda')\,d\lambda' , \tag{5.15}$$

or if the latter condition

$$\gamma(\lambda) = 2\,\mathrm{Re}\,[c_0 f'(\lambda)] + \oint_{P_0} \mathrm{Re}[K(\lambda,\lambda')]\,\gamma(\lambda')\,d\lambda' , \tag{5.16}$$

the function $K(\lambda,\lambda')$ being given by

$$K(\lambda,\lambda') = \frac{1}{is} f'(\lambda) \coth\frac{\pi}{s}(f(\lambda') - f(\lambda)) .$$

In principle either of the above equations could be used. Equation (5.15) was used by Isay[5.58]. Unfortunately he was only partially successful because the numerical treatment of his first order integral equation was quite tedious, the kernel being singular. This also led to limitations on the blade thickness. The credit for the solution of the cascade flow problem must really go to Martensen who, in 1959, used equation (5.16).

If

$$f(\lambda') = m(\lambda') + in(\lambda') ;$$

equation (5.16) then has the real part

$$\gamma(\lambda) = 2c_{xo}\dot{m}(\lambda) + 2c_{yo}\dot{n}(\lambda) + \oint_{P_0} K(\lambda,\lambda')\gamma(\lambda')\,d\lambda' \qquad (5.17)$$

where

$$K(\lambda,\lambda') = \frac{1}{s}\cdot\frac{\dot{n}(\lambda)\sinh\frac{2\pi}{s}\{m(\lambda)-m(\lambda')\} - \dot{m}(\lambda)\sin\frac{2\pi}{s}\{n(\lambda)-n(\lambda')\}}{\cosh\frac{2\pi}{s}\{m(\lambda)-m(\lambda')\} - \cos\frac{2\pi}{s}\{n(\lambda)-n(\lambda')\}} .$$

Now

$$\gamma(\lambda) = c(\lambda)\sqrt{\dot{m}^2 + \dot{n}^2}$$

$$\therefore\; c(\lambda)\sqrt{\dot{m}(\lambda)^2 + \dot{n}(\lambda)^2} - \oint_{P_0} K(\lambda,\lambda')c(\lambda')\sqrt{\dot{m}(\lambda)^2 + \dot{n}(\lambda)^2}\,d\lambda' = 2c_x\dot{m}(\lambda) + 2\left(\frac{c_{y1}+c_{y2}}{2}\right)\dot{n}(\lambda) \qquad (5.18)$$

which is regular and is a Fredholm Integral Equation of the second kind. Martensen was able to demonstrate the existence of a solution quite generally.

Although their presentation was less rigorous, Bueckner and Schnackel [5.59] also gave a solution in 1959, complete with worked examples for turbine blading. The procedure of Giesing[5.60] is similar to that of Martensen although the theory is strictly equivalent to a surface source distribution rather than a distribution of vortices. This conceptual difference in no way affects the result.

If a smooth contour is assumed then

$$\oint_{P_0} K(\lambda,\lambda')\,d\lambda' = 1 . \qquad (5.19)$$

Equation (5.18) is linear so that Martensen was able to synthesize the flow from a combination of three basic solutions

$$c(\lambda) = \frac{\left[c_x a(\lambda) + \tfrac{1}{2}(c_{y1}+c_{y2})b(\lambda) + \Gamma\gamma(\lambda)\right]}{\sqrt{\dot{m}(\lambda)^2 + \dot{n}(\lambda)^2}} \qquad (5.20)$$

where $a(\lambda)$ corresponds to a flow at zero inlet angle,
$b(\lambda)$ corresponds to a flow at 90° inlet angle and
$\gamma(\lambda)$ corresponds to purely circulatory flow.

Figure 5.4 depicts these three basic flows.

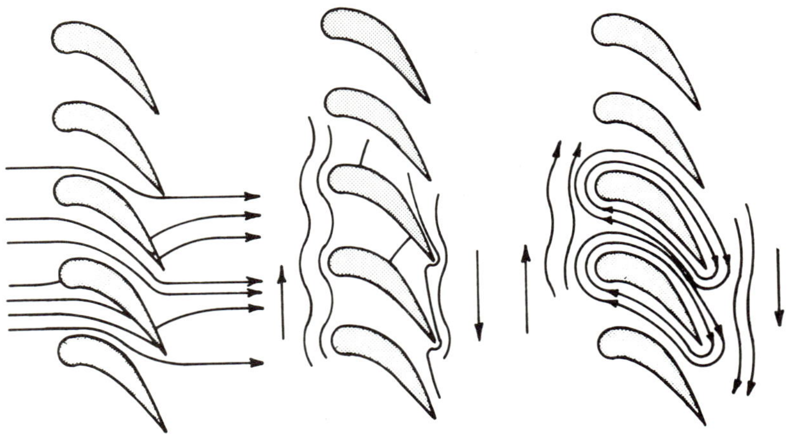

Fig. 5.4. Superposition of flow solutions in Martensen theory.

The equations for solution are then

$$a(\lambda) - \oint_{P_0} K(\lambda,\lambda')\, a(\lambda')\, d\lambda' = 2\dot{m}(\lambda) \tag{5.21 a}$$

$$b(\lambda) - \oint_{P_0} K(\lambda,\lambda')\, b(\lambda')\, d\lambda' = 2\dot{n}(\lambda) \tag{5.21 b}$$

$$\gamma(\lambda) - \oint_{P_0} K(\lambda,\lambda')\, \gamma(\lambda')\, d\lambda' = 0 \tag{5.21 c}$$

with

$$\oint_{P_0} a(\lambda')\, d\lambda' = \oint_{P_0} b(\lambda')\, d\lambda' = 0\ , \qquad 2\pi \oint_{P_0} \gamma(\lambda')\, d\lambda' = 1\ .$$

A numerical solution to equation (5.21) is obtained as follows:-
The blade profile is divided into 2N parts over the range $0 < \lambda' < 2\pi$. The equations are then expressed as a series of simultaneous linear equations for $\gamma(\lambda_i)$

$$\gamma(\lambda_i) - \frac{1}{2\pi}\sum_j \gamma(\lambda_i) K(\lambda_i,\lambda_j)(\lambda_{j+1}-\lambda_j) = 2c_x\dot{m}(\lambda_i) + 2c_y\dot{n}(\lambda_i)\ . \tag{5.22}$$

Since equations (5.22) are linearly dependent, equation (5.19) can be written

$$\sum_{p=1}^{2N} K_{pp} = 2N \tag{5.23}$$

and equations (5.22) can be written as a set of simultaneous equations for a_p, b_p, and γ_p where p = 1,2............2N giving

$$\frac{1}{2N}\begin{bmatrix} 2N-K_{11} & -K_{12} & \cdots \\ -K_{21} & 2N-K_{22} & \\ \cdot & \cdot & \\ \cdot & \cdot & \\ \cdot & \cdot & \\ -K_{2N,1} & -K_{2N,2} & \\ 2\pi & 2\pi & \end{bmatrix} \cdot \begin{bmatrix} a_1 & b_1 & \gamma_1 \\ a_2 & b_2 & \gamma_2 \\ \cdot & \cdot & \cdot \\ \cdot & \cdot & \cdot \\ \cdot & \cdot & \cdot \\ a_{2N} & b_{2N} & \gamma_{2N} \end{bmatrix} = \begin{bmatrix} 2x_1 & 2y_1 & 0 \\ 2x_2 & 2y_2 & 0 \\ \cdot & \cdot & \cdot \\ \cdot & \cdot & \cdot \\ \cdot & \cdot & \cdot \\ 2x_{2N} & 2y_{2N} & 0 \\ 0 & 0 & 1 \end{bmatrix}. \tag{5.24}$$

Price improved the integration of the equations by postulating $\gamma(\lambda')$, $\dot{x}(\lambda')$, $\dot{y}(\lambda')$, to be invariant between points, rather than the previous case in which $K(\lambda, \lambda')$ had been assumed constant[5.61]. Giesing[5.60] also did this, approximating the profile by straight line elements. This technique is easy to apply and avoids possible troubles, especially near thin trailing edges.

Experience shows that unless the quality of input data is very good it is worthwhile taking a fair amount of time and trouble to input smooth curves into the program. Martensen had used a Fourier Series representation of the 2N data points, giving a convenient computed value of $(\dot{x}_1, \dot{y}_1)\ldots(\dot{x}_{2N}, \dot{y}_{2N})$. Martensen admits that this can give trouble, especially with fixed-point computers. Payne[5.62] shows that these difficulties can be overcome by using the blade surface length s as a parameter instead of λ'. Other possibilities include using curve fitting of the points with cubics in $y=f(\lambda)$ or with an appropriate spline fit.

If the blade edges are not well defined the use of a rounding technique might be considered. Bueckner and Schnackel[5.59] gave such a technique, in which the edge curve is defined by

$$y(x) = y_m(x) \pm \sqrt{a_0 + a_1x + a_2x^2 + a_3x^3} \qquad x \geqslant 0 \tag{5.25}$$

y_m is located on the camber line and the coefficients a_0, a_1, a_2 and a_3 are so determined that the first and second derivatives of the profile are continuous at the intersection between edge and main profiles. The result is a curve which resembles an ellipse with a major to minor axis ratio of two. The use of a polynomial fit for the edge is not recommended and it should be stressed that discontinuities in first or second derivative will disrupt the program.

The number of points which must be taken on the profile depends on the accuracy of the data. For very accurate data good results have been obtained with less than 30 points but in general 100 points or more should be used; as a rough guide some 15 of these should be placed on each blade edge. Where the blade is thin it is good practice to place pairs of points opposite each other with respect to the camber line. As a general rule the spacing along the profile between adjacent points should not increase by more than 50% as one moves from one point to the next.

5.2 An Exact Solution for Cascades

Results have been published, based upon one or other of the above potential flow methods, for many different aerofoil profiles and blade configurations. However, due to the possibility of error in the lengthy computations and to the differing assumptions made, discrepancies are noticeable when two or more of the methods are applied to the same blade profile. As an example of these errors, Pollard and Wordsworth [5.40] compared pressure distributions calculated for a similar cascade configuration using the Schlichting and Howell methods and found sufficient disagreement near the peak of suction pressure to make subsequent boundary layer calculations useless. It was clear that an absolute standard was required in order to check the validity of potential flow solutions; the author has provided such a standard[5.63] using as a basis the early work of Merchant and Collar[5.4].

Collar[5.64] produced an analysis giving a transformation linking the known potential flow around a cascade of ovals to that around a cascade of staggered flat plates. During the years 1941-44 Merchant and Collar used a similar procedure on a physically simpler series of ovals and gave suggestions for extending the theory to a cascade of aerofoil profiles in an analogous manner to the theory of isolated Joukowsky transform aerofoils. As a result of this they were able to give useful plots of outlet angle and lift coefficient, assuming application of the Kutta-Joukowsky condition, for a range of cascade configurations. The theory provided the basis of a standard for evaluating the accuracy of pressure and velocity distributions obtained from approximate solutions.

It became evident, as a result of later development of the exact theory, that the Kutta-Joukowsky condition has no meaning with respect to the general case of the potential field around aerofoils in cascade; the outlet angle itself becomes an independent variable. This general case of profiles with a rounded trailing edge will be discussed in section 5.2.2. It was obvious that an indeterminate state does not exist in real fluid flow and that therefore the condition for determining the outlet flow angle (or rear stagnation point) must be obtainable from a study of the effect of viscosity. The search for such a condition constitutes the main thrust of Chapter 7.

5.2.1 The exact solution to the flow through a derived cascade. The procedure for evaluation of the blade profile shape and cascade configuration is as follows:

(i) The normal flow (i.e. flow perpendicular to the imaginary axis) past a series of ovals on the imaginary axis has been given by Lamb[5.65].

(ii) The normal flow past a series of laminae lying along the imaginary axis is also known and a transformation can be obtained which converts the laminae into the ovals of (i) (Fig. 5.5(a)).

(iii) In a similar way the known general flow (i.e. the flow with inlet angle and circulation) around the laminae gives the general flow around the ovals.

(iv) A particular case of the general flow around the ovals is that for which the flow at infinity is inclined to the axis but for which there is no circulation. In this case the ovals can immediately be transformed into a cascade of flat plates parallel to the direction of flow at infinity (Fig. 5.5(b)).

(v) Application of this transformation to ovals which are offset from the origin produces a cascade of aerofoil shapes. This is the class of aerofoil for which the profile shape, and subsequently the aerodynamic characteristics, will be obtained.

The procedure is similar to the usual Joukowsky process for an isolated aerofoil and, if the axes of each oval are small compared with the spacing, these ovals tend to become circles and the transformation used becomes the Joukowsky transformation. The full analysis is given below; the scope and limitations of the theory have been discussed in the author's papers [5.66], [5.67].

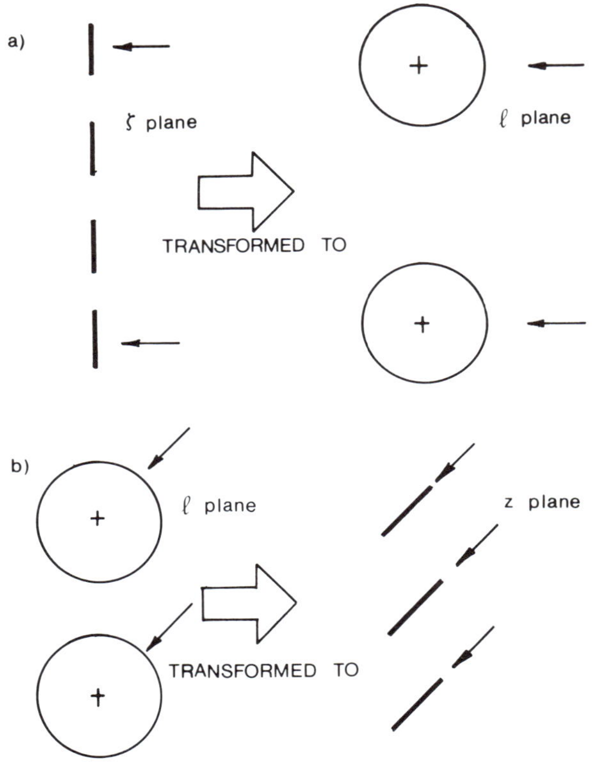

Fig. 5.5. Stages in the transformation.

5.2.2 Derivation of aerofoils. The potential field due to normal flow of a uniform stream past a series of uniform doublets lying along the imaginary ℓ plane axis is given by Lamb[5.65] :-

$$\omega = U(\ell + \sinh^2\beta \coth\ell). \tag{5.26}$$

The streamline $\psi = 0$ marks part of the $n = 0$ axis and the closed oval

$$\cosh 2m = \cos 2n + \frac{\sinh^2\beta \sin 2n}{n} \tag{5.27}$$

β is now seen to be the length of the semi-major oval axis.

Considering a series of laminae distributed along the η axis of the ζ plane with a period of π, the stagnation points of the flow around these laminae may be made to correspond to those in the ℓ plane. We can thus put $\ell = \pm\beta$, as the stagnation point, in equation (5.26).

For simplicity, following Merchant and Collar, λ and γ are defined as

$$\lambda = \ell + \sinh^2\beta \coth \ell , \tag{5.28}$$

$$\gamma = \beta + \sinh^2\beta \coth \beta . \tag{5.29}$$

The flow around the laminae is given by

$$\omega = U \cosh^{-1}(\cosh \gamma \cosh \zeta) \tag{5.30}$$

and since, from (5.26), $\omega = U\lambda$ for the ovals

$$\cosh \lambda = \cosh\gamma \cosh \zeta . \tag{5.31}$$

This is thus the required transformation connecting the ℓ plane ovals and the ζ plane laminae.

The general flow past the ζ plane laminae is known to be

$$\frac{d\omega}{d\zeta} = \frac{U \sinh\zeta + iW \cosh \zeta}{\sqrt{\sinh^2\zeta + \tanh^2\zeta}} - iV \tag{5.32}$$

where V is the component of velocity parallel to the η axis, and there is a circulation $\Gamma = -2\pi W$ around each lamina.

If transformation (5.31) is applied to the general flow past the laminae, the general flow past the ovals is obtained:-

$$u_\ell - iv_\ell = \frac{d\omega}{d\ell} = \left[U + i\left(\frac{W\cosh \lambda - V\sinh\lambda}{\sinh^2\lambda - \sinh^2\gamma}\right)\right] \cdot \left[1 - \frac{\sinh^2\beta}{\sinh^2\ell}\right] . \tag{5.33}$$

The ovals are next transformed into inclined flat plates and aerofoils.

Considering the particular case in which $W = 0$, $V = U\tan\chi$, equation (5.33) becomes

$$u_\ell - iv_\ell = \frac{d\omega}{d\ell} = U\left[1 - \frac{i \tan\chi \sinh \lambda}{\sinh^2\lambda - \sinh^2\gamma}\right] \cdot \left[1 - \frac{\sinh^2\beta}{\sinh^2\ell}\right] . \tag{5.34}$$

Also, considering the flow past a cascade of inclined flat plates, as shown in Fig. 5.6

$$\frac{d\omega}{dz} = U(1 - i\tan\chi) . \tag{5.35}$$

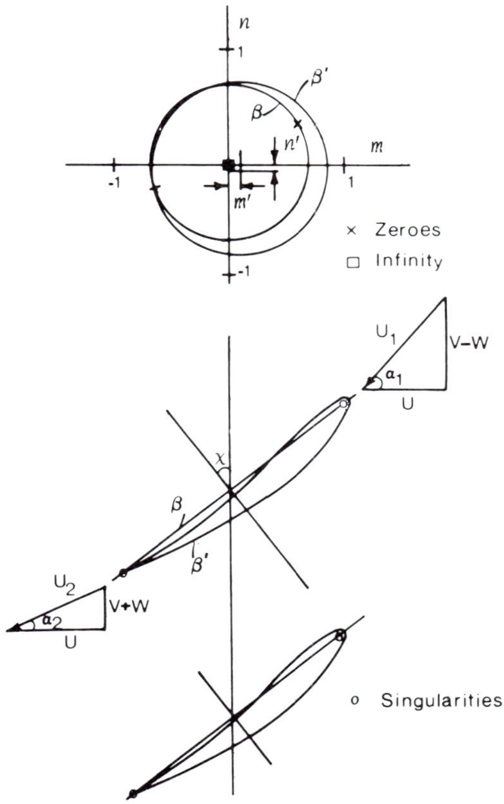

Fig. 5.6. The ℓ plane ovals and the z plane cascade.

Elimination of w in (5.24) and (5.35) gives

$$\frac{dz}{d\ell} = e^{i\chi}\left[\cos\chi - \frac{i\sin\chi\ \sinh\lambda}{\sinh^2\lambda - \sinh^2\gamma}\right]\times\left[1 - \frac{\sinh^2\beta}{\sinh^2\ell}\right]. \tag{5.36}$$

Hence the transformation connecting the oval and cascade planes is given by

$$z = e^{i\chi}(\lambda\cos\chi - i\sin\chi\ \text{arccosh}(\text{sech}\,\gamma\ \cosh\lambda)) . \tag{5.37}$$

The procedure for the derivation of a cascade of aerofoils is thus to select a suitable set of ℓ plane ovals, postulate a set of larger ovals with offset centres, and apply transformation (5.37) to these offset ovals.

Experience enables the required type of cascade to be generated. For example, β should be around 0.725 to give a space-chord ratio of unity and β' should be approximately 10% larger than β to give a maximum thickness of 10% of the chord. Variation of camber and position of maximum thickness is obtained by variation of m' and n' (the coordinates of the offset oval centre); the proviso is that the transformation singularities must be enclosed within the offset oval, or for a cusped trailing edge, lie on this oval.

5.2.3 The flow around the aerofoils. The relationship between velocities in the ℓ and z planes is given by

$$u_z - i v_z = \frac{u_\ell - i v_\ell}{\left|\frac{dz}{d\ell}\right|} \tag{5.38}$$

where subscript z refers to the local velocity on the z plane cascade profile, subscript ℓ refers to the local velocity on the ℓ plane profile.

From equations (5.34) and (5.36) the velocity in the z plane is given by

$$u_z - i v_z = \frac{\left[U + i\,\frac{W\cosh\lambda - V\sinh\lambda}{\sqrt{\sinh^2\lambda - \sinh^2\gamma}}\right]\cdot\left[1 - \frac{\sinh^2\beta}{\sinh^2\ell}\right]}{\left|\left[\cos\chi - \frac{i\sin\chi\sinh\lambda}{\sqrt{\sinh^2\lambda - \sinh^2\gamma}}\right]\cdot\left[1 - \frac{\sinh^2\beta}{\sinh^2\ell}\right]\right|} . \tag{5.39}$$

To obtain W, the stagnation point value of ℓ is substituted into equation (5.34). For a cusped trailing edge the Kutta-Joukowsky condition must be satisfied. Since $d\omega/dz = (d\omega/d\ell)/(dz/d\ell)$ and both $(d\omega/d\ell)$ and $(dz/d\ell)$ become zero if the stagnation point is at the point of the cusp, the condition is satisfied.

From equation (5.34), at the rear stagnation point

$$W = V\tanh\lambda_T + i\,U\sqrt{1 - \frac{\cosh^2\gamma}{\cosh^2\lambda_T}} , \tag{5.40}$$

where the subscript T refers to trailing edge conditions.

Now the circulation is taken as $\Gamma = -2\pi W$, giving the following values for air angles α_1 and α_2

$$\tan\alpha_1 = \frac{V - W}{U} , \qquad \tan\alpha_2 = \frac{V + W}{U} .$$

Hence

$$V = U\tan\alpha_1 + W$$

and from equation (5.40)

$$W = U\left[\frac{\tan\alpha_1\tanh\lambda_T + \sqrt{\frac{\cosh^2\gamma}{\cosh^2\lambda_T} - 1}}{1 - \tanh\lambda_T}\right] ,$$

$$V = U\left[\frac{\tan\alpha_1 + \sqrt{\dfrac{\cosh^2\gamma}{\cosh^2\lambda_T} - 1}}{1 - \tanh\lambda_T}\right].$$

Substituting these values for W and V into equation (5.39) we obtain

$$\frac{u_Z - i v_Z}{U} = \frac{(1 + iA)\cdot\left(1 - \dfrac{\sinh^2\beta}{\sinh^2\ell}\right)}{\left|\left[\cos\chi - \dfrac{i\sin\chi\sinh\lambda}{\sinh^2\lambda - \sinh^2\gamma}\right]\cdot\left[1 - \dfrac{\sinh^2\beta}{\sinh^2\ell}\right]\right|} \tag{5.41}$$

where

$$A = \frac{\left\{\tan\alpha_1\tanh\lambda_T + \sqrt{\dfrac{\cosh^2\gamma}{\cosh^2\lambda_T} - 1}\right\}\cosh\lambda - \left\{\tan\alpha_1 + \sqrt{\dfrac{\cosh^2\gamma}{\cosh^2\lambda_T} - 1}\right\}\sinh\lambda}{(1 - \tanh\lambda)\sqrt{\sinh^2\lambda_T - \sinh^2\gamma}}.$$

For the potential flow around the β' ovals to be determined, these ovals must be relocated with their centres at the points (0,0), $(0,\pi)$, $(0,2\pi)$, in the ℓ plane.

If $\quad \ell_{T2} = \ell_T - \ell' \quad$ where $\quad \ell' = m' + in'$

then

$$\lambda_{T2} = \ell_{T2} + \sinh^2\beta'\coth\ell_{T2}.$$

Now

$$\tan\alpha_1 = \frac{V - W}{U} \qquad \text{and} \qquad \tan\alpha_2 = \frac{V + W}{U}$$

thus

$$\tan\alpha_2 = \frac{\tan\alpha_1(1 + \tanh\lambda_{T2}) + 2\sqrt{\dfrac{\cosh^2\gamma}{\cosh^2\lambda_{T2}} - 1}}{1 - \tanh\lambda_{T2}}. \tag{5.42}$$

The complete flow around the relocated β' ovals will be needed, for use in obtaining the cascade profile pressure distribution; from

$$\frac{u_Z - i v_Z}{U} = \left[1 - \frac{\sinh^2\beta'}{\sinh^2\ell'}\right]\times \tag{5.43}$$

$$\left[1 + i\,\frac{\left\{\tan\alpha_1\tanh\lambda_T + \sqrt{\dfrac{\cosh^2\gamma}{\cosh^2\lambda_T} - 1}\right\}\cosh\lambda - \left\{\tan\alpha_1 + \sqrt{\dfrac{\cosh^2\gamma}{\cosh^2\lambda_T} - 1}\right\}\sinh\lambda}{(1 - \tanh\lambda_T)\sqrt{\sinh^2\lambda - \sinh^2\gamma}}\right].$$

In equation (5.43) all λ's and ℓ's refer to conditions around the β' oval relocated and centred at the origin.

Making use of equation (5.36)

$$\left|\frac{dz}{d\ell}\right| = \left|\left(\cos\chi - \frac{i\sin\chi\sinh\lambda}{\sqrt{\sinh^2\lambda - \sinh^2\gamma}}\right)\times\left(1 - \frac{\sinh^2\beta}{\sinh^2\ell}\right)\right| \qquad (5.44)$$

Here, as in the basic transformation used to determine the aerofoil profile, the β of the smaller oval is employed.

It is now possible to evaluate the $(u_z - iv_z)/U$ of equation (5.41) using the right hand side of equation (5.43) as the numerator, and that of (5.44) as the denominator. The value of the pressure coefficient for the corresponding point on the aerofoil surface is given by application of the Bernoulli equation.

Preliminary calculations and exact results for a cascade of cusped compressor blades are provided in Appendix A.

5.2.4 Extension to blades with rounded trailing edges. In the previous section the particular case was considered in which m' and n' were chosen such that the two ovals intersected at one singularity of the transformation and such that the larger (β') oval enclosed the other singularity. The transformation of the flow about the ovals into the flow about the aerofoils in the z plane was thus conformal everywhere and a cascade of profiles with cusped trailing edges was derived, as in Fig. 5.7.

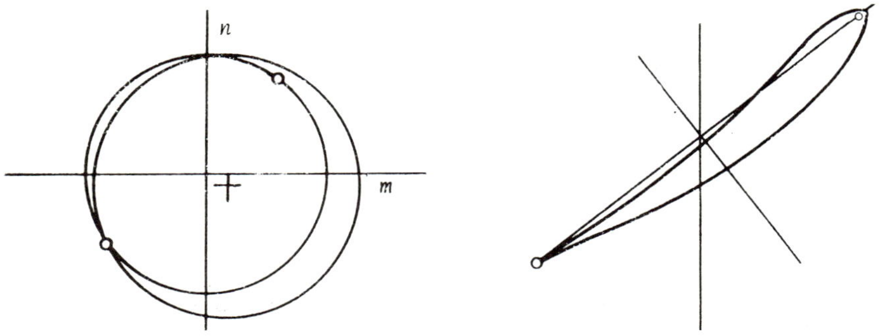

Fig. 5. 7. Profile with cusped trailing edge.

This case may be regarded as a limiting case of the shortest distance, between the β' oval and one singularity, tending to zero. At the other limit the larger oval is concentric with the β oval, thus implying $\ell' = m' + in' = 0$. Upon transformation to the z plane a cascade of symmetrical uncambered aerofoils which tend towards an elliptic shape is obtained, as in Fig. 5.8. The significance of the relationship between the distance in the ℓ plane from the singularity to the nearest point of the β' oval and the radius of curvature on the nose or tail in the z plane thus becomes apparent.

The general case is intermediate between the two extremes mentioned above. As long as both singularities in the ℓ plane are enclosed by the β' oval, any position of ℓ' will give a transform which is conformal. It is thus possible, by correct selection of the normal distance in the ℓ plane between the singularity and the basic oval, to obtain any value of radius of curvature at the nose or tail in the z plane, from zero to a high positive value.

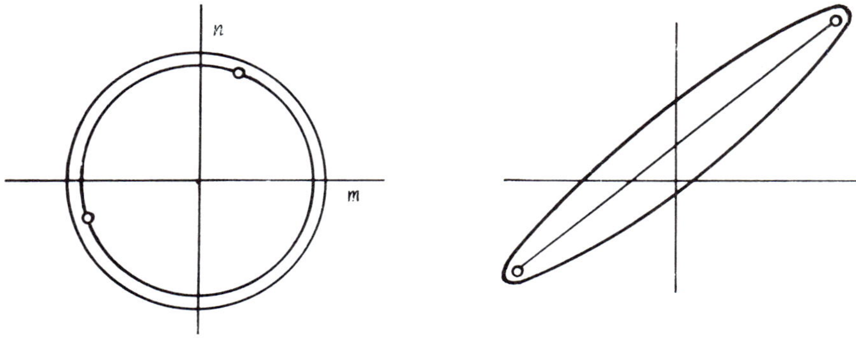

Fig. 5.8. Elliptical profile.

By a "trial and error" alteration of the variables β', m' and n' it is possible to approximate to practical compressor cascade profiles, which have rounded leading and trailing edges (Fig. 5.9).

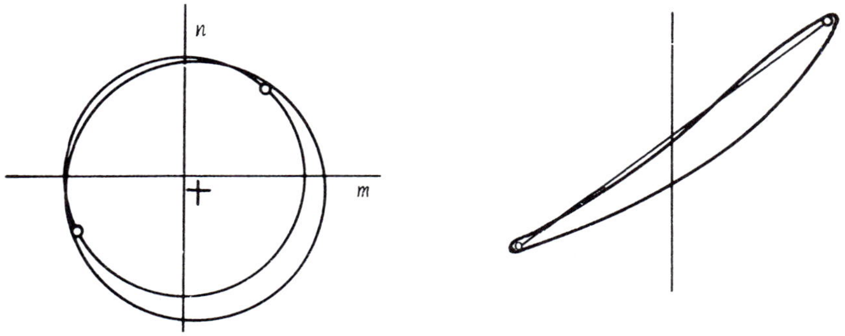

Fig. 5.9. Practical compressor blade.

An aid to the convergence of the iterative process is given by a knowledge of the position of maximum thickness in the cusped profile of the previous section (at 23% of the chord length) and the knowledge that for the ovals in a concentric configuration the position of the maximum thickness is at 50% chord. With these extremes fixed any desired position of maximum thickness can be obtained by an approximate interpolation for $\ell' = m' + in'$.

The procedure for calculation of the pressure distribution and outlet angle for a profile with a rounded trailing edge follows that of the previous section. With this new type of profile, however, there is an additional complication. For profiles with a cusped trailing edge the selection of the position of rear stagnation point, and hence the outlet angle, followed automatically from the Kutta-Joukowsky condition[5.1,5.2] which avoided infinite velocities at the trailing edge by placing the rear stagnation point at the transformation singularity, on the point of the cusp. This condition does not apply to potential flows which have no singularity and which do not tend to give rise to infinite velocities on the profile. The Kutta-Joukowsky condition cannot, therefore, be applied to the usual type of compressor or turbine blade which has a rounded trailing edge.

No alternative condition which could be applied to the potential flow around a cascade of aerofoils with rounded trailing edges was known. All previous investigators have either:-

(a) replaced the aerofoil by an equivalent one with a cusp at the trailing edge (e.g. the Schlichting method).

(b) placed the stagnation point either at the intersection of the profile with the line connecting the trailing edges of each aerofoil in the cascade or at the rear end of the camber line.

(c) specified the outlet angle as well as the inlet angle (e.g. the Martensen method).

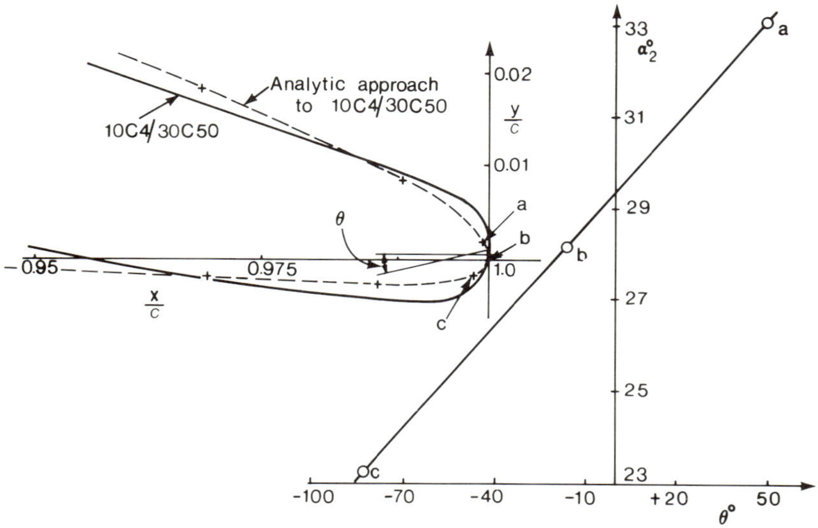

Fig. 5.10. Outlet angle as a function of stagnation point location.

Figure 5.10 gives the variation of outlet angle with the position of the stagnation point for a typical compressor cascade profile. Such results as these, predicting a large variation in outlet angle for a comparatively small alteration in the position of the stagnation point, emphasize the fact that the potential flow around a conventional aerofoil in cascade is not completely "determined" by specification of the cascade configuration and the inlet angle but that the position of the rear stagnation point is a further variable. Figures 5.11 and 5.12 show that the effect of the position of the rear stagnation point upon the pressure distribution for a compressor cascade profile is also large.

Since no condition has been discovered which gives a unique solution to the potential flow around a cascade, the quest for such a condition will be postponed until the effect of viscosity on the flow is considered. In this chapter the author will confine the investigation to obtaining outlet angles and pressure distributions for clearly stated and arbitrarily selected positions of rear stagnation point. This procedure, whilst introducing a further variable, will provide the maximum amount of information for use when considering viscosity effects.

5.2.5 <u>Potential flow near the rear stagnation point.</u> In order to obtain a qualitative understanding of the pressure distribution of Fig. 5.12 the conditions are depicted in an exaggerated fashion in Fig. 5.13.

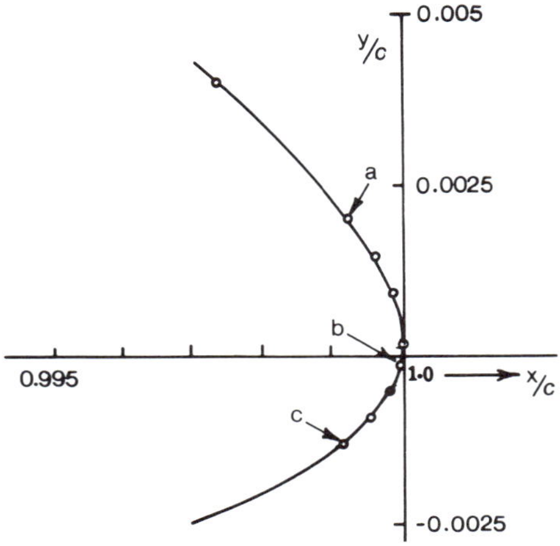

Fig. 5.11. Trailing edge detail for analytic profile with alternative locations of rear stagnation point.

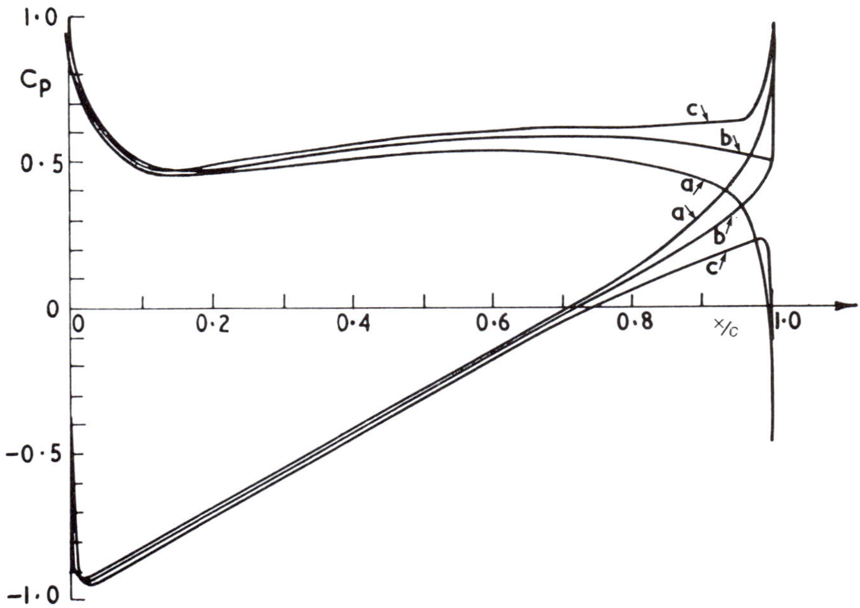

Fig. 5.12. Effect of location of rear stagnation point on pressure distribution for analytic profile.

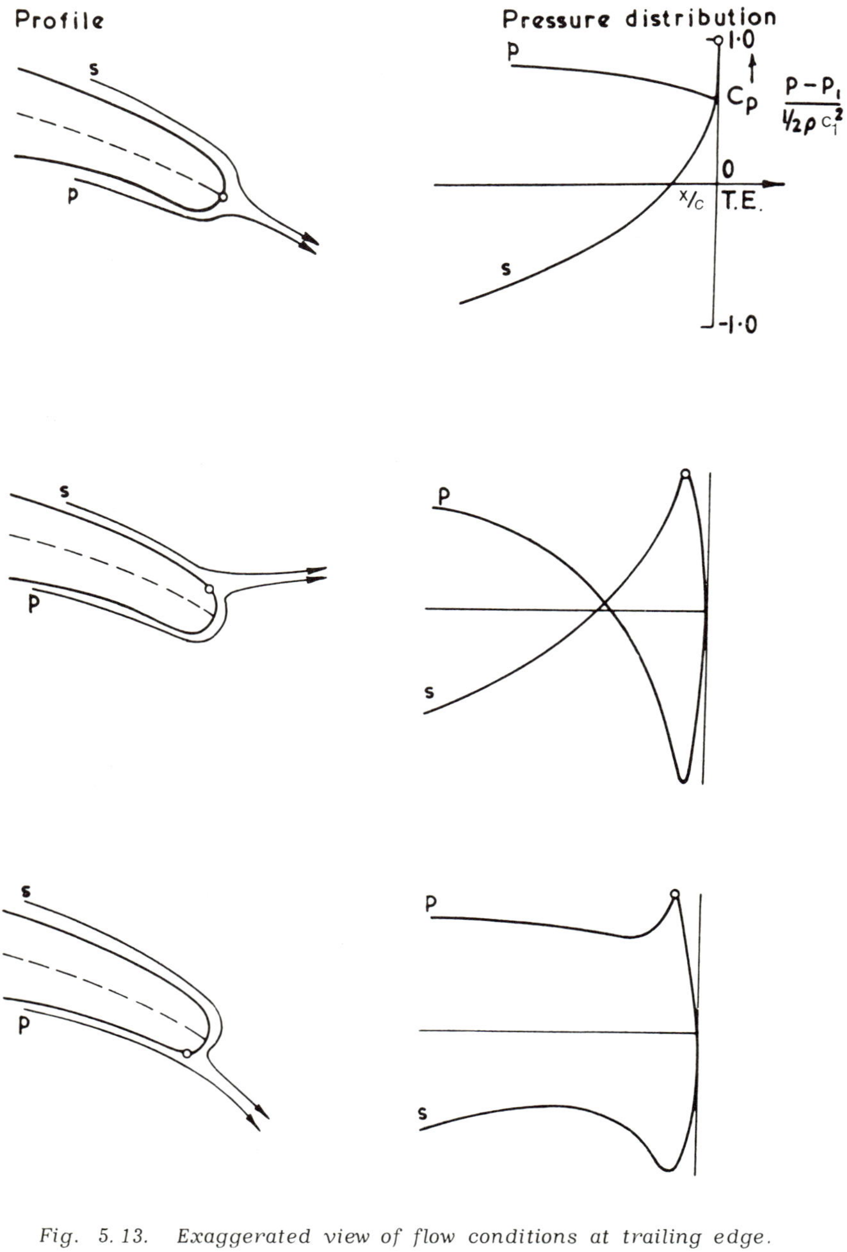

Fig. 5.13. Exaggerated view of flow conditions at trailing edge.

Figure 5.13(a) shows the stagnation point located at the extreme rear of the aerofoil. Since the flow does not attain very high velocities anywhere, the suction surface pressure rises smoothly to the stagnation point, at exactly $x/c = 1$. Since the path of a particle travelling along the pressure surface towards the stagnation point would follow similar curvatures, the velocity attained is similar and the pressure distribution approaches the stagnation point in a similar way.

In Fig. 5.13(b) the stagnation point is well up on the suction surface. Consequently, a fluid particle travelling along this surface is soon decelerated to approach the stagnation condition before the end of the chord. On the other hand the pressure surface route entails a path of high curvature and an associated high velocity (which would be infinite in the limiting case of a cusped trailing edge). After negotiation of this region of high curvature, a rapid deceleration would bring the particle to the stagnation point. The upper and lower surface pressure distributions intersect somewhat upstream of the stagnation point position.

Figure 5.13(c) shows the stagnation point on the pressure surface. Such configurations have possible parallels in jet flap work. In this case it is the pressure surface which gives low velocities and an early stagnation point. The suction surface velocity reaches high values in the region of high curvature. As in the case of the pressure surface velocity of Fig. 5.13(b), deceleration to the stagnation velocity is then rapid.

This exaggerated flow model should explain the pressure distributions of Fig. 5.12 and provide a qualitative picture of the potential flow conditions at the trailing edge.

5.3 Use of Exact Theory to Assess Accuracy of Existing Methods

5.3.1 Comparisons for cusped blade. Many of the more recent methods have been applied to the cascade of cusped blades given in Appendix A with the object of checking their predictions of pressure distribution and outlet angle. Figure 5.14 shows the blade profile and results of various programs for that cascade.

Difficulties with the Schlichting singularity method have been discussed in section 5.4. The predicted outlet angle is 0.7° too high. Discrepancies are seen to occur in the pressure distribution near the suction peak. Results provided by Hall and Thwaites [5.41], based on the Garrick method, show an accuracy in outlet angle of almost four decimal places and very good agreement in pressure distribution.

CABIS is G.E.'s Martensen-type program and the agreement shown is very good except in the trailing edge region. Another G.E. program is that of Novak and Haymann-Haber[5.68]. This uses a power series expansion for the passage flow and the accuracy is seen to be surprisingly good.

A finite element solution has been reported by O'Carroll and Morgan [5.69]. It gives good agreement but, despite local mesh refinement, accuracy drops off in the leading-edge region. A corresponding finite difference scheme was less accurate. Mesh refinement was also used in finite element solutions by Ecer and Akay [5.70] and Baskharone and Hamed [5.29]. These have also been compared with the test case and gave excellent agreement with the analysis. Caspar et al [5.71] have produced a finite area technique for high speed flows. This gave near-perfect agreement for the cusped cascade although it was reported that when the Kutta-Joukowsky condition was employed the predicted exit angle was in error by 0.77° and the pressure was unloaded slightly near the trailing edge.

A rapid (4 secs.) panel method has been demonstrated by McFarland [5.72] with near-perfect results for the pressure distribution and outlet angle of the cusped blade.

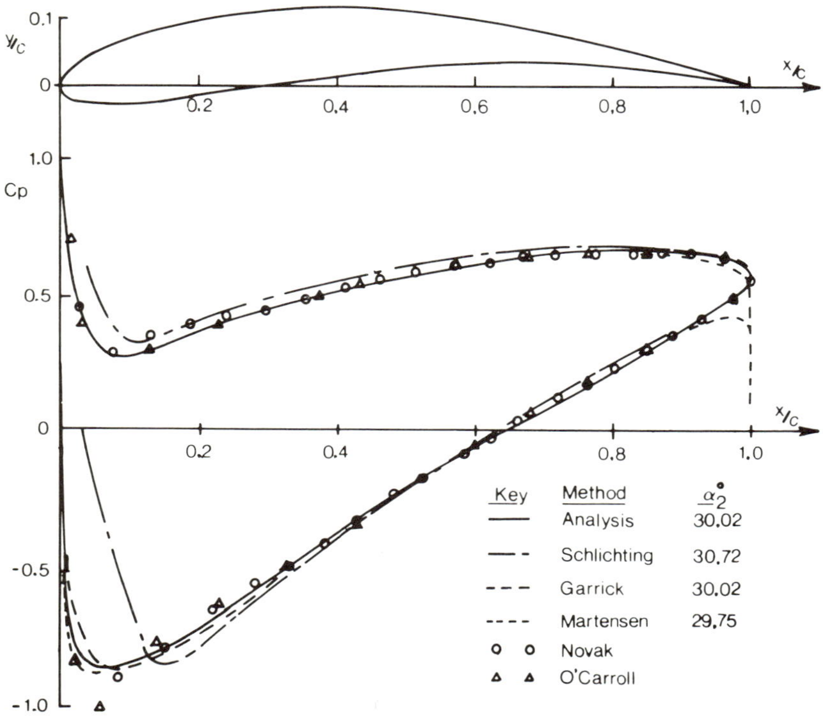

Fig. 5.14. Pressure distributions and blade profile for cusped blade cascade.

5.3.2 Blading having rounded trailing edges. The previous analysis for establishing an analytic profile with a rounded trailing edge has been used to obtain an approximation to a typical compressor cascade profile. The profile chosen was a 10C4/30C50 section set at a stagger of 36° with a space-chord ratio of unity.

It was found that, following the procedure of sub-section 5.2.4, only three iterations were necessary to give very good agreement with the standard C4 form. The maximum difference between the 10C4/30C50 section and the analytic approach to it is 0.3% of the chord length. The difference is most marked at the trailing edge. Fig. 5.10 gave a magnified view of the trailing edge. In this region, although the ordinates are still accurate to within 0.3% of the chord, the radius of curvature is only 40% of the given C4 trailing edge radius.

Having established a blade profile similar to those used in industrial applications, the author checked the results of several well-known potential flow methods. To eliminate one variable the position of the rear stagnation point was arbitrarily selected to be at the end of the camber line, position 'b', the co-ordinates of this position being (1.0) in the z(x,y) plane. For the used inlet angle of 51° this resulted in an outlet angle given by $\tan \alpha_2 = 0.5359714$, according to the analysis.

Methods of solution used were those of Garrick[5.37], Martensen[3.27], Howell[5.36] and Schlichting[3.25].

The computer programs for the Garrick, Howell and Schlichting methods were not available to the author at the time of comparison; however, these computations had been carried out previously for a 10C4/30C50 profile with the same stagger of 36°, the same inlet angle of 51° and a slightly different space chord ratio of unity (as compared with 0.9901985). As a result of the assumptions implicit in the Schlichting method, the trailing edge became cusped and the Kutta-Joukowsky condition was applied. As mentioned previously, the discrepancey between the 10C4/30C50 profile and the ordinates for the analytical approach to this was nowhere greater than 0.3% of the chord. The previously computed results from the Garrick, Howell and Schlichting methods are thus compared directly with the results of the Martensen method and of the exact solution in Fig. 5.15.

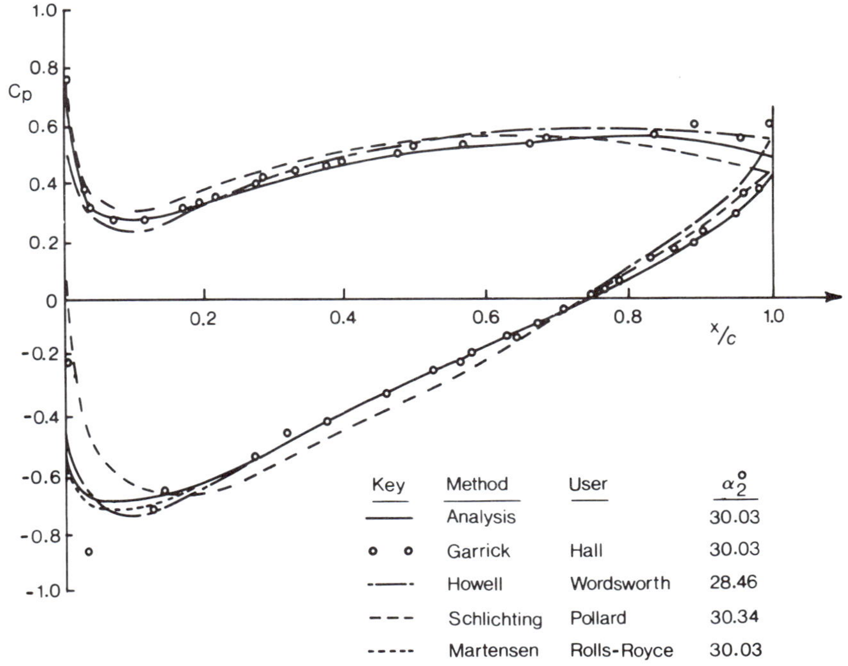

Fig. 5.15. Pressure distributions for C4-type profile.

The Garrick method gives good agreement, apart from one point near the suction peak. The Martensen method is seen to give complete agreement apart from a very slight error at the suction peak. The Howell method gives reasonable agreement, the maximum error in C_p being of the order of 10%. The Schlichting method in general gives a similar order of accuracy but misplaces the suction peak, just as it did for the profile with a cusped trailing edge, and has a maximum error in C_p of 15%.

5.3.3 70° camber profile. A need for an analytical solution to a highly cambered blade arose from doubts as to the applicability of the Schlichting method at high cambers. In the Schlichting theory the profile is generated by a source distribution. Ideally the sources would be distributed along the camber line, but due to the complexity of the mathematics involved the sources are usually distributed along the chord line, (this difficulty is partially overcome in refs. 5.51, 5.73). As a result of this simplification, errors were thought to be present in the prediction of the performance of highly cambered profiles. The following comparison is an attempt to assess the magnitude of these errors.

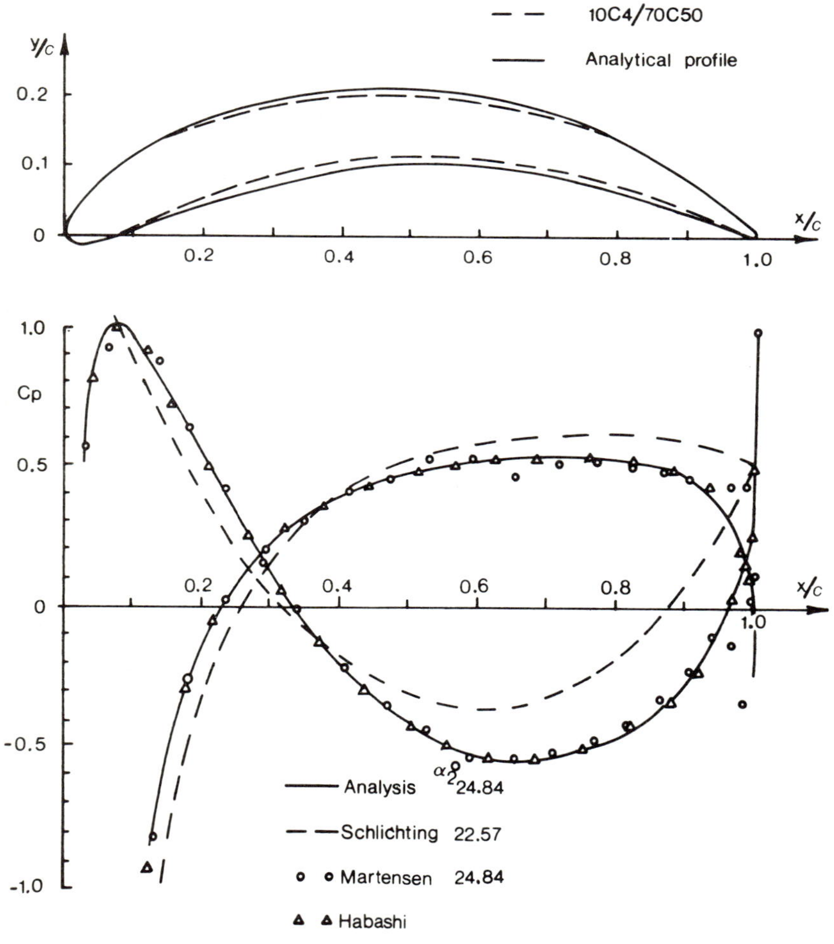

Fig. 5.16. Pressure distributions and blade profile for 70° camber cascade.

Figure 5.16 demonstrates that it was possible to obtain a highly cambered profile from the analysis. In this instance the stagger was zero, the camber approximately 70°, and the space-chord ratio was 0.9003643 - fairly typical for an impulse cascade.

Again for this comparison the rear stagnation point was arbitrarily selected at point (1,0) in the z plane. Exact solutions were run for incidences of zero and -70°. The latter case is a very stringent test of potential flow theories because of the severe pressure gradients present.

In fact over twenty different programs have been compared with the test case of Fig. 5.16 at -70° incidence and it is possible to draw some general conclusions.

At least seven different Martensen method decks have been compared with the analytic results of Fig. 5.16 and these all give good agreement. A typical set of Martensen results is given as an example.

Two workers have compared their Katsanis method[5.33] decks and these both gave excellent agreement as did the singularity method of Ribaut[5.74].

The finite element methods of refs.[5.30] and [5.31] give excellent agreement with the exact theory and results from the Habashi[5.30] method are plotted.

Matrix and relaxation methods in general did not give a very accurate solution.

Schlichting-type singularity methods, which were the original object of the investigation, gave surprisingly good agreement with the analysis considering that the singularities are distributed on the chord line. A sample Schlichting calculation is presented in Fig. 5.16.

5.3.4 <u>112° camber profile.</u> Since the theories of Garrick, Martensen and Schlichting were found to be applicable at 70° camber it was decided to obtain an exact solution for a more highly cambered blade. This was achieved, although the blade is thick and has a large trailing edge radius. This profile provides a stringent test for the accuracy of potential flow methods.

The co-ordinates and pressure coefficients are tabulated in Appendix B for the chosen position of rear stagnation point and are plotted in Fig. 5.17.

The flow computation methods used for this profile were the Schlichting method, as developed by Lewis [5.73], the Martensen method program of Payne and Latimer, and the streamline curvature code of Stuart.

The Schlichting theory gives a large error over the latter half of the profile. The error is probably due to the fact that Schlichting's theory gives the flow around a similar but cusped profile and also due to the placing of singularities on the chord line.

The results of the Martensen method for this profile, with the given values of inlet and outlet angle, indicate that if care is taken with the accuracy of input data then excellent results are possible for a very wide range of camber angles.

This is also true of the streamline curvature technique which previously suffered from errors in the blade edge regions.

5.3.5 <u>Discussion on the accuracy of different methods.</u> The accuracy of a wide range of potential theories has been checked using four exact solutions, ranging from a cusped compressor blade in cascade to a 112° impulse blade having a rounded trailing edge.

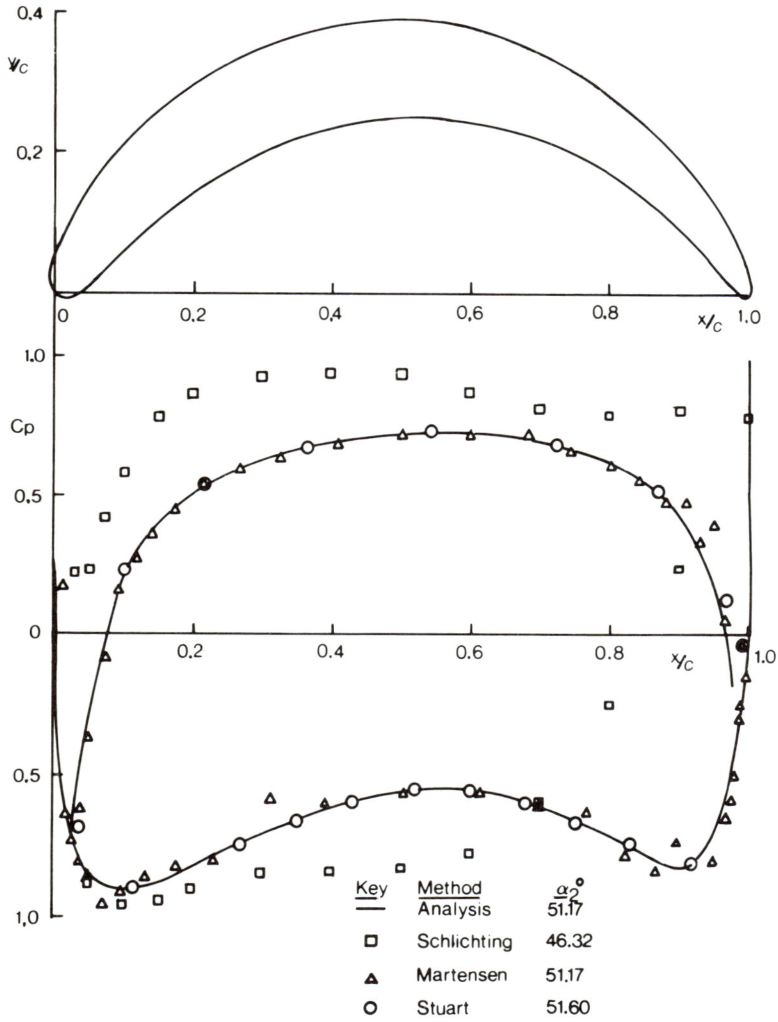

Fig. 5.17. Pressure distributions and blade profile for 112° camber cascade.

For some methods, such as the singularity method of Schlichting, the comparisons have served to identify limitations and areas of potential inaccuracy. These concern the suction peak region for some profiles, the treatment of rounded trailing edges and the use of the method for profiles having more than 70° camber.

A wide range of methods has been explored for conventional camber angles and have shown good accuracy under these conditions. These include the conformal transformation methods of Garrick and Howell, various finite element analyses and panel methods.

Two classes of method gave excellent accuracy for the whole range of profiles considered; these were solutions based on the Martensen method and streamline curvature solutions. There are, however, minor residual doubts on the applicability of these techniques. The Martensen theory is weakest when the blade profile becomes thin, as at the cusped trailing edge. Work on trailing edge representation by Wilkinson [5.75] seems to have cured this problem. Streamline curvature methods are weakest in the opposite case, for thick blade edges and sudden changes in curvature. The results of Fig. 5.17 suggest that this difficulty may have been overcome.

No unique solution was found within potential flow theory to the problem of locating the rear stagnation point; the position was arbitrarily specified in all computations and the search for a uniqueness condition was carried into the field of viscous flow theory.

5.4 Solutions to the Indirect Problem for Incompressible Potential Flow

In the indirect problem, sometimes called the "inverse" or "design" problem, the distribution of velocity over the surface length is prescribed and it is required to find the cascade geometry which will support this distribution. The leading and trailing edge conditions have been treated variously by different authors; the velocity distribution has also been specified in different ways including use of a hodograph representation, to be described in sub-section 5.4.2.

5.4.1 Development of the problem. Early work on singularity theories by Betz [5.76] for isolated aerofoils and by Pistolesi [5.18] for cascades, was useful in the first solutions to the indirect cascade flow problem.

In Ackeret's singularity theory [5.77] the interference effects of other blades upon the flow around a typical aerofoil were approximated by a simple superposition method. Ackeret himself considered that his method, although only approximate, was sufficiently flexible and useful for design purposes and that it showed promise of eventual application to annular cascades. This early assessment is probably still valid.

The data required for the computation are;

- (i) Velocity triangles upstream and downstream
- (ii) Blade spacing, s
- (iii) Maximum thickness, t, and its location
- (iv) Vorticity distribution
- (v) Source/sink distribution

The result of the calculation will be the camber line and thickness distribution of a cascade blade.

Ackeret firstly considered that the cascade produced the desired velocity distribution by a continuous vorticity distribution within the blade row. To this was added a continuous distribution of sources (with equal sink distribution) to allow representation of a finite blade thickness. The vortex distribution was evaluated using equation (5.14) and the source distribution correspondingly.

For the placing of a periodic vorticity distribution to be possible a zeroth-order approximation to the camber line is firstly computed assuming an infinite number of thin blades. The next step is to move to consideration of a finite number of blades by introducing the appropriate vortex and source distributions, periodic along the cascade, in such a way that vorticity and source strength are cancelled out in the region between the blades. The circulation distribution will have strong maxima at the blades and the thickness of the circulation distribution (representing the thickness of the blade) will typically be of the order of 10-20% of the blade spacing. Sawyer[5.78], using Ackeret's method, suggested vortex and source distributions for the typical aerofoil which could be approximated in Fourier series for the entire cascade.

The major approximations in the Ackeret method are the use of the zeroth order camber line as a base for the vorticity distribution and the assumption that the singularity distributions lie within the profile. In evaluating the velocity distribution it is also assumed that the aerofoil surface is parallel to the camber line.

In 1947, Goldstein and Jerison[5.79] developed a singularity method which, because of difficulties in regions of large surface curvature, was only useful in the design of thin blades. Woolard[5.80] and Scholz[5.55] have also given approximate singularity solutions to the 'indirect' problem. Diesendruck[5.81] was able to apply the iterative interference method for direct problem solutions, as given by Katzoff, Finn and Laurence[5.49], to the indirect problem but only for thin blades. Diesendruck obtained a solution in terms of integrals of vorticity; these had to be evaluated numerically and as an initial approximation the blade was assumed to have a circular-arc shape. Alpert[5.82] developed a procedure for the design of blade passages when the shape of one wall is known and the velocity distribution is prescribed on that wall.

Many proposed solutions to the indirect problem are of the conformal transformation type. Traupel[5.45] solved the indirect problem by considering transformations between the cascade plane and a unit circle and between this unit circle and a complete half plane. Finally an auxiliary plane was introduced with infinity points symmetrical about the imaginary axis. The flow was easily determined in this plane and Traupel was able to work back to obtain the blade shape in the cascade plane. Another method utilising the mapping into a unit circle plane was given by Goldstein; the computational procedure for this method is described in refs. 5.83 and 5.84. In Otsuka's approximate method[5.85] points of similar potential were located in the cascade (z) and circle (ζ) planes. The transformation was then established by expressing $(d\zeta/dz)$ as a series in z .

The theories that have been most useful in turbine design work are probably those of Stanitz. Stanitz developed an approximate theory and a more exact one, both of which are suitable for use at any subsonic Mach Number. His approximate design method[5.86] is applicable to blading of low space/chord ratio and rests on assumptions of linear flow variation over the blade passage and of a mass-weighted average velocity half-way between the prescribed velocities on the blade surfaces, having an identical direction to that of the camber line. The method is well described in refs. 3.10 and 5.86.

5.4.2 Solutions using the hodograph plane. It is often convenient to represent cascade flows in the hodograph plane. Both the complex potential function $\omega(z) = \varphi + i\psi$ and its derivative, the conjugate velocity $\omega(z) = d\omega/dz$, are functions of z . The elimination of z gives $\Omega = \Omega(\omega)$ and Ω can be mapped onto the hodograph plane. $\Omega = \Omega(\zeta)$ and its complex velocity $d\Omega/d\zeta$ can be examined and the derivative of the transformation represented by

$$\frac{dz}{d\zeta} = \frac{d\omega}{d\zeta} \Bigg/ \frac{d\omega}{dz} \quad . \tag{5.45}$$

Another possibility is to take the logarithm of the conjugate velocity and enter a new plane $\Omega = \Omega \ \ln(\omega)$. This is termed the 'logarithmic hodograph plane'. Many workers have used the hodograph plane representation which is even more helpful for compressible flows since the equations relating stream and potential functions in compressible flow become linear when expressed in the hodograph plane. Problems specified in the hodograph plane are strictly neither 'direct' or 'indirect' but provided the obvious difficulties of specification and identification can be overcome then useful blade shapes can certainly be derived. A further difficulty is that the hodograph of a cascade flow is generally many-valued and must be scrutinised on two or more Riemann surfaces.

In the method of Simonov[5.9] the hodograph (ζ) plane contains a vortex source and vortex sink and has two infinity points which correspond to the two stagnation points on the aerofoil surface. Simonov mapped the flow onto the interior of a circle in a further auxiliary plane in which the potential function can easily be found. The transformation between the circle plane and the hodograph plane may be established numerically; thus a selected ζ plane hodograph has a corresponding cascade plane profile. Simonov's method allows analytical integration of the transformation to produce aerofoils in the cascade plane having rounded leading edges and finite wedge-shaped trailing edges.

Chang[5.87] expressed $d\Omega/d\zeta$ and $d\Omega/dz$ as products or quotients of linear factors, thus avoiding tedious numerical integration. Unfortunately this was only possible for profiles with cusped trailing edges; indeed, the aerofoil thickness sometimes became negative. Chang used a unit circle for his auxiliary (ζ) plane, as did Lin[5.88] and Yeh[5.89], who specified the velocity directly along the blade surface, as a function of arc length. Accordingly the potential variation in the cascade plane, and thus in the hodograph plane, was prescribed.

Lighthill[5.90] initially used the transformation $Z = \exp(2\pi z/s)$ to move from the cascade plane into a Z plane having only one aerofoil. He used a hodograph plane in which the aerofoil is represented by a unit circle. By choosing the velocity on the surface of the aerofoil as a function of points on the unit circle, the corresponding points on the aerofoil were found.

Legendre[5.91] worked with hodograph boundaries formed using various hypergeometric equations. A logarithmic transformation converted these boundaries into straight lines and the enclosed area was then mapped into a half-plane using the Schwarz-Christoffel transformation. Having found the potential function in terms of the hodograph coordinates it was necessary to construct the profile by integration. Lecomte[5.92] has programmed the method for rapid design use with an interactive graphic display.

In 1958 Kraft[5.93] published his method of 'hodograph synthesis' for the design of laminar flow impulse buckets. This was followed a year later by Cantrell and Fowler's publication[5.94] of the more sophisticated 'schlicht' method favoured by General Electric designers.

Cantrell and Fowler utilized a mapping from the logarithmic hodograph plane to the physical plane. The coordinates of the logarithmic hodograph plane are related to the physical flow by the expression

$$\zeta = \xi + i\eta = \ln(u - iv) = (\ln\omega) - i\theta \ . \tag{5.46}$$

if the physical flow is described by the complex potential function one then has

$$\zeta = \ln(u - iv) = \ln\left(\frac{d\varphi}{dx} + i\frac{d\psi}{dx}\right) = \ln\left(\frac{d\Omega}{dz}\right) \tag{5.47}$$

and a conformal mapping exists between the two planes.

For a hodograph solution

$$\Omega(\zeta) = \varphi(\xi,\eta) + i\psi(\xi,\eta) ;$$

the transformation from the ζ plane to the z plane is given by the inverse of equation (5.47) as

$$dz = e^{-\zeta} \frac{d\Omega}{d\zeta} d\zeta .$$

The use of such a mapping permits precise control over the suction surface velocities which can be constant.

Cantrell and Fowler idealized the log-hodograph representation of a turbine bucket by a polygon. The interior of this polygon was transformed to the upper half of an auxiliary $R = P + iQ$ plane by the Schwarz-Christoffel transformation

$$\zeta = \int C \frac{[(R - P_6)(R - P_7)]^{\frac{1}{2}}}{[(R - P_4)(R - P_5)]^{\frac{1}{2}}} \frac{dR}{R} + \zeta_0 . \qquad (5.48)$$

The corners of the polygon are rounded since it would be difficult to numerically integrate through sharp corners. This was achieved by mapping two radial lines from the R plane into the log-hodograph plane. An auxiliary transformation was required between the R plane and a further auxiliary plane.

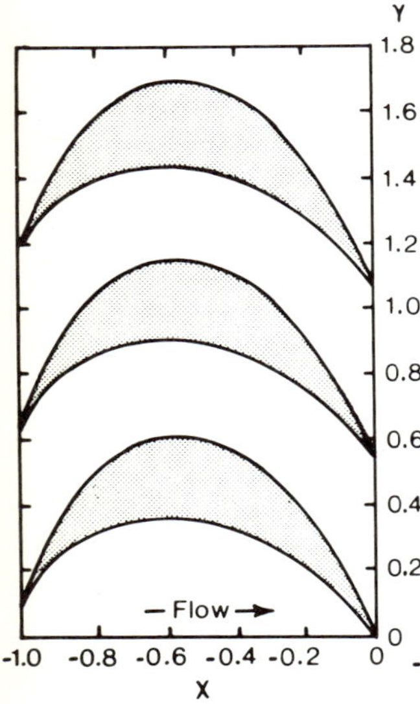

Fig. 5.18. Typical impulse bucket.

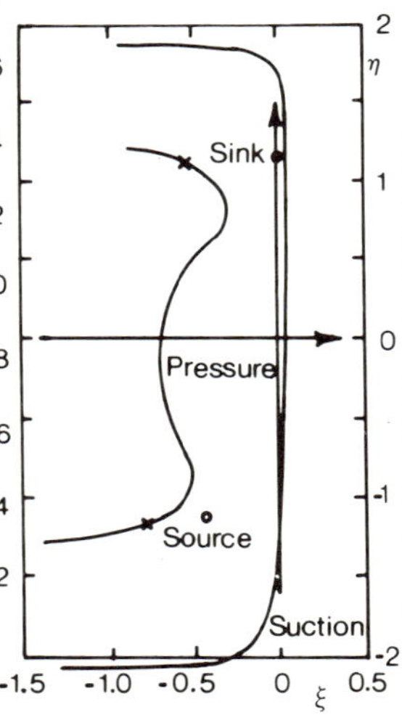

Fig. 5.19 Velocity specification in hodograph plane.

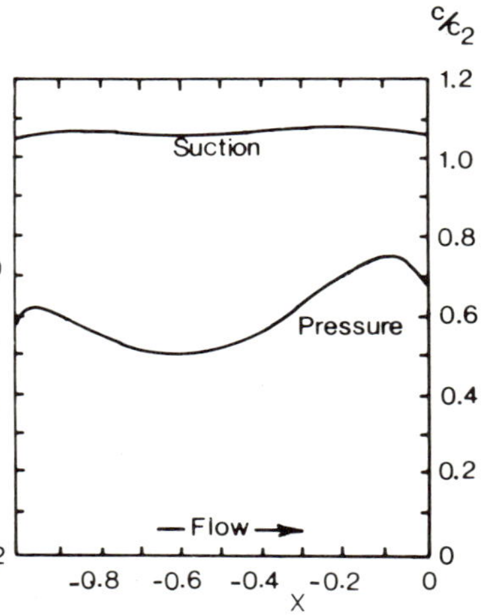

Fig. 5.20. Calculated velocity distribution.

The blade spacing was left as very much an independent variable and for this reason the method is a 'trial-and-error' procedure as indeed any design method must be. In practice use of the transformation is iterative, the first run producing blading of often-unconventional shape. Scrutiny of early results does, however, give useful information to assist subsequent iterations. Many successful gas and steam turbine rotors have been designed using the above procedure, including turbine buckets for the General Electric J79 engine.

Figure 5.18 gives results for a typical impulse bucket, showing in Fig. 5.19 a practical velocity specification in the log-hodograph plane, and in Fig. 5.20 the calculated velocity distribution. The points marked 'x' in Fig. 5.19 correspond to points of the blade surface at a distance of 0.1 percent of the blade width from the actual leading and trailing edges. Closure of the final profile is no problem with this method, an accuracy of 0.1 percent of blade width being easily attainable.

5.4.3 Indirect use of the Martensen method. The Martensen method has exhibited good accuracy and reliability for solutions to the direct problem and this gave an incentive to explore ways of using it in solutions to the design problem.

Wilkinson[5.75] firstly produced an approximate method in which corrections were made to vorticity on the camber line rather than on the blade surface.

Murugesan and Railly[5.95] derived an equation giving the normal velocity distribution around the blade for a given tangential velocity distribution. From this normal velocity an initial arbitrarily chosen blade shape was successively modified until a blade was found having the prescribed velocity distribution. Reasonable accuracy was usually obtained after a few iterations.

Schwering[5.96] subsequently produced a solution quite similar to the Martensen method but formulated in terms of stream function and leading to a Fredholm integral equation of the first kind. He also defined the blade surface in terms of analytic functions, giving improved accuracy in the solution of the integrals.

Lewis[5.97] has recently published a truly inverse procedure. The blade profile is outlined by elements and the velocity at the midpoint of each element is calculated from the prescribed velocity distribution. Blade streamlines are reconstructed from the resulting velocity vectors; a profile closure condition is then imposed and iteration applied until convergence. Lewis provided examples for various blade shapes and it appears that this method should prove to be a versatile computer graphic design tool.

REFERENCES

5.1 KUTTA, W.M. Auftriebskräfte in strömenden Flüssigkeiten. Mus. Aero. Mitt. (1902)

5.2 JOUKOWSKY, N.E. Collected works. (Vol.6) (in Russian) OGIZ, State Publ. House of Tech. Theo. Lit., Moscow (1950)

5.3 CHAPLYGIN, S.A. Theory of the cascade wing. Math. Journ., Vol. XXIX (1913)

5.4 MERCHANT, W. and COLLAR, A.R. Flow of an ideal fluid past a cascade of blades. (Part II). ARC R&M 1893 (1941)

5.5 JOUKOWSKY, N.E. Aerodynamique. Gauthier-Villars, Paris (1916)

5.6 GRAMMEL, R. Die hydrodynamischen Grundlagen des Fluges. F. Vieweg, Braunschweig. (1917)

5.7 MCWOOD, F. and GLAUERT, H. Preliminary investigation of multiplane interference applied to propeller theory. Adv. Cttee. Aero, TR620 (1918-1919)

5.8 KOENIG, E. Potential Strömung durch Gitter. ZAMM vol.2, pg.422 (1922)

5.9 SIMONOV, L.A. Construction of profiles by means of a velocity hodograph. Proc. USSR Acad. Sci. vol.5, no.2. Trans. by S. Bergman (1941)

5.10 SCHILHANSL, M. Näherungsweise Berechnung von Auftrieb und Druckverteilung in Flügelgittern. Jahrb. d. Wiss. Ges. f. Luftfahrt 151-167 (1927)
5.11 SORENSEN, E. Potentialstromung durch rotierende Kreiselrader. ZAMM Vol. 7, pg. 89 (1927)
5.12 BRAUN, E. Fur Gittertheorie der Propeller rader. Escher Wyss Mitt, Vol. 5, pg. 64 (1932)
5.13 WEINEL, E. Beiträge zur rationellen Hydrodynamik der Gitterströmung. Ing. Archiv. 5, pg. 91 (1934)
5.14 NUMACHI, F. Aerofoil theory of propeller pumps with special reference to the effects of blade interference upon the lift and cavitation. Tohoku Imp. Univ., Tech. Rep. 8, 3 (1929)
5.15 WEINIG, F. Die Strömung um die Schaufeln von Turbomaschinen. Joh. Ambr. Barth., Liepzig (1935)
5.16 BETZ, A. Diagramme zur Berechnung von Flügelreihen. Ing. Archiv. 2, pg. 359 (1931) NACA TM 1032 (1942)
5.17 VON KARMAN, T. and BURGERS, J.M. Flow through a lattice composed of airfoils. Vol. 2 of Aerodynamic Theory. Div. E, ch. 2. pt.b sec. 23 pg. 91 W.F. Durand ed., Julius Springer Berlin (1935)
5.18 PISTOLESI, E. On the calculation of an infinite cascade of thin aerofoils. L'Aerotechnica V17 (1937)
5.19 LIEBLEIN, V. The calculation of the lift characteristics of an aerofoil section in a cascade. M.O.S. Volkenrode, Rep. and Trans. 442 (W. Bente, trans.) (1947)
5.20 ERWIN, J.H. and YACOBI, L.A. Method of estimating the incompressible-flow pressure distribution of compressor blade sections at design angle of attack. NACA R&M L53F17 (1953)
5.21 WU. CHUNG HUA and BROWN, C.A. Method of analysis for compressible flow past arbitrary turbomachine blades on general surfaces of revolution. NACA TN 2407 (1959)
5.22 FANTI, R.A., KEMP, N.H. and NILSON, E.N. A theory of thin airfoils isolated and in cascade, yielding finite pressure at smooth leading edges. J. Aero. Sci., 25 409 (1958)
5.23 FERRARI, C. Un metodo rapido approssimatocalcolo delle caratteristicho aerodinamiche delle chiere di profili alari. Atti della Accad. delle Sci. di Torino. B1 264 (1947)
5.24 HLAVKA, G.E. An approximate theory for potential flow through cascades of airfoils. Ph. D. thesis, Caltech (1954)
5.25 HUMIMOTO, B., KAMIMOTO, G. and HIROSE, K. Theory of wing lattice. Kyoto Univ. Eng. Res. Inst., Tech. Rep. (1952)
5.26 KRUGER, H. Investigation of surface velocity and pressure distribution along cascade profiles. NACA TR 193
5.27 MURAI, H. Theory on the mutual interference of hydro-or aerofoil profiles arranged in a straight grate. Trans. JSME vol. 17 (1951)
5.28 TANABE, K. and HORLOCK, J.H. A simple method for predicting the performance of cascades of low solidity. Aero. Quart. pg. 225 (1967)
5.29 BASKHARONE, E., and HAMED, A. A new approach in cascade flow analysis using the finite element method. AIAA J. 19, 1 pg. 65 (1981)
5.30 HABASHI, W.G., DUECK, E.G. and KENNY, D.P. Finite-element approach to compressor blade-to-blade cascade analysis. AIAA J, 17, 7 pg. 693 (1979)
5.31 WHITEHEAD, D.S. and GRANT, R.J. Force and moment coefficients for high-deflection cascades. Cambridge Univ. CUED/A - Turbo/TR98 (1980)
5.32 KAVANAGH, P. and SEROVY, G.K. Through-flow solution for axial-flow turbomachine blade rows. NASA CR-363 (1966)
5.33 KATSANIS, T. A computer program for calculating velocities and streamlines for two-dimensional, incompressible flow in axial blades rows. NASA TN D-3762 (1967)
5.34 DAIGUJI, H. A simplified analysis of cascade flow in mixed-flow pumps and turbines. Bull. JSME 10,39 pg. 496 (1967)
5.35 THEODORSEN, T. Theory of wing sections of arbitrary shape. NACA Rep. 411 (1931)

5.36 HOWELL, A.R. The theory of arbitrary aerofoils in cascades. Phil. Mag. 39 299 (1948)
5.37 GARRICK, I.E. On the plane potential flow past a lattice of arbitrary aerofoils. NACA Rep. 788 (1944)
5.38 CARTER, A.D.S. and HUGHES, H.P. A theoretical investigation into the effect of profile shape on the performance of aerofoils in cascade. ARC R&M 2384 (1950)
5.39 VAZSONYI, A. On the aerodynamic design of axial-flow compressors and turbines. Trans. ASME, J. App. Mech., 15, 1, pg. 53 (1948)
5.40 POLLARD, D. and WORDSWORTH, J. A comparison of two methods for predicting the potential flow around arbitrary aerofoils in cascade. ARC CP 618 (1962)
5.41 HALL, W.S. and THWAITES, B. On the calculation of cascade flows. ARC 25,350 (1963)
5.42 THWAITES, B. On the numerical calculation of Theodorsen's transformation. ARC 24,367 (1963)
5.43 WATSON, E.J. Formulae for the computation of the functions employed for calculating the velocity distribution about a given aerofoil. ARC R&M 2176 (1945)
5.44 MUTTERPERL, W. A solution of the direct and inverse potential problems for arbitrary cascades of airfoils. NACA WR L-81 (1944)
5.45 TRAUPEL, W. Die Berechnung der Strömung durch Schaufelgitter. Sulzer Tech. Rundschau, 1, 25 (1945)
5.46 BIRNBAUM, W. Die Tragende Wirbelfläche als Hilfsmittel zur Behandlung des ebenen Problems der Tragflügeltheorie. ZAMM, 3,4. pg. 290 (1923)
5.47 GLAUERT, H. A theory of thin aerofoils. ARC R&M 910 (1924)
5.48 ALLAN, W.K. A theory of cascade interference. ARC 23,214
5.49 KATZOFF, S., FINN, R.S. and LAURENCE, J.C. Interference method for obtaining potential flow past an arbitrary cascade of airfoils. NACA Rep. 879 (1947)
5.50 CHAUVIN, J. and BREUGELMANS, F.A.E. Private communications (1964)
5.51 RICHTER, W. Berechnung der Druckverteilung von ebenen Schaufelgittern mit stark gewölbten dicken Profilen bei inkompressibler Strömung. Ing. Archiv. 29. S.351 (1960)
5.52 LEWIS, R.I. and PENNINGTON, G.A. Theoretical investigation of some basic assumptions of Schlichting's singularity method of cascade analysis. ARC 26,159 (1964)
5.53 RIEGELS, F.W. Aerofoil Sections. Butterworths (1961)
5.54 MELLOR, G.L. An analysis of axial compressor cascade aerodynamics. I and II Trans. ASME, J. Basic Eng. 81,3 pg. 362 (1959)
5.55 SCHOLZ, N. Ein Berechnungsverfahren zum Entwurf von Schaufelgitterprofilen. VDI Forschungsheft 442, 25-48 (1954)
5.56 JORGE, J. Potential flow through cascades with blades of finite camber and thickness. Ph.D. thesis, Princeton Univ.
5.57 MANI, R. and ACOSTA, A.J. Quasi two-dimensional flow through a cascade. ASME Paper 67WA/FE-9 (1967)
5.58 ISAY, W.H. Zur Berechnung der Strömung durch axiale Schaufelgitter. ZAMM 37,5,321 (1957)
5.59 BUECKNER, H.F. and SCHNACKEL, H.C. The calculation of incompressible flow through turbine cascades. ASME Annual Meeting (1959)
5.60 GIESING, J.P. Extension of the Douglas Neumann program to problems of lifting infinite cascades. Douglas Rep. LB31653 (1964)
5.61 PRICE, D.W. Calculation of the velocity distribution around profiles in cascade in two dimensional potential flow. (unpublished Rolls-Royce Rep.) (1963)
5.62 PAYNE, D. Isolated and cascade aerofoils. Dissertation, London Univ. (1964)
5.63 GOSTELOW, J.P. Potential flow through cascades. A comparison between exact and approxiamate solutions. ARC CP 807 (1964)
5.64 COLLAR, A.R. The flow of a perfect fluid through cascades of aerofoils. J. Roy. Aero. Soc. (May, 1941)

5.65 LAMB, H. Hydrodynamics. 6th ed. Cambridge Univ. Press (1932)
5.66 GOSTELOW, J.P. Potential flow through cascades. Extensions to an exact theory. ARC CP 808 (1964)
5.67 GOSTELOW, J.P. The calculation of incompressible flow through cascades of highly cambered blades. Adv. Probs. in Turbo. VKIFD (1965)
5.68 NOVAK, R.A. and HAYMANN-HABER, G. A mixed-flow cascade passage design procedure based on a power series expension. ASME Paper 82-GT-121 (1982)
5.69 O'CARROLL, M.J. and MORGAN, L.A. Finite element and difference methods for cascades. In "Computational Methods and Problems in Aeronautical Fluid Dynamics" ed. B.C. Hewitt et al., Academic Press (1976)
5.70 ECER, A. and AKAY, H.U. Investigation of transonic flow in a cascade using the finite element method. AIAA J. 19,9 pg. 1174 (1981)
5.71 CASPAR, J.R., HOBBS, D.E. and DAVIS, R.L. The calculation of two-dimensional compressible flow in cascades using finite area techniques. AIAA J. 18,1, pg. 103 (1980)
5.72 MCFARLAND, E.R. Solution of plane cascade flow using improved surface singularity methods. Trans. ASME, J. Eng. Power 104, 3 pg. 668 (1982)
5.73 LEWIS, R.I. Restrictive assumptions and range of validity of Schlichting's cascade analysis. Adv. Probs. in Turbo. VKIFD (1965)
5.74 RIBAUT, M. Three-dimensional calculation of flow in turbomachines with the aid of singularities. Trans. ASME, J. Eng. Power, pg. 258 (1968)
5.75 WILKINSON, D.H. A numerical solution of the analysis and design problems for the flow past one or more aerofoils in cascade. ARC R&M 3545 (1967)
5.76 BETZ, A. Untersuchung einer Joukowskychen Tragflüche. ZFM6 173-179 (1915)
5.77 ACKERET, J. The design of closely spaced blade grids. British M.A.P. R.T.P. Trans. 2007 (1942)
5.78 SAWYER, W.T. Experimental investigation of a stationary cascade of aerodynamic profiles. Mitt. aus d. Inst. f. Aero., Heft 17, Zurich (1949)
5.79 GOLDSTEIN, A.W. and JERISON, M. Isolated and cascade airfoils with prescribed velocity distribution. NACA Rep. 869 (1947)
5.80 WOOLARD, H.W. The incompressible flow about a cascade of airfoils. Cornell Aero. Lab., Rep. AF-734-A-1 (1950)
5.81 DIESENDRUCK, L. Iterative interference methods in the design of thin cascade blades. NACA TN 1254 (1947)
5.82 ALPERT, S. Design method for two-dimensional channels for compressible flow with application to high-solidity cascades. NACA TN 1931 (1949)
5.83 HANSEN, A.G. and YOHNER, P.L. A numerical procedure for designing cascade blades with prescribed velocity distributions in incompressible potential flow. NACA TN 2101 (1950)
5.84 Aerodynamic design of axial-flow compressors. NASA SP-36 (1965)
5.85 OTSUKA, J.S. Method of designing lattice wing profile with prescribed velocity distribution. Transp. Tech. Res. Inst., Tokyo. Rep. 16 (1955)
5.86 STANITZ, J.D. Approximate design method for high-solidity blade elements in compressors and turbines. NACA TN 2408 (1952)
5.87 CHANG, Y.S. Anwendung der Blasiusschen Satze für Auftrieb und Moment auf Profile in Gitterstellung. Thesis T.H. Berlin (1938)
5.88 LIN, C.C. On the subsonic flow through circular and straight lattices of airfoils. J. Math. Phys., 38, 2, pg. 117 (1949)
5.89 YEH, H. The development of cascade profiles for high subsonic potential flow. J. Aero. Sci. 19 630 (1952)
5.90 LIGHTHILL, M.J. A mathematical method of cascade design. ARC R&M 2104 (1945)
5.91 LEGENDRE, R.G. Calcul d'un profil ou d'une grille de profils pour une distribution de vitesse donnee en fonction du potentiel. La Recherche Aerospatiale, pg. 231 (1972-1974)
5.92 LECOMTE, C. Calculation of cascade profiles from the velocity distribution. ASME Paper 74-GT-70 (1974)
5.93 KRAFT, H. The development of a laminar wing type turbine bucket. ZAMP, 9 404 (1958)

5.94 CANTRELL, H.H. and FOWLER, J.E. The aerodynamic design of two-dimensional turbine cascades for incompressible flow with a high-speed computer. Trans. ASME, J. Basic Eng., 81,3,349 (1959)

5.95 MURUGESAN, K. and RAILLY, J.W. Pure design method for aerofoils in cascade. J. Mech. Eng. Sci., 11,5 (1969)

5.96 SCHWERING. W. Design of cascades for incompressible plane potential flows with prescribed velocity distribution. ASME Paper 70-GT-87 (1970)

5.97 LEWIS, R.I. A method for inverse aerofoil and cascade design by surface vorticity. ASME Paper 82-GT-154 (1982)

Appendix A. Parameters, co-ordinates and pressure distribution for cusped blade.

Parameters were selected as follows:- Set $\chi = 37.5^{o}$, $\alpha_1 = 53.5^{o}$ and $\beta = 0.725$. From equation (5.29) this gives $\cosh \gamma = 2.91481083$. For λ at the zero positions $\sinh \lambda = \pm \cos \chi \sinh \gamma$ (from equation (5.36)) is used. The two values obtained are $\pm (0.632248112 + 0.357257149i)$. Set $\beta' = 0.8$ for larger ovals; position these to include zeroes and infinities or, for a cusped blade, the larger oval should pass through the zero.

In this example β' was displaced to have its centre at $\zeta' = m' + in' = 0.112512215 - 0.0632i$. The blade profile (from equation (5.37)) and pressure distribution (from equation (5.41)) were then computed. The results also gave $s/c = 0.9901573$ and $\tan\alpha_2 = 0.57793012$.

Suction Surface			Pressure Surface		
x/c	y/c	Cp	x/c	y/c	Cp
-0.0001414	+0.0021437	+0.4363445	+0.0000631	-0.0003936	+0.6827467
+0.0011316	+0.0086575	-0.1328097	+0.0012478	-0.0041009	+0.9449901
+0.0046149	+0.0153348	-0.4766543	+0.0059857	-0.0099254	+0.9211494
+0.0099973	+0.0220979	-0.6573834	+0.0151904	-0.0150978	+0.6346141
+0.0170584	+0.0288868	-0.7538457	+0.0308941	-0.0192226	+0.4121105
+0.0256397	+0.0356536	-0.8072119	+0.0575823	-0.0214915	+0.3034295
+0.0356275	+0.0423583	-0.8368069	+0.1109821	-0.0193545	+0.2977494
+0.0469424	+0.0489665	-0.8519049	+0.1451326	-0.0159400	+0.3248899
+0.0595318	+0.0554474	-0.8571795	+0.1464393	-0.0157915	+0.3260522
+0.0733664	+0.0617726	-0.8550898	+0.1653660	-0.0135341	+0.3432965
+0.0884371	+0.0679148	-0.8470333	+0.1784929	-0.0118722	+0.3554600
+0.1047549	+0.0738467	-0.8337270	+0.2320319	-0.0046258	+0.4037846
+0.1223507	+0.0795404	-0.8156043	+0.2472125	-0.0025069	+0.4167019
+0.1412773	+0.0849661	-0.7928747	+0.2708527	+0.0007947	+0.4359517
+0.1616129	+0.0900909	-0.7656136	+0.2725677	+0.0010332	+0.4373066
+0.1834664	+0.0948774	-0.7337957	+0.3219801	+0.0077840	+0.4738663
+0.1840367	+0.0949930	-0.7329363	+0.4251358	+0.0203300	+0.5360268
+0.2069867	+0.0992816	-0.6973123	+0.4999570	+0.0274401	+0.5710162
+0.2322761	+0.1032498	-0.6559725	+0.5622544	+0.0317420	+0.5947677
+0.2458548	+0.1050500	-0.6333965	+0.6165678	+0.0341203	+0.6119034
+0.2599128	+0.1067143	-0.6094902	+0.6649646	+0.0350486	+0.6245159
+0.2746029	+0.1082308	-0.5841975	+0.6916590	+0.0350432	+0.6304108
+0.2899889	+0.1095855	-0.5574506	+0.7085867	+0.0348389	+0.6337517
+0.3061483	+0.1107616	-0.5291659	+0.7481384	+0.0337241	+0.6403130
+0.3231776	+0.1117387	-0.4992400	+0.7840864	+0.0318926	+0.6446481
+0.3411983	+0.1124916	-0.4675408	+0.8167564	+0.0295050	+0.6471121
+0.3603669	+0.1129881	-0.4338966	+0.8463846	+0.0267036	+0.6478887
+0.3808910	+0.1131861	-0.3980766	+0.8731469	+0.0236169	+0.6471476
+0.4030553	+0.1130286	-0.3597584	+0.8971765	+0.0203631	+0.6449974
+0.4272691	+0.1124343	-0.3184689	+0.9185750	+0.0170523	+0.6415090
+0.4541581	+0.1112801	-0.2734666	+0.9374189	+0.0137879	+0.6367226

Appendix A (continued).

Suction Surface x/c	Suction Surface y/c	Suction Surface Cp	Pressure Surface x/c	Pressure Surface y/c	Pressure Surface Cp
+0.4847715	+0.1093610	-0.2234749	+0.9537636	+0.0106676	+0.6306511
+0.5211449	+0.1062824	-0.1659479	+0.9676454	+0.0077845	+0.6232786
+0.5685042	+0.1010401	-0.0941639	+0.9790821	+0.0052283	+0.6145736
+0.5941925	+0.0976375	-0.0566467	+0.9880724	+0.0030860	+0.6044715
+0.6500792	+0.0889340	+0.0219360	+0.9945930	+0.0014433	+0.5928825
+0.7105268	+0.0775910	+0.1033092	+0.9985953	+0.0003856	+0.5796842
+0.7630253	+0.0661807	+0.1722613	+0.9999912	+0.0000000	+0.5647053
+0.7860596	+0.0607335	+0.2023781	+1.0000000	+0.0000000	+1.0000000
+0.8266043	+0.0505188	+0.2557337			
+0.8559117	+0.0426614	+0.2950046			
+0.8792122	+0.0361504	+0.3269870			
+0.8985445	+0.0305851	+0.3542904			
+0.9149604	+0.0257549	+0.3782450			
+0.9290874	+0.0215307	+0.3996382			
+0.9413357	+0.0178259	+0.4189820			
+0.9519903	+0.0145782	+0.4366321			
+0.9612585	+0.0117402	+0.4528476			
+0.9692963	+0.0092748	+0.4678245			
+0.9821363	+0.0053448	+0.4946432			
+0.9871093	+0.0038330	+0.5067059			
+0.9912043	+0.0025968	+0.5179857			
+0.9944718	+0.0016191	+0.5285513			
+0.9986835	+0.0003785	+0.5478059			

Appendix B. Co-ordinates and pressure distribution for 112° camber blade.

Suction Surface x/c	Suction Surface y/c	Suction Surface Cp	Pressure Surface x/c	Pressure Surface y/c	Pressure Surface Cp
-0.00352110	+0.00923877	+0.82211789	+0.00000000	+0.00000000	+0.99115082
-0.00416633	+0.03418151	+0.19303330	+0.00631790	-0.00588533	+0.20927346
+0.00226933	+0.06470695	-0.25466742	+0.03286025	+0.00103115	-0.63938475
+0.01529799	+0.09905553	-0.56120251	+0.08598451	+0.04805304	+0.09227916
+0.03535433	+0.13654865	-0.76505931	+0.12668707	+0.08629330	+0.32567045
+0.06389262	+0.17722395	-0.88100575	+0.21407971	+0.15390654	+0.54935134
+0.10429606	+0.22198988	-0.91332828	+0.39228241	+0.23125038	+0.69055688
+0.16549821	+0.27368098	-0.86322892	+0.53900388	+0.24535753	+0.71546992
+0.21215796	+0.30450654	-0.80623787	+0.74604444	+0.19334415	+0.66544259
+0.28660146	+0.34220073	-0.71846674	+0.85691462	+0.12335896	+0.54732966
+0.45195727	+0.38307185	-0.57979164	+0.90807529	+0.07792880	+0.41580197
+0.51896138	+0.38370916	-0.55705256	+0.96952949	+0.01444430	-0.03605225
+0.68884350	+0.34397029	-0.60545506	+0.99508686	-0.00299272	-0.09077782
+0.76647825	+0.30492329	-0.68184755	+1.00000000	+0.00000000	+1.00000000
+0.81521102	+0.27273354	-0.74463437			
+0.87930408	+0.21893512	-0.81802829			
+0.92202323	+0.17275505	-0.81606584			
+0.95276306	+0.13120458	-0.73969601			
+0.97509409	+0.09334566	-0.58974412			
+0.99057698	+0.05922620	-0.35784959			
+0.99971382	+0.02975946	-0.00545534			
+1.00201575	+0.00721731	+0.62637309			

CHAPTER 6

Compressible Flow Theories

The inlet Mach Number for turbine and compressor blade rows usually has a high subsonic value and therefore a theoretical treatment of subsonic compressible flow through cascades is essential. Much of this chapter will therefore be devoted to the direct and design problems in subsonic flow.

With the widespread use of high rotational speeds in fans and compressors, supersonic inlet Mach Numbers are also becoming commonplace. In turbine blade rows the discharge flows are often supersonic, especially in contemporary steam turbine practice. The latter part of the chapter therefore concerns the problems of supersonic and transonic flow analysis in cascades.

6.1 Subsonic Flow Calculations

The potential equation in two-dimensional compressible flow is

$$\varphi_{xx} + \varphi_{yy} = \frac{1}{a^2}\left[\varphi_x^2\varphi_{xx} + \varphi_y^2\varphi_{yy} + 2\varphi_x\varphi_y\varphi_{xy}\right] \tag{6.1}$$

where

$$a^2 = a_0^2 - \tfrac{1}{2}(\gamma - 1)\cdot(\varphi_x^2 + \varphi_y^2) \quad . \tag{6.2}$$

No universally valid technique for solving this equation has yet been achieved and it is therefore helpful to classify cascade configurations into three types for the purpose of solving the above equation:-

(a) Cascades of low space/chord ratio (e.g. turbine hub and some guide vane cascades).

(b) Cascades of high space/chord ratio which avoid the occurrence of high local velocities (e.g. some fan blades).

(c) The general case of cascades intermediate between the two extremes, including most compressor, fan and turbine cascades.

6.1.1 Closely-spaced cascades. The flow through such cascades can be analysed approximately by the use of channel flow techniques. Provided that the space/chord ratio is not greater than about 0.7 finite difference and relaxation computer methods are relatively straightforward to implement.

Early solutions to the problem were given by Stanitz and Prian[6.1] (a rapid approximate solution), Huppert and McGregor[6.2] (a stream filament method in which a linear variation of streamline curvature is assumed) and by Wu and Brown[5.21].

Wu and Brown treated the right hand side of equation (6.1) as a known function of x and y, thus linearising the equation. Novak[5.68] has recently improved the Wu and Brown formulation to provide an effective design procedure for shock-free cascades. The range of applicability of mean-streamline methods has been explored by Cai[6.3]. A widely used approximate design method is that of Stanitz[6.4]. The governing equations are written in terms of potential and stream function since the velocity distribution is prescribed. The blade surfaces in the physical (z = x + iy) plane are streamlines and are therefore lines of constant ψ in the potential plane. In ref. 3.15 the continuity and irrotationality conditions are written as

$$\frac{1}{\rho}\left(\frac{\partial \log \rho}{\partial \varphi} + \frac{\partial \log q}{\partial \varphi}\right) + \frac{\partial \theta}{\partial \psi} = 0 \tag{6.3}$$

$$\frac{\rho\, \partial \log q}{\partial \psi} - \frac{\partial \theta}{\partial \varphi} = 0 \, . \tag{6.4}$$

Differentiating equations (6.3) and (6.4) with respect to φ and ψ, respectively, and combining to eliminate $(\partial^2\theta / \partial\varphi\, \partial\psi)$ yields

$$\frac{\partial^2 \log \rho}{\partial \varphi^2} + \frac{\partial^2 \log q}{\partial \varphi^2} - \frac{\partial \log \rho}{\partial \varphi}\left(\frac{\partial \log \rho}{\partial \varphi} + \frac{\partial \log q}{\partial \varphi}\right) + \rho^2 \frac{\partial \log q}{\partial \psi} - \frac{\partial \log \rho}{\partial \psi} + \rho^2 \frac{\partial^2 \log q}{\partial \psi^2} = 0 \, . \tag{6.5}$$

The distribution of log q in the potential plane is determined from equation (6.5) by either relaxation, a matrix method or a Green's function approach. Having found log q throughout a blade passage the flow angle distribution $\theta(\varphi, \psi)$ can be obtained along the streamlines (and hence blade surfaces) by integrating equation (6.4) keeping ψ constant. Similarly the distribution along each potential line is found by integrating equation (6.3) keeping φ constant.

It is then a straightforward matter, using the geometric conditions

$$dx = ds \cos\theta \tag{6.6}$$

$$dy = ds \sin\theta \tag{6.7}$$

and the stream and potential function definitions

$$d\psi = \rho q \, dn \tag{6.8}$$

$$d\varphi = q \, ds \tag{6.9}$$

to arrive at the physical-plane blade geometry by integrating from the leading edge

$$x = \int_{\psi} \frac{\cos\theta}{q}\, d\varphi, \qquad y = \int_{\psi} \frac{\sin\theta}{q}\, d\varphi \, .$$

The method was mainly used by Stanitz for channel flows, with and without the simplifying assumption of a linear pressure-density isentrope. An example from ref. 3.15 shows that there is little difference between the predicted geometries for values of γ of 1.4 and -1 (Fig. 6.1). The Green's function solution for the Stanitz method has been further developed by Payne[5.62] and has been widely used in industry.

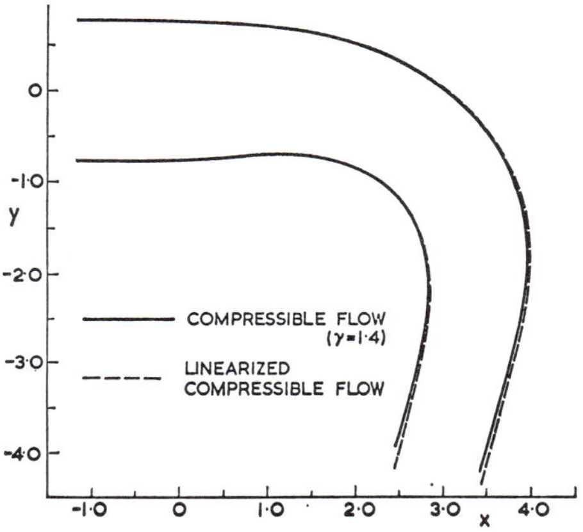

Fig. 6.1. Stanitz results showing effect of changes in γ.

The Cantrell and Fowler solution in the hodograph plane[5.94] has been generalised for compressible flows by Uenishi[6.5]. Reasonable agreement between theory and experiment was obtained for a wide variety of impulse and reaction blades at moderate subsonic velocities. Uenishi corroborated the finding of Fig. 6.1 and obtained a maximum closure error of 0.6 percent of chord length. Cohen[6.6] has obtained an interesting design solution by representing the blade passage by a rectangular profile in the log-hodograph plane, although accurate solutions were only demonstrated for the special case of free-vortex flow between concentric constant velocity walls.

Most solutions proposed for the channel-flow approach share the limitations that no accurate results are possible in the vicinity of leading or trailing edges and that accuracy is poor for blade spacings used in axial flow compressors.

6.1.2 Widely-spaced cascades. For such blading the solution can be conveniently approached by linearisation in either the physical or hodograph planes.

Linearisation in the physical plane is attained by making certain restrictive assumptions with respect to equation (6.1). The main assumption is that all boundary gradients and perturbations on the inlet velocity vector are small. It follows that errors will be present in the solution for highly-cambered cascades and for any blade near the leading and trailing edges. Despite this limitation and certain further approximations, the simple second Prandtl-Glauert rule[5.47],

$$Cp_c = Cp_{inc}/\sqrt{1 - M^2} \tag{6.10}$$

is obtained and has been used estensively, especially in Germany. Moreover, Lakomy[6.7] has shown that the Prandtl-Glauert law can be applied to most cascades in general use if an empirical correction factor (a function of angle of attack) is included. Zhou et al[6.8] have advocated the use of five transonic similarity parameters for cascades.

Since linearisation in the physical plane may introduce inaccuracies which limit its application, the more general case of linearisation in the hodograph plane will be considered. In order to render (6.1) linear without neglect of any terms it is necessary to introduce new variables and to transform the equation into the hodograph plane.

Putting

$$q = \sqrt{u^2 + v^2} = \sqrt{\varphi_x^2 + \varphi_y^2} \tag{6.11}$$

and

$$\theta = \tan^{-1}(v/u) = \tan^{-1}(\varphi_y/\varphi_x) \,. \tag{6.12}$$

The following linear equations are easily obtained,

$$\varphi_\theta = (q/\rho)\,\psi_q \tag{6.13}$$

$$\varphi_q = -\left(1 - \frac{q^2}{a^2}\right)\frac{1}{\rho q}\,\psi_\theta \,. \tag{6.14}$$

To achieve integration more conveniently we put

$$\varphi_\theta = \psi_\lambda \frac{q\,\lambda'(q)}{\rho} \qquad \text{and} \qquad \varphi_\lambda = -\left(1 - \frac{q^2}{a^2}\right)\frac{1}{\rho\, q\, \lambda(q)}\,\psi_\theta$$

$$\lambda'(q) = \frac{1}{q}\sqrt{\left(1 - \frac{q^2}{a^2}\right)} \qquad \lambda = \int \frac{\sqrt{1 - (q^2/a^2)}}{q}\,dq$$

$$\varphi_\theta = (\lambda)\,\psi_\lambda \tag{6.15}$$

$$\varphi_\lambda = -\,l(\lambda)\,\psi_\theta \tag{6.16}$$

where

$$l(\lambda) = \frac{1}{\rho}\sqrt{1 - \frac{q^2}{a^2}} \,.$$

The above transformations are real if the flow is subsonic. Equations (6.15) and (6.16) are still perfectly general in that no assumptions concerning the value of γ have been made.

It should be noted that if $l(\lambda) = 1$ then Laplace equations with respect to θ and λ are obtained. The Mach Number can now be chosen to correspond to the flow under consideration. Von Karman and Tsien chose the Mach Number, and hence the point at which the tangent gas line intersects the real gas relationship, to correspond to the free stream condition. A suitable basis for cascades is the inlet Mach Number although some workers prefer a mean flow Mach Number. Another possibility is to arrange intersections at both the upstream and downstream flow conditions.

Having chosen a linear pressure volume relationship, it is possible to relate compressible flow equations

$$\varphi_\theta = \psi_\lambda \tag{6.17}$$

and

$$\varphi_\lambda = -\psi_\theta \tag{6.18}$$

to the equivalent equations in incompressible flow, and thus to determine a relationship between the compressible and incompressible flow pressure coefficients.

Such a relationship has been given by Von Karman[6.9] and Tsien[6.10],

$$Cp_c = \frac{Cp_{inc}}{\sqrt{1 - M^2} + \dfrac{M}{\sqrt{1 - M^2} + 1}\cdot\dfrac{Cp_{inc}}{2}} \,. \tag{6.19}$$

For a thin profile at low Mach Numbers this method provides dependable results.

Reverting to equations (6.15) and (6.16), which are as general as equation (6.1), it is possible to eliminate φ between these equations.

$$l(\lambda)\nabla\psi + l'(\lambda)\psi_\lambda = 0 \tag{6.20}$$

where

$$l'(\lambda) = dl/d\lambda \quad \text{and} \quad \nabla = \frac{\partial^2}{\partial\lambda^2} + \frac{\partial^2}{\partial\theta^2} .$$

Defining

$$\Psi = \sqrt{l}\,\psi \tag{6.21}$$

we obtain

$$\nabla\Psi = L(\lambda)\Psi_\lambda \tag{6.22}$$

where

$$L(\lambda) = \nabla\sqrt{l}/\sqrt{l} . \tag{6.23}$$

Equation (6.22) has been used by Chaplygin[6.11], Von Mises[6.12], Bergman[6.13] and Bers[6.14] to obtain solutions to the compressible flow equations.

A method of designing cascade blades with a prescribed velocity distribution in compressible potential flow using equations (6.17) and (6.18) was given by Costello[6.15]. The velocity distribution was used to select a suitable incompressible potential flow about the unit circle. The Lin[5.88] transformation was then used to transform this incompressible flow into a linearised compressible flow about a cascade. Unfortunately the conditions imposed upon the prescribed velocity distribution are rather stringent.

6.1.3 <u>The general case.</u> Whereas theoretical simplifications facilitate solution of the compressible flow for the extremes of blade spacing, neither channel flow treatment nor linearisation are possible for most cascades. In the general case at least five basic approaches are available. Following early work the solution can be expanded in series form; alternatively fixed point iterative schemes of computation with arithmetical determination of convergence may be used; or, more recently, finite difference, finite element and streamline curvature solutions have been used with success.

6.1.4 <u>Series solutions.</u> Since linearisation would give unacceptable inaccuracies no attempt is made in the series solutions to either transform to a hodograph plane or change the form of the gas laws. Thus equation (6.1) is used directly in methods given by Poggi[6.16] and by Janzen[6.17] and Rayleigh[6.18]. In both methods equation (6.1) was rendered integrable by obtaining a series solution of the form

$$\Phi = \Phi_0 + \Phi_1 M^2 + \Phi_2 M^4 + \ldots + \Phi_n M^{2n} . \tag{6.24}$$

The Mach Number can have any reference value but is usually the inlet free stream Mach Number.

Poggi regarded a compressible fluid as an incompressible fluid having a continuous source distribution in the region external to the body. If, in equation (6.24), Φ_0 is a known incompressible flow obtained from $\nabla\varphi_0 = 0$ then φ is obtained by successively calculating $\nabla\varphi_1 = f_1(\varphi_0, x, y)$, $\nabla\varphi_2 = f_2(\varphi_0, \varphi_1, x, y)$ etc, where f_1 and f_2 are functions obtained from equation (6.1).

The method of Poggi was improved by Kaplan[6.19] who introduced the conjugate complex notation of Milne Thompson[6.20]

$$\omega\bar{\omega} = (-u + iv)\cdot(-u - iv) = u^2 + v^2 \tag{6.25}$$

and obtained

$$\nabla\varphi_0 = 0 \tag{6.26}$$

$$\nabla\varphi_1 = \frac{1}{2}\left[\frac{\partial\varphi_0}{\partial x}\cdot\frac{\partial}{\partial x}\left(\frac{\omega_0\bar{\omega}_0}{u^2}\right) + \frac{\partial\varphi_0}{\partial y}\cdot\frac{\partial}{\partial y}\left(\frac{\omega_0\bar{\omega}_0}{u^2}\right)\right] \tag{6.27}$$

and similarly for $\nabla\varphi_2$, $\nabla\varphi_3$ etc.

Kaplan was then able to relate the right-hand sides to successive approximations to a source-sink distribution. Using the conformal transformation of the obstacle to a circle and the known image system within the circle from the sources outside, Kaplan expressed all velocities in terms of contour integrals. If the integration is achieved analytically, the accuracy of compressible flow prediction is limited only by the number of terms taken.

D.C. Prince[6.21] has used the Janzen-Rayleigh approach but replaced the potential functions of (6.24) by stream functions to simplify satisfaction of the boundary conditions.

Prince considered only the first term

$$\psi = \psi_0 + \psi_1 M^2 \tag{6.28}$$

but allowed ψ_1 to be a function of Mach Number. The flow field is specified in terms of three stream-function components

(a) the incompressible solution, ψ_0

(b) a low Mach number compressibility disturbance, ψ_1, and

(c) an iterated solution, ψ_2, for the compressibility disturbances at a high subsonic reference Mach Number, M. Application of the component solutions to other reference Mach Numbers is achieved by treating the difference between components (b) and (c) as the M^4 component of the Janzen-Rayleigh theory, viz:-

$$\psi = \psi_0 + M_\infty^2\left[\psi_1 + \frac{M_\infty^2}{M_{ref}^2}(\psi_2 - \psi_1)\right]. \tag{6.29}$$

The above approach appears to give reasonable approximations to local Mach Numbers as high as 1.1, provided the dominant feature controlling the pressure distribution is the transverse pressure gradient due to streamline curvature rather than the one-dimensional area variation.

Cascade compressibility effects are conveniently handled by calculation of the compressibility disturbance, (c), at a reference upstream Mach Number of 0.6.

Figure 6.2 gives an example of the type of improvement possible when a method such as Prince's is used. The original NACA 65 series blading was showing high losses due to the presence of shocks. A revised blade was then specified with a slightly longer chord and less curvature in the throat region. The analysis shows that nowhere on the new blade is the sonic condition reached.

6.1.5 Iterative solutions. The right hand side of equation (6.1) may be simulated by a source-sink distribution within the field of incompressible flow around the profile. The difference between this concept and that of Poggi is that whereas in the Poggi and Janzen and Rayleigh theories the right hand side was expanded into a Mach Number power series, in this model the right hand side remains intact and is calculated in iterative steps from the basic flow.

The procedure is that the source distribution corresponding to the right hand side of equation (6.1) is firstly obtained from the incompressible flow solution. This is then used to obtain an equation of the form

$$\Phi_{xx} + \Phi_{yy} = f(x,y). \tag{6.30}$$

Perturbation velocity components due to each element of source distribution are obtained and the total additional velocity at any point resulting from compressibility effects is obtained by integration over the field.

This additional velocity is incomplete, however, since in this form the final solution would not converge to the flow around the original boundaries. The nett induced velocity will in general have a component normal to the body surface. This normal component must be counteracted by placing a line source around the body. The line source should be exactly sufficient to cancel all normal velocities and is incorporated into the equation in a similar way to the point sources representing density variation.

Perturbations on the original velocities are obtained and hence a new flow is established. This new flow is then treated on the same basis and the procedure is repeated until convergence is obtained.

Iterative solutions based upon the Martensen[3.27] potential flow theory have been proposed by Imbach[6.22] and by Price[6.23], who suggested the use of a line source to counteract the normal velocity component. Imbach used the vortex and source boundary conditions to derive two integral equations which were solved by a method of successive approximations.

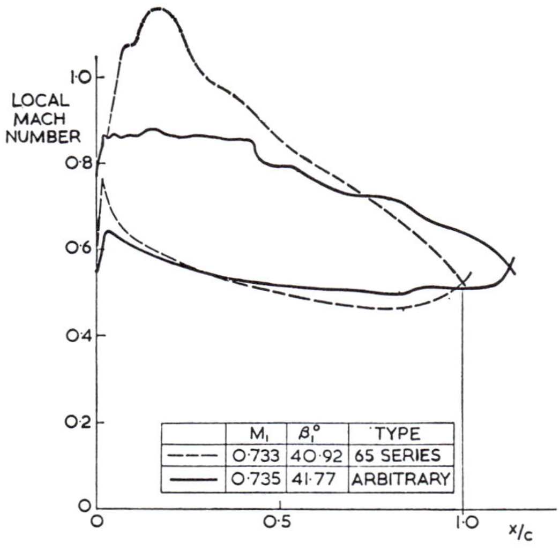

	M_1	$\beta_1°$	TYPE
– – –	0·733	40·92	65 SERIES
——	0·735	41·77	ARBITRARY

Fig. 6.2. Use of compressible flow calculations to eliminate a supersonic region.

If the source distribution is expanded by means of the binomial theorem the series obtained is identical to that of Poggi's method. Furthermore the description is simplified if the complex notation used by Kaplan is introduced. This has been achieved by Payne[5.62] also using the Martensen theory as a foundation.

Unfortunately the iterative methods converge slowly as the sonic condition is approached. This is because the source distribution is by no means a small perturbation. Thus important compressibility terms are left 'trailing' one cycle behind in the iterative procedure.

6.1.6 Finite difference solutions. Finite difference solutions have been developed by Smith and Frost[6.24] who also derived a Poisson-type differential equation. This was then solved using finite difference techniques with a ten-point star. A band matrix solution was chosen, being efficient and more stable numerically than a relaxation method which would depend excessively on user input of an over-relaxation factor. If a relaxation method is chosen then under-relaxation must get stronger as Mach Numbers approach unity.

As with the iterative procedures it is good practice to increase the inlet Mach Number in gradual steps to the desired value.

Ref. 6.24 suggested that the calculation of density should be allowed to lag the stream function calculation by one iteration, thus increasing stability for compressible flows.

This recommendation is at variance with the reasoning of Sylvester and Fitch[6.25] who express density gradients in terms of a dependent variable and collect these together in such a way that density derivatives do not need to be specified 'a priori' or guessed. Smith's is undoubtedly a safe approach and locally sonic conditions have been handled without difficulty.

Davis and Millar[6.26] used 10, and in some cases 12, grid points in an elaborate finite difference mesh. Their report was a useful and detailed description of the techniques adopted for the complex boundary conditions and matrix-manipulation techniques.

6.1.7 Finite-element solutions. Finite-element methods have advantages in flexibility compared with finite-difference methods. Irregular grids can most readily be developed using finite elements and higher order elements may be imposed locally to improve accuracy. Where boundaries are irregular finite elements may be chosen such that grid points coincide with the boundary, since the computation locations are also the corners of the elements.

The finite element method proceeds by the formation of integrals of approximation over the domain under consideration. This is most readily achieved by identification of a variational principle requiring maximisation of a scalar function over the domain.

Although the identification of such a variational principle is not difficult for incompressible flows[5.69,6.27] it is not generally possible for the non-linear equations of subsonic and transonic compressible flows.

One technique which appears to work well for subsonic and low transonic flows is the local linearisation approach of Habashi et al[5.30]. Since no variational principle has been found for the full compressible equations Habashi chooses to linearise the equations with regard to the local Mach Number throughout the flow field. A separate linearisation is therefore performed for each element, the variational principle for that element being based on the Laplace equation for the (locally) linearised compressible flow. This local linearisation is deployed within an iterative scheme and provides stability and rapid convergence for all subsonic Mach Numbers.

A different approach is taken by Adler and Krimerman[3.32] who rewrite the governing equations in the form of Poisson equations having right hand sides composed of 'known' functions. The variational principle is known and the Poisson equations are readily solved by conventional finite-element techniques.

An alternative pseudo-variational approach is by the application of weighting functions. In the weighted residual method the compressible flow equations are replaced by weighted integral statements or shape functions. The most common solution technique involves application of the Galerkin method; the shape function is substituted into the governing equation and the resulting residual is required to be orthogonal to the shape functions.

Ecer and Akay[5.70] argue that consistent use of a suitable variational function gives equivalent elements to those obtained using the Galerkin approach. Nevertheless the Galerkin approximation is well-suited to the non-linear problems of flows for which identification of a variational principle is no longer possible.

Because the Galerkin approximation is essentially elliptic in nature it does not cope well with transonic and supersonic flows where discrimination between the upstream and downstream transmission of information is essential. The introduction of artificial compressibility is a means of overcoming this limitation and is advocated by Deconinck and Hirsch[6.28] who demonstrate high quality solutions for compressible flows in cascades using a Galerkin-based finite element procedure.

Other finite element programs based on the Galerkin approximation include the Newton iterative scheme of Prince[6.29] and the unsteady flow theory of Whitehead and Grant[5.31].

6.1.8 Streamline curvature solutions. Into this category fall the powerful numerical treatments of Katsanis[6.30], Wilkinson[6.31] and Stuart[6.32]. The objectives of these methods appear to be efficiency of operation and the capability to consider velocities above the critical. The methods have originated using streamlines and 'quasi-orthogonals', which are simply lines passing from one channel wall to the other, in an arbitrary (usually pitchwise) direction. It appears to be advantageous to switch to full orthogonal grids for transonic cascades.

Reference 6.30 gives a description and listing of the Katsanis code. The program is versatile and allows for quasi three-dimensionality, losses, rotation and mixed flows. The solution is based on a finite difference technique with the streamline curvature routine taking over for transonic flows.

Rolls-Royce are using a method developed by Stuart, which uses a full orthogonal grid. The program is fast, as is the Wilkinson program, and has been integrated into a complete direct/indirect design system including boundary layer calculation. The method of Bindon and Carmichael[6.33] also uses an orthogonal grid but the program is slower than the Wilkinson and Stuart programs by a factor of ten.

In a typical streamline curvature program the input data are differentiated twice and the values obtained are used in the velocity gradient equation. A start is made with approximate velocities, along the mid-streamline, obtained from the previous iteration. A numerical integration is then performed in both directions to find the blade surface velocities. Velocity levels are then adjusted to satisfy continuity.

In order to ensure rapid convergence, the appropriate damping factor must be chosen. The change in velocity gradient appears to be a sensitive convergence criterion and this is reduced between iterations to prevent divergence.

The least satisfactory feature of streamline curvature methods is their limited ability to give an accurate potential flow solution in the vicinity of the blade edges. The stagnation streamline has a right-angled kink at the blade edge which cannot be properly represented; consequently, errors must be present in the velocity distribution.

With the object of assessing the magnitude of any such errors, results from three streamline curvature programs, together with corresponding results from Smith's matrix program have been compared in Fig. 6.3.

All four calculation methods had been used to predict the velocity distribution around the stator mean section of a cold-air turbine test, by Whitney et al[6.34], and an instructive comparison is possible.

The overall impression is that all four methods give quite acceptable accuracy. Agreement for most of the pressure surface is excellent; for the high Mach Numbers of the suction surface it is quite good, although the Katsanis method exhibits wayward tendencies in places. It appears that the Stuart method displays good accuracy everywhere. The other three methods all have difficulty in the blade edge regions, either putting in unnecessary velocity peaks (Katsanis and Smith) or failing to predict a genuine peak (Wilkinson). It should be noted that streamline curvature techniques are not alone in having problems of blade edge definition; methods using a finite difference approach can also encounter these difficulties.

The test case chosen was not particularly difficult, having only medium velocities and a reasonably low space/chord ratio, but the four methods used were reassuringly successful. Had other methods also been run for this case it is anticipated that agreement would have been reasonable. It therefore appears that, apart from the additional work needed to secure good accuracy in the blade edge regions, the direct problem for subsonic flows is essentially solved.

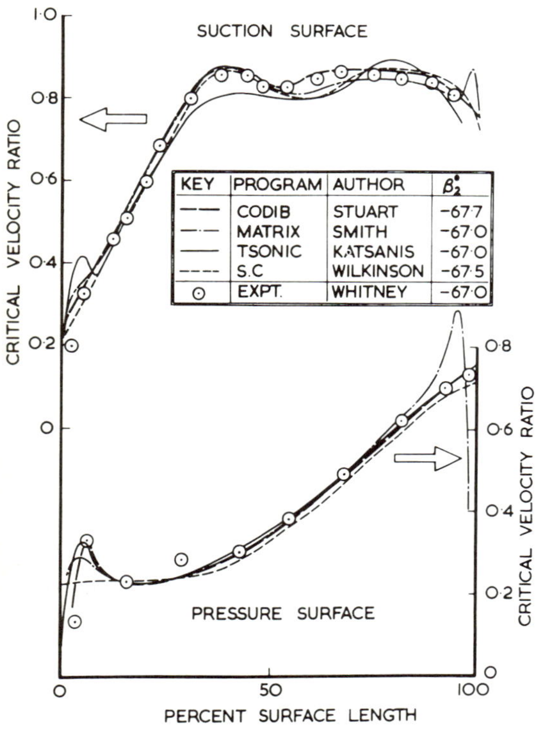

Fig. 6.3. Comparison of four compressible flow theories with experimental results.

6.2 Supersonic Flow Calculations

Most streamline curvature methods are stable at supersonic speeds. Isentropic flow is assumed and prediction of sonic regions terminating in a weak shock can be attempted. Figure 6.4 gives an example of the accuracy obtained. The pressure distributions, computed for the steam turbine blade of ref. 6.35 using the programs of refs. 6.32,6.33 give reasonable agreement with the experimental results except in the peak velocity region. The characteristics solution assumes coincidence of a straight sonic line and the geometric throat.

In cases where purely supersonic flows exist the method of characteristics is employed. The term "characteristics" is used to denote the real solutions to any hyperbolic equa-

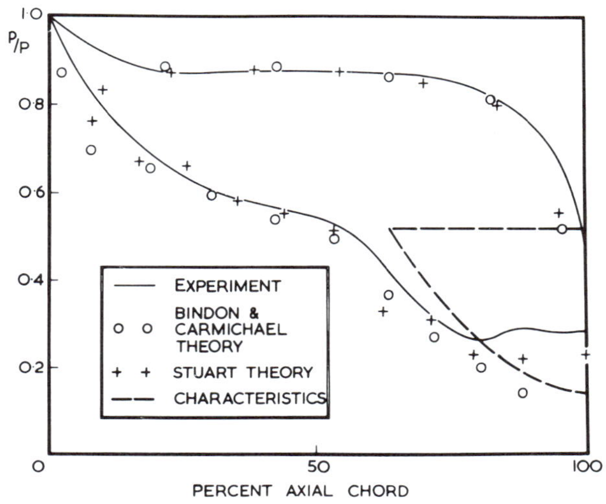

Fig. 6.4. Comparison of theory and experiment for supercritical steam turbine blade.

tion. These solutions take the form of lines along which information may be transmitted. No information is obtained concerning the derivatives of dependent variables normal to those lines.

For two-dimensional, irrotational, isentropic flow the characteristics consitute two families of simple Mach waves. In supersonic flows disturbances are propagated along these Mach waves, which are of two families I and II drawn at $\mp \mu^0$ to the stream direction in Fig. 6.5.

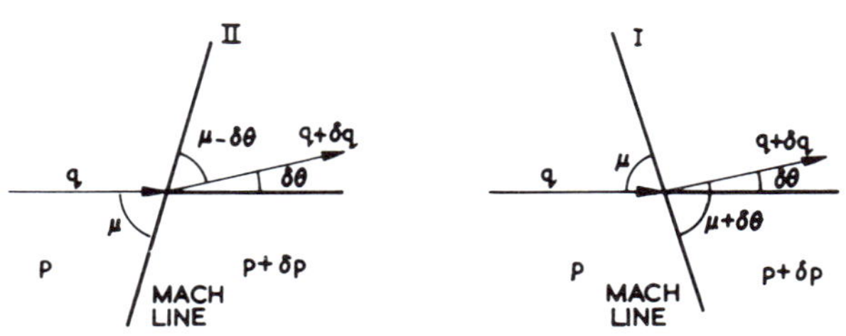

Fig. 6.5. Disturbance propagation along Mach waves.

Using family II as an example the velocity vector q changes in magnitude by δq and in direction by $\delta\theta$ across the Mach line. Now across the wave tangential momentum is conserved,

$$q \cos \mu = (q + \delta q) \cos (\mu - \delta\theta) \tag{6.31}$$

and

$$\frac{1}{q} \cdot \frac{dq}{d\theta} = - \tan \mu = - \frac{1}{\sqrt{M^2 - 1}} . \tag{6.32}$$

Since, from the adiabatic relations

$$\frac{q}{\sqrt{T}} = \frac{M\sqrt{\gamma R}}{\sqrt{1 + \frac{1}{2}(\gamma - 1) M^2}} \tag{6.33}$$

and T is constant we have

$$\frac{dq}{q} = \frac{d M^2}{2 M^2 [1 + \frac{1}{2}(\gamma - 1) M^2]} \tag{6.34}$$

therefore

$$d\theta = \frac{\mp \sqrt{M^2 - 1}\ dM}{M [1 + \frac{1}{2}(\gamma - 1) M^2]} . \tag{6.35}$$

This is integrated along the streamline to give

$$\theta + \omega(M) = \text{constant} = 2 I \tag{6.36}$$

where

$$\omega (M) = - \sqrt{\frac{\gamma + 1}{\gamma - 1}} \tan^{-1} \sqrt{\left\{\frac{\gamma - 1}{\gamma + 1}\right\} . (M^2 - 1)} + \tan^{-1} \sqrt{M^2 - 1} \tag{6.37}$$

is the Prandtl-Meyer expansion angle.

ω has the significance of being the deflection angle required to accelerate the flow from the sonic condition to the desired Mach Number.

The above result, equation (6.37), will also apply if equation (6.35) is integrated along the expansion Mach Line. Alternatively, if (6.35) be integrated along Mach Line II there results

$$\theta - \omega (M) = \text{constant} = 2 II . \tag{6.38}$$

The initial information will usually be given along a curve such as L, in Fig. 6.6, which is not generally a characteristic. The "triangular" network of characteristic lines and their characteristic relationships are built up in a step-by-step manner working outwards from the given curve.

Consider one step of the process - a move from cell A to cell B. the change is governed by equation (6.32). This also applies for the move from B to C and we deduce that the relationships hold all along the Mach wave.

Equations (6.36) and (6.38) hold for all wave crossings (I_m, II_n) and hence the flow field is determined by the conditions (namely angle of inclination and Mach Number) of the physical boundary from which the waves originate. Any intersection in the flow field can thus be characterised by the numbers I and II.

For a more rigorous description, with applications, the reader is referred to the works of Shapiro[1.13] and Courant and Friedrichs[6.36]. In the following section its application to compressor cascades and to impulse and reaction turbine cascades is briefly considered.

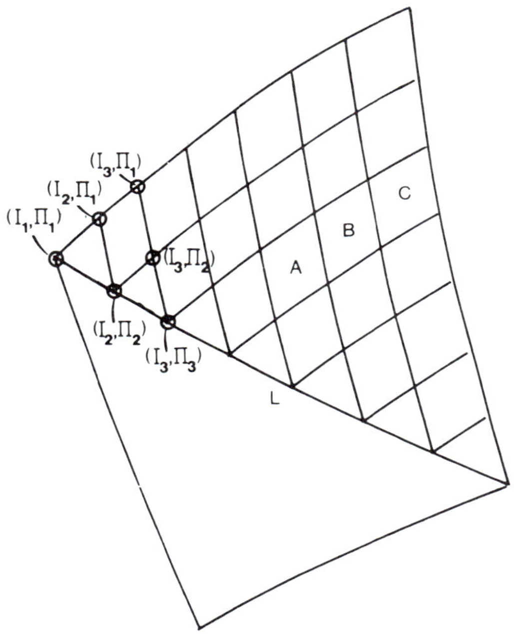

Fig. 6.6. Construction of a network of characteristics.

6.2.1 Compressor cascades. The inlet Mach Number to a compressor blade row needs to be quite high (usually at least 1.5) before worthwhile areas of purely supersonic flow are present around the blade. Some current fan rotor tip sections exhibit supersonic flow over all of the suction surface and much of the pressure surface. Such a case needs analysis by the method of characteristics for optimum design.

Such techniques are not particularly new. In 1951 Costilow[6.37] was able to apply the method of characteristics to the tip section of a supersonic impeller using hand computation.

Other interesting applications include the analysis of radial diffuser cascade for high-performance radial outflow compressors. Kenny[6.38] shows that rather involved configuration assuming a uniform inlet circle source/vortex flow with sharp leading edge vanes. A strong, normal, passage shock occurs at the throat entry.

6.2.2 Impulse cascades. The use of purely supersonic impulse cascades is more widespread, applications having arisen very naturally. For some time the designers of highly loaded turbine stages for marine astern propulsion and for rocket pumps had been using high pressure ratios. Extrapolation of old subsonic bucket data was clearly resulting in bad designs and work was needed to properly prescribe passage geometries.

These cascades employ supersonic free-vortex flow in the main portion of the passage using a vortex net derived by Busemann[6.39]. Since the streamlines of such a flow are concentric circles the problem is one of designing suitable entrance and exit transition regions. If the blade edges are cusped, the supersonic flow pattern will be as shown in Fig. 6.7. The flow in region I will be identical to the flow far upstream. In region II, the flow is expanding, through waves generated by the convex transition arc, and in region III the flow is compressed, through waves generated by the concave transition arc. In region IV the desired supersonic free vortex pattern is established. The problem is thus to generate suitable transition arcs using the method of characteristics. An illustration of this is given in ref. 6.40.

Stratford and Sansome[6.41] gave a very simple early example of the above procedure; Goldman and Scullin[6.42] published a program for the example shown in Fig. 6.7 which includes boundary layer displacement thickness effects.

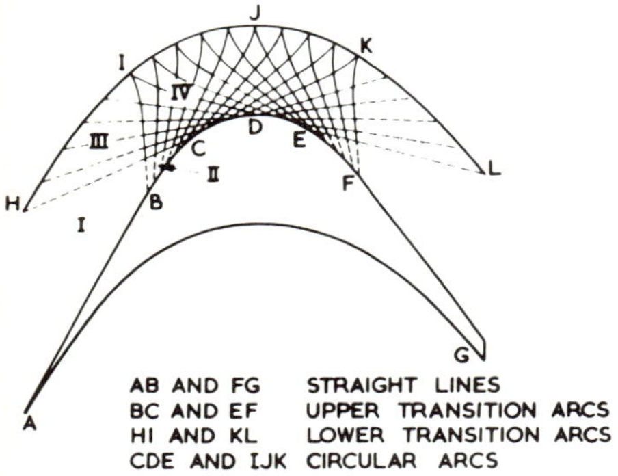

Fig. 6.7. Impulse passage designed for supersonic vortex flow.

6.2.3 Turbine cascades. In general for a mixed supersonic flow field it is reasonable to divide the field into distinct areas. Even in the case of a supercritical turbine nozzle or other blade with subsonic inlet flow there would be a definite sonic line and a new supersonic calculation could be started from that line. Refs. 6.35, 6.43 give good applications of this procedure.

There are important cases, especially for steam turbine rotor tips, where both inlet and outlet flows are supersonic. Lawaczeck[6.44] has classified the various flow configurations for such cases. These mainly depend on whether the flow is choked or not. The mode of solution for such a cascade having a supersonic discharge axial Mach Number has been given by Lichtfuss and Starken[6.45]. In the entrance region the flow is irrotational and isentropic. The flow within the entrance region (Fig. 6.8) has one family of simple Mach waves coming from far upstream. These are straight lines and the flow conditions along the wave do not vary.

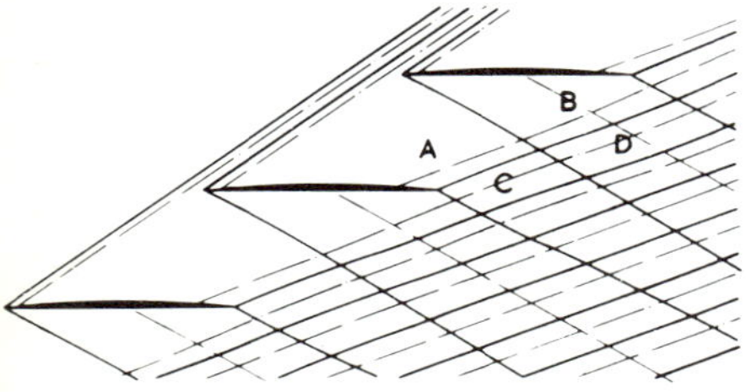

Fig. 6.8. Low turning turbine cascade flow with $M_1 = 2.1$ and $M_2 = 2.43$.

Having determined this region of the flow (marked A in Fig. 6.8) the next region B has to be calculated. This area is not a simple wave flow and the method of characteristics has to be applied. On complete coverage of area B the flow angle and Mach Number at either side of the trailing edge are known. This information then fixes the inclination of the two shocks which spring from the trailing edge (assumed sharp in ref. 6.44) by means of the shock relationships.

Having found this information for the two shocks, the area between these C, can be tackled. Since one set of waves emanates from upstream, this area is one of simple wave flow. However, for region D this is not the case and again the method of characteristics must be employed.

Since the entire downstream region is covered by waves emanating from the blades the back pressure is fixed and cannot be varied by downstream throttling.

6.3 Transonic Flow Calculations

Transonic flows, within a two-dimensional cascade context are taken to imply flow patterns in which supersonic and subsonic flow regimes co-exist.

6.3.1 Flow patterns and equation types. Blading having mixed subsonic and supersonic flows is often encountered in practice. Theories based on the transonic small disturbance assumption, successful for conditions very close to sonic, are inadequate for flows containing strong discontinuities.

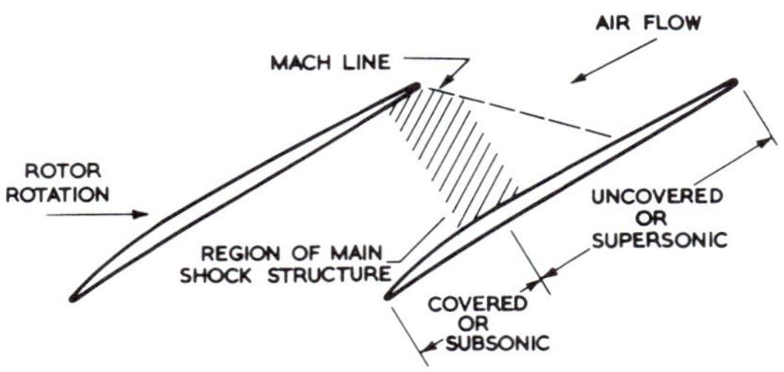

Fig. 6.9. Typical transonic compressor rotor passage.

A typical transonic compressor rotor passage is depicted in Fig. 6.9. The inlet Mach Number might be around 1.4; the position of the main shock structure is of considerable significance for the machine. Obviously no subsonic diffusion can be attempted upstream of the shock's impingement on the suction surface. To achieve this diffusion efficiently the available subsonic chordwise distance must be maximised. At the same time the shock losses, which are an important fraction of the total loss, must be reduced. The shock strength, and therefore the shock loss, is related to the supersonic expansion in the uncovered region of the passage (that region ahead of the first Mach wave from the adjacent blade). The shape of the forward portion therefore determines the shock loss. In achieving optimum performance the designer is therefore constrained to balance the shock and diffusion losses.

The shock configuration for a supercritical turbine blade passage is shown in Fig. 6.10. Such blading is typical of the last rotor row of a large steam turbine where good performance is desired over a wide range of discharge conditions.

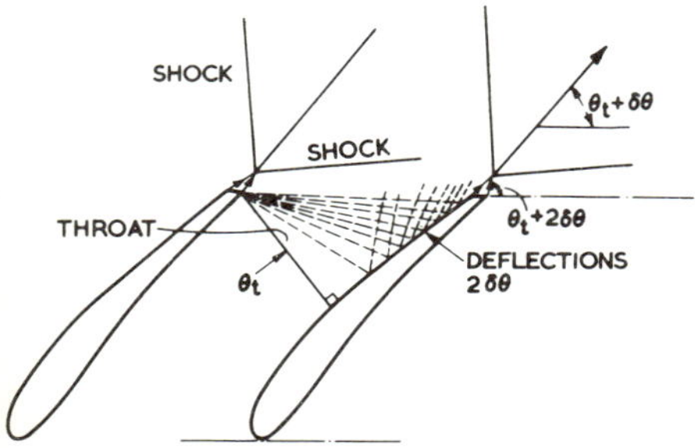

Fig. 6.10. Supercritical turbine nozzle passage.

Blading which has considerable areas of supersonic flow can be analysed by the method of characteristics. If equation (6.1) be rewritten as

$$A\varphi_{xx} + 2B\varphi_{xy} + C\varphi_{yy} = 0,$$

where

$$A = 1 - \frac{\varphi_x^2}{a^2}, \qquad B = -\frac{\varphi_x\varphi_y}{a^2} \quad \text{and} \quad C = 1 - \frac{\varphi_y^2}{a^2}$$

then a necessary and sufficient condition for the existence of real characteristics is

$$D = B^2 - AC > 0 \quad .$$

The differential equation is elliptic if D<O, hyperbolic if D>O. In the elliptic (subsonic) case no characteristics or discontinuities in the higher derivatives of φ are possible. Hyperbolic equations represent supersonic flows for which different solutions may join with a continuous derivative along the characteristic curve. Representation of a steady flow field having both subsonic and supersonic regions entails solution of both hyperbolic and elliptic equations.

6.3.2 Test cases for shock-free flow. The designer will not, in general, have 'a priori' knowledge of the shape and position of the sonic line, which must therefore be predicted. The sonic line often has a marked 'S' shape both in compressor cascades [6.46] and in turbine cascades [6.47]. This line is the boundary at which the elliptic and hyperbolic solutions are matched and this description gives the clue to some recent prediction techniques.

The description of the sonic line in the hodograph plane is trivial. Nieuwland[6.48], relying on the earlier work of Lighthill[6.49], was able to use this fact to produce some remarkable results. Basing his work on the full compressible equations, Nieuwland evolved a method for deriving aerofoil shapes for which waves reflected at the sonic line did not develop into a shock. These cases derive from exceptional hodograph solutions. It follows that, for a given shape of blade, even a small change in inlet Mach Number could disrupt the shock-free flow. Nevertheless the shock-free flow patterns have been observed experimentally, for isolated aerofoils, and are the subject of considerable design effort.

Bauer, Garabedian and Korn[6.50] have formulated a characteristic initial value problem amenable to solution by finite difference procedures. The use of one of Korn's cascades in the development of supercritical compressor blade technology has been described in refs.6.51 and 6.52.

Shock-free impulse turbine cascades developed by Hobson[6.47] using a hodograph approach provide excellent test cases for transonic numerical methods.

Hobson used different coordinate systems for the elliptic and hyperbolic domains.

In the elliptic domain the hodograph plane was divided by constant M and θ lines into four-pointed finite difference stars. For each mesh point the finite difference equations were written as

$$[F]\bar{\psi} = \bar{b} \tag{6.39}$$

where $[F]$ is a matrix whose elements depend on the local value of M and the star line length. Inversion by standard methods yielded

$$\bar{\psi} = [F]^{-1}\bar{b} \ . \tag{6.40}$$

The derivative $(\partial\psi/\partial M)$ was then calculated at the assumed sonic line and stored.

The procedure for the hyperbolic domain was broadly similar. However in the supersonic region the characteristic directions are preferred as coordinates since derivatives of stream function in any other direction are likely to be discontinuous.

The finite difference scheme was therefore set up in the I, II plane as defined in section 6.2.

Across the sonic line the derivatives $(\partial\psi/\partial M)$ and $(\partial\psi/\partial\theta)$ are finite and continuous and

$$\psi = \psi(M,\theta) = \psi(I,II) \ . \tag{6.41}$$

Solutions obtained for these quantities from the elliptic and hyperbolic domains were therefore compared and an iterative scheme adopted to convergence to better than one part in 10,000 which gave a corresponding accuracy in physical shape.

Suction Surface			Pressure Surface		
x/c	y/c	M*	x/c	y/c	M*
0.0043	0.0074	0.3250	0	0	0
0.0283	0.0481	0.4250	0.0282	0.0112	0.3500
0.0883	0.1487	0.5250	0.1343	0.0434	0.4250
0.1206	0.2018	0.5750	0.2103	0.0608	0.4500
0.1469	0.2440	0.6250	0.3312	0.0803	0.4720
0.1747	0.2869	0.7000	0.3612	0.0838	0.4750
0.1980	0.3193	0.7750	0.4179	0.0873	0.4770
0.2158	0.3414	0.8250	0.5000	0.0909	0.4800
0.2560	0.3828	0.9000			
0.2918	0.4124	0.9500			
0.3310	0.4387	1.0000			
0.3750	0.4618	1.0777			
0.3937	0.4695	1.1267			
0.4072	0.4741	1.1871			
0.4167	0.4767	1.2523			
0.4318	0.4800	1.3007			
0.4503	0.4827	1.3380			
0.4843	0.4847	1.3658			
0.5000	0.4854	1.3772			

Coordinate and critical Mach Number distributions are symmetrical about the mid-chord.

$s/c = 1.0121$
$M_1^* = M_2^* = 0.5100$
$\beta_1 = \beta_2 = 0.7600$ radians.

Table 6.1. Coordinates and critical Mach Number distribution for Hobson's shock-free impulse cascade.

The coordinates of one of the resulting profiles are given in Table 6.1. This cascade will be used for evaluation purposes in Chapter 10. Attempts to predict this accurate solution have been made using at least twenty different numerical methods. Most attempts at this very stringent test case resulted in the termination of the supersonic zone by a shock. The methods of refs. 6.52, 6.53 and 3.86, which have proved most successful to date, are relaxation, singularity and time marching solutions respectively.

6.3.3 Solutions to the potential equation. In 1970 Murman and Cole[6.54] developed the first successive line over-relaxation method for transonic flows. Their finite difference solution to the transonic small perturbation equation employed central differencing in supersonic regions. The directional bias thus introduced corresponds to the upwind region of dependence in supersonic flows.

The simplicity and reliability of the method guaranteed its use in external aerodynamics. However the convergence rate was somewhat slow for practical use. Furthermore the full potential equation is usually required for real configurations and especially for cascades.

Jameson[6.55] was able to develop solutions to the full potential equation using a rotated difference approach. The treatment of supersonic flows in complex geometries proceeds most smoothly if the direction of upwind difference is rotated to confirm with the local flow direction. Ives and Liutermoza[6.52] have used the rotated difference approach in their nonconservative method for cascades. Dodge[1.5] applied an upwind difference approach with a characteristic-related grid.

If an upwind difference scheme is interpreted in terms of central differences a third order term is effectively introduced which is equivalent to an 'artificial viscosity'. Jameson followed Murman and Cole in using upwind differencing for the non conservative form of the equations or by explicitly introducing artificial viscosity into the conservative schemes.

When a suitable trailing edge condition is imposed a continuous and unique solution exists to the potential equation (6.1) for the subsonic flow past a cascade. This is not the case when the flow becomes supersonic and weak solutions propagate. The hyperbolic boundary value problem is ill-posed under these conditions and non-physical discrete solutions emerge, including expansion shocks, which prevent the iterative scheme from converging onto the correct solution. Some criterion, such as the entropy condition, is required for selection of the physically significant solution.

In principle the introduction of viscous dissipation is equivalent to the stipulation of an entropy condition and should permit the selection of the correct weak solution to (6.1). In realisation of this Murman and Cole had introduced upwind differencing as a form of viscous dissipation. Jameson used a switching function to add a metred dissipational correction which was zero for subsonic flows and increased strongly with supersonic Mach Numbers.

The whole scheme of solution, having the density modified according to an artificial viscosity term, was labelled the artificial compressibility method by Hafez et al[6.56]. Ref.6.56 presented a new and simple approach to the solution of the potential equation in conservation form. It was shown that the magnitude and form of the modified density were critical; there was a preference for upwind evaluations of density.

The introduction of artificial compressibility has permitted the formulation of the transonic cascade problem as fully elliptic in character. For example the elliptic Galerkin approximation may be applied in finite element solutions. The upwind-differenced artificial viscosity terms are included in non-linear artificial compressibility coefficients which are re-evaluated each iteration.

Artificial compressibility is now widely used for cascade flows in finite difference, finite area and finite element formulations. Examples are given in refs. 6.28, 6.57, 5.30 6.58 , 6.59.

Despite the success of artificial compressibility in selecting the correct solution, practical difficulties with the peakiness of shocks remain. These non-physical peaks, which are accentuated by mesh refinement, are a numerical result of attempts to represent a discontinuous function. Many users have amplified the artificial compressibility function to cope with this and it has appeared that such additional viscosity was necessary for calculation stability. However this has often resulted in the weakening and misplacing of the shock. Habishi and Hafez[6.60] have overcome the problem by taking into account upstream as well as local Mach Numbers in the switching function. This had the effect of eliminating peaks without smearing shocks.

After careful studies of its effects Caspar[6.57] counsels that the role of the artificial compressibility[6.56] method (or artificial viscosity in general) is to eliminate non-physical discrete solutions to the potential equation. Artificial compressibility should not be used to enhance convergence.

6.3.4 Solutions to the Euler equations. The methods developed for external aerodynamics often have difficulty coping with the repeating boundary conditions imposed by cascade flows. Furthermore solutions to the potential equation have not demonstrated the ability to capture strong shocks.

The difficulty of mixed elliptic and hyperbolic domains with strong discontinuities has been overcome by use of 'time marching' techniques, treating the governing time-dependent equations for compressible flow. These are hyperbolic for all Mach Numbers and it is premised that the asymptotic solution reached after an iteration process is the physical steady solution.

The two-dimensional equations for conservation of mass, momentum in each direction and energy are:-

$$\frac{\partial \rho}{\partial t} + \frac{\partial}{\partial x}(\rho u) + \frac{\partial}{\partial y}(\rho v) = 0 \; . \tag{6.42}$$

$$\frac{\partial}{\partial t}(\rho u) + \frac{\partial}{\partial x}(p + \rho u^2) + \frac{\partial}{\partial y}(\rho uv) = 0 \; , \tag{6.43}$$

$$\frac{\partial}{\partial t}(\rho v) + \frac{\partial}{\partial x}(\rho uv) + \frac{\partial}{\partial y}(p + \rho v^2) = 0 \; , \tag{6.44}$$

$$\frac{\partial}{\partial t}(\rho E) + \frac{\partial}{\partial x}(\rho uH) + \frac{\partial}{\partial y}(\rho vH) = 0 \; . \tag{6.45}$$

These equations are to be solved in conjunction with the perfect gas relationships:-

$$p = \rho RT \; , \tag{6.46}$$

$$H = c_p T + \tfrac{1}{2}q^2 \; , \tag{6.47}$$

$$E = c_v T + \tfrac{1}{2}q^2 \; . \tag{6.48}$$

In solving the above two-dimensional Euler equations some fundamental choices must be made.

Firstly the equations may be treated in either integral form with finite volume methods, such as those of McDonald[6.61] or Denton[3.86], or differential form using entirely finite difference techniques as in the method of Gopalakrishnan and Bozzola[6.62] or Veuillot and Viviand[6.63]. Although this question is not resolved the finite volume techniques have advantages in simplicity of operation and interpretation.

Secondly there is the question of whether, in combination with the momentum equations, isentropic flow is to be assumed or the energy equation is to be solved. In the former case stagnation enthalpy will not be conserved and in the latter case entropy will not be conserved.

In his finite volume time marching method McDonald[6.61] chose to assume isentropic flow. Surprisingly good agreement was demonstrated between isentropic time marching and experiment for cases having local Mach Numbers as high as 1.43. In McDonald's scheme the conservation laws are automatically satisfied between two adjacent elements, giving the desired accuracy with a coarser mesh than earlier time-dependent finite difference solutions, such as that of ref.6.62 in which a second-order McCormack scheme was used.

In an improved finite volume scheme Denton[3.86] chose to solve the energy equation, thus permitting entropy variations in the flow field. In practice the solution of the energy equation can be a trivial assumption of constant stagnation enthalpy for steady two-dimensional flows. This assumption is incorrect for unsteady or three-dimensional flows in which case the energy equation may require solution in full.

Denton used a simple four-sided grid and a simple and fast first-order difference scheme with the progressive addition of higher order corrections as the calculation converges. This produced a fast and robust method which has been widely used for design and analysis[6.64].

Methods involving solution of the energy equation often produce strong entropy changes which influence the entire downstream flow field. It has been found that these are particularly prevalent in the leading edge region where attempts to represent sharp changes in gradient and curvature are frequently unsuccessful.

Although mesh refinement shows some promise of increasing precision in the blade edge regions transformation of the blade geometry from a physical to a computational plane[6.65] and the use of special fitted grids [6.58,6.65] are more elegant and potentially more efficient and accurate methods of blade edge treatment.

The time step in explicit time integration is limited by the stability bounds of the Courant-Friedrichs-Lewy criterion which ensures that the numerical domain of dependence encloses the domain of dependence determined by the characteristic cone. It is this criterion which limits convergence speed and may dictate the use of several hundred time steps. Even when the criterion is satisfied oscillations and lack of convergence may still be caused by the amplification of non-physical weak solutions.

There is, in the Euler equations, no strong dissipation source which would lead to effective elimination of spurious unsteady perturbation waves. Some additional means of damping is therefore required and this may be provided by various techniques such as spatial smoothing[6.61] or explicit artificial viscosity[6.63]. Essers[6.66] uses completely artificial time dependent equations chosen to give strong damping and improved convergence. Kurzrock and Novick[6.67] produced a predictor-corrector time marching solution to the Navier Stokes equations. Good agreement with experimental results for a transonic compressor cascade was demonstrated.

In Denton's scheme[6.68] the derived values of pressure are used in the subsequent momentum equations and this makes the scheme partially implicit. Denton also demonstrates the value of multigrid techniques in enhancing convergence. Whereas in an explicit scheme the variables at any location are expressed directly, in terms of known information from the previous time step, an implicit scheme calls upon undetermined variables from other locations.

Implicit methods for solving the Euler equations, although more complex in formulation, are stable regardless of time step size. Steger et al[6.69] have used implicit approximate factorization to predict the transonic flow through a compressor cascade. Delaney [6.70] has described the use of a hopscotch method combining explicit and implicit finite difference formulae at alternate grid points. This method optimises the pseudo-viscous damping.

Two other related methods which have been demonstrated convincingly for turbine cascades are the time marching method of Nagayama et al [6.71] and the partially parabolic method of Singhal and Spalding[6.72].

6.3.5 Shock boundary layer interactions. At present the only reliable way to predict passage flows having shock-boundary layer interactions is to use empirical data, as plotted in Fig. 6.11. The graph shows peak pressure rise across the shock as a function of upstream Mach Number.

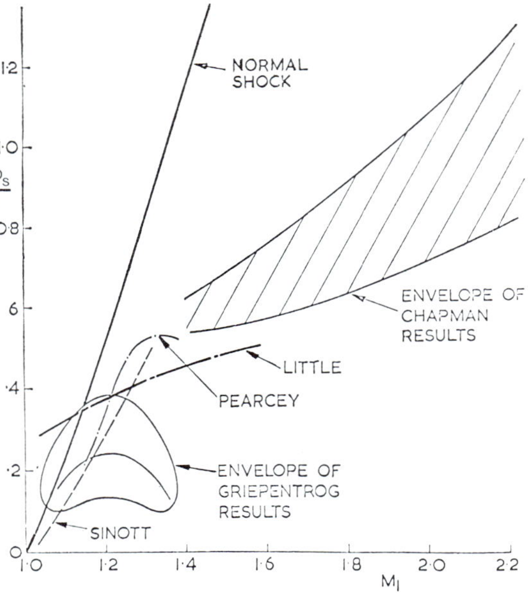

Fig. 6.11. Data for static pressure rise across passage shocks.

The overall trends are established by Chapman's data[6.73] obtained from a wide variety of Reynolds' Numbers. These data, extrapolated back, agree roughly with the trends of the Sinnott[6.74] and Pearcey[6.75] data and have a higher pressure rise at around Mach 1.5 than those quoted in ref. 6.76. At Mach Numbers around two the Chapman data are confirmed by the annular cascade tests of Paulon et al[3.89]. The data of Griepentrog[6.77] tend to be open to wide interpretation being derived from a variety of boundary conditions.

Sinnott's results are obtained from tests on isolated aerofoils. In predicting pressure distributions the subsonic and supersonic regimes were treated separately and then linked by an empirical shock-pressure rise relation. Sinnott's results suggest that the shock appears at that point on the aerofoil at which an equivalent shock-free flow would have its down-stream sonic point.

The rule thus derived was used with the theory of Spreiter and Alkesne[6.78] to give good agreement with subsequent experiments for both position and strength of shock. In common with some previously described theoretical methods, Sinnott's approach is limited to non-peaky profiles.

REFERENCES

6.1 STANITZ, J.D. and PRIAN, V.D. A rapid approximate method for determining velocity distribution on impeller blades of centrifugal compressors. NACA TN 2421 (1951)

6.2 HUPPERT, M.C. and MCGREGOR, C. Comparison between predicted and observed performance of gas-turbine stator blade designed for free-vortex flow. NACA TN 1810 (1949)

6.3 CAI, RUI-XIAN. The accuracy of mean-stream-line method in subsonic and transonic regions. 5th ISABE, Bangalore (1981)

6.4 STANITZ, J.D. Approximate design method for high-solidity blade elements in compressors and turbines. NACA TN 2408 (1951)

6.5 UENISHI, A. A design method and the performance of two-dimensional turbine cascades for high subsonic flows. ASME Paper 71-GT-34 (1971)

6.6 COHEN, M.J. A hodograph design method for compressible flow problems. Trans. ASME, J. App. Mech. 29 (1962)

6.7 LAKOMY, C. The calculation of the flow of the compressor cascade at high subsonic speeds. Strojnicky Casopis XVII c. 1 (1965)

6.8 ZHOU, S., SHEN, M.Y. and LIN, B.Z. Similarity law of transonic flow fields around cascades. ASME Paper 81-GT-36 (1981)

6.9 VON KARMAN, T. Compressibility effects in aerodynamics. J. Aero. Sci., 8;9 (1941)

6.10 TSIEN, H.S. Two-dimensional subsonic flow of compressible fluids. J. Aero. Sci., 6;10 (1939)

6.11 CHAPLYGIN, S.A. On gas jets. NACA TM 1063 (1944)

6.12 VON MISES, R. Mathematical theory of compressible fluid flow. Academic Press (1958)

6.13 BERGMAN, S. Methods for determination and computation of flow patterns of a compressible fluid. NACA TN 1019 (1946)

6.14 BERS, L. Velocity distribution on wing sections of arbitrary shape in compressible potential flow. III: Circulating flows obeying the simplified density speed relation. NACA TN 2056 (1950)

6.15 COSTELLO, G.R. Method of designing cascade blades with prescribed velocity distributions in compressible potential flow. NACA Rep. 978 (1949)

6.16 POGGI, L. Campo di velocita in una corrente piana di fluida compressibile. L'Aerotecnica 12, 1519 (1932)

6.17 JANZEN, O. Beitrag zu einer Theorie der stationären strömung kompressibler Flussigkeiten. Phys. Zeits. 14, 639 (1913)

6.18 RAYLEIGH, Lord. On the flow of a compressible fluid past an obstacle. Phil. Mag. 32, 1 (1916)

6.19 KAPLAN, C. On the use of residue theory for treating the subsonic flow of a compressible fluid. NACA Rep. 728 (1941)

6.20 MILNE-THOMPSON, L.M. Theoretical Hydrodynamics. MacMillan (1938)

6.21 PRINCE, D.C. Jr. Evaluation of compressibility effects by the compressible flux plot computer program. GE R66FPD (1966)

6.22 IMBACH, H.E. Calculation of the compressible, frictionless subsonic flow through a plane blade cascade. Brown Boveri Rev. 51, 12 pg. 752 (1964)

6.23 PRICE, D. Two dimensional compressible potential flow around profiles in cascade. GTCC Aero Rep. 547

6.24 SMITH, D.J.L. and FROST, D.H. Calculation of the flow past turbomachine blades. Proc. I. Mech. E. 184, 3G(II) (1969-70)

6.25 SYLVESTER, M.E. and FITCH, C.M. Matrix methods for the design of cascades to prescribed surface velocity distributions and for fully compressible flow. Conf. at Penn. State Univ. (1970)

6.26 DAVIS, W.R. and MILLAR, D.A.J. A matrix method applied to the analysis of the flow past turbomachine blades. Carleton Univ. Rep. ME/A 72-7 (1972)

6.27 THOMPSON, D.S. Finite element anlaysis of the flow through a cascade of aerofoils. Cambridge Univ. CUED Turbo /TR 45 (1973)

6.28 DECONINCK, H. and HIRSCH, CH. Finite element methods for transonic blade-to-blade calculation in turbomachines. Trans. ASME, J. Eng. Power 103, 4 pg. 665 (1981)

6.29 PRINCE, T.C. Prediction of steady inviscid compressible flow on a blade-to-blade surface by finite element method. AIAA Paper 78-244 (1978)

6.30 KATSANIS, T. Fortran program for calculating transonic velocities on a blade-to-blade stream surface of a turbomachine. NASA TN D-5427 (1969)

6.31 WILKINSON, D.H. Calculation of blade-to-blade flow in a turbomachine by streamline curvature. ARC R&M 3704 (1970)

6.32 STUART, A.R. Private communication (1972)

6.33 BINDON, J.P. and CARMICHAEL, A.D. Streamline curvature analysis of compressible and high Mach Number cascade flow. J. Mech. Eng. Sci. 13,5(1971)

6.34 WHITNEY, W.J., SZANCA, E.M., MOFFITT, T.P. and MONROE, D. Cold-air investigation of a turbine for high-temperature engine application. NASA TN D-3751 (1967)

6.35 FORSTER, V.T. Turbine-blading development using a transonic variable-density cascade wind tunnel. Proc. I. Mech. E. 179, 1, 6 (1964-65)

6.36 COURANT, R. and FRIEDRICHS, K.O. Supersonic Flow and Shock Waves. Interscience, N.Y. (1949)

6.37 COSTILOW, E.L. Application of a characteristic blade solution to flow in a supersonic rotor with varying stream filament thickness. NACA TN 2992 (1955)

6.38 KENNY, D.P. Supersonic radial diffusers. AGARD Lect. Ser. 39 (1970)

6.39 BUSEMANN, A. Unpublished work described in Section G of "The Aerodynamics of Turbines and Compressors" (W.R. Hawthorne ed.) Princeton Univ. Press (1964)

6.40 ERWIN, J.R. and VITALE, N.G. The radial outflow compressor. Advanced Centrifugal Compressors. ASME (1971)

6.41 STRATFORD, B.S. and SANSOME, G.E. Theory and tunnel tests of rotor blades for supersonic turbines. ARC R&M 3725 (1960)

6.42 GOLDMAN, L.J. and SCULLIN, F.J. Computer program for design of two-dimensional supersonic turbine rotor blades with boundary-layer correction. NASA TM X-2434 (1971)

6.43 CURTIS, E.M., HUTTON, M.F. and WILKINSON, D.H. Theoretical and experimental work on losses in 2-D turbine cascades with supersonic outlet flow. I. Mech. E. Conf. Pub. 3, pg. 19 (1973)

6.44 LAWACZECK, O.K. Calculation of the flow properties up- and downstream of and within a supersonic turbine cascade. ASME Paper 72-GT-47 (1972)

6.45 LICHTFUSS, H.J. and STARKEN, H. Supersonic exit flow of two-dimensional cascades. ASME Paper 72-GT-49 (1972)

6.46 DVORAK, R. On the development and structure of transonic flow in cascades. in 'Symposium Transsonicum II', Springer Verlag (1976)

6.47 HOBSON, D.E. Shock-free transonic flow in turbomachinery cascades. Ph.D. thesis, Cambridge Univ. (1974)

6.48 NIEWLAND, G.Y. Theoretical design of shock free transonic flow around aerofoil sections. 5th ICAS Cong., London (1966)

6.49 LIGHTHILL, M.J. The hodograph transformation in transonic flow, II and III. Proc. R. Soc., 1-191 (1947)

6.50 BAUER, F., GARABEDIAN, P. and KORN, D. Supercritical wing sections III. Lecture Notes in Economics and Mathematical Systems, Vol. 150, Springer-Verlag, N.Y. (1977)

6.51 STEPHENS, H.E. Application of supercritical airfoil technology to compressor cascades. Comparison of theoretical and experimental results. AIAA J. 17 pg. 594 (1979)
6.52 IVES, D.C. and LIUTERMOZA, J.F. Second-order accurate calculation of transonic flow over turbomachinery cascades. AIAA J. 17, pg. 870 (1979)
6.53 RIBAUT, M. and VAINIO, R. On the calculation of two-dimensional subsonic and shock-free transonic flow. Trans. ASME, J. Eng. Power 97, pg. 603 (1975)
6.54 MURMAN, E.M. and COLE, J.D. Calculation of plane steady transonic flows. AIAA J. 9 1 pg. 114 (1971)
6.55 JAMESON, A. Iterative solution of transonic flows over airfoils and wings, including flows at Mach 1. Comm. on Pure and App. Math. 27;283 (1974)
6.56 HAFEZ, M.M., SOUTH, J.C. and MURMAN, E.M. Artificial compressibility methods for numerical solutions of transonic full potential equation. AIAA J. 17 8 pg. 838 (1979)
6.57 CASPAR, J.R. Unconditionally stable calculation of transonic potential flow through cascades using an adaptive mesh for shock capture. ASME Paper 82-GT-238 (1982)
6.58 FARRELL, C. and ADAMCZYK, J. Full potential solution of transonic quasi-3D flow through a cascade using artificial compressibility. Trans. ASME, J. Eng. Power 104, 1 pg. 143 (1982)
6.59 AKAY, H.U. and ECER, A. Transonic flow computations in cascades using finite element method. Trans. ASME, J. Eng. Power 103, pg. 657 (1981)
6.60 HABASHI, W.G. and HAFEZ, M.M. Finite element solutions of transonic flow problems. AIAA J. 20 10 pg. 1368 (1982)
6.61 MCDONALD, P.W. The computation of transonic flow through two-dimensional gas turbine cascades. ASME Paper 71-GT-89 (1971)
6.62 GOPALAKRISHNAN, S. and BOZZOLA, R. Computation of shocked flows in compressor cascades. ASME Paper 72-GT-31 (1972)
6.63 VEUILLOT, J.P. and VIVIAND, H. Pseudo-unsteady method for the computation of transonic potential flows. AIAA J. 17 7 (1979)
6.64 SARATHY, K.P. Computation of three-dimensional flow fields through rotating blade rows and comparison with experiment. Trans. ASME, J. Eng. Power 104, 2 pg. 394 (1982)
6.65 SOBIECZKY, H. and DULIKRAVICH, D.S. A computational design method for transonic turbomachinery cascades. ASME Paper 82-GT-117 (1982)
6.66 ESSERS, J.A. and KAFYEKE, F. Application of a fast pseudo-unsteady method to steady transonic flows in turbine cascades. Trans. ASME, J. Eng. Power 104, 2 pg. 420 (1982)
6.67 KURZROCK, J.W. and NOVICK, A.S. Transonic flow around rotor blade elements. ASME Paper 75-FE-24 (1975)
6.68 DENTON, J.D. An improved time marching method for turbomachinery flow calculation. ASME Paper 82-GT-239 (1982)
6.70 DELANEY, R.A. Time-marching analysis of steady transonic flow in turbo-machinery cascades using the hopscotch method. ASME Paper 82-GT-152 (1982)
6.71 NAGAYAMA, T., ADACHI, T., HAYASHI, Y. and MIYATA, E. Numerical analysis of flow through turbine cascades by the modified FLIC method. Mitsubishi Tech. Rev. (Oct. 1978)
6.72 SINGHAL, A.K. and SPALDING, D.B. A 2D partially-parabolic procedure for axial-flow turbomachinery cascades. ARC R&M 3807 (1976)
6.73 CHAPMAN, D.R., KUEHN, D.M. and LARSON, H.K. Investigation of separated flows in supersonic and subsonic streams with emphasis on the effect of transition. NACA Rep. 1356 (1957)
6.74 SINNOTT, C.S. On the prediction of mixed subsonic-supersonic pressure distributions. J. Aero. Sci. 27, 767 (1960)
6.75 PEARCEY, H.H. Some effects of shock-induced separation of turbulent boundary layers in transonic flow past aerofoils. ARC R&M 3108 (1959)
6.76 LITTLE, B.H.Jr. Effects of initial turbulent boundary layer on shock induced separation in transonic flow. VKI TN 39 (1967)

6.77 GRIEPENTROG, H. Transonic compressor cascades. AGARD Lect. Ser. 39 Advanced Compressors (1970)

6.78 SPREITER, J.R. and ALKESNE, A.Y. Thin airfoil theory based on approximate selection of the transonic flow equation. NACA Rep. 1359 (1958)

CHAPTER 7

Viscous Flows

For the whole of the last century a schism existed between the elegant science of hydrodynamics and the practical investigations of hydraulics. This division has existed for much of recorded history and its legacy is still with us in the various ways in which fluid mechanics is taught.

However in 1904 Prandtl's boundary layer theory [7.1] provided a basis for reconciliation. Prandtl observed that where inertia effects are dominant over viscous effects the latter are confined to a thin layer of fluid adjacent to the body and to the ensuing wake. The resulting paradigm was one in which the hydrodynamic field calculations are matched to those of the viscous boundary layer; this has rendered most steady flows calculable including parameters such as drag, hitherto empirically derived.

In most turbomachinery applications the Reynolds' Number is sufficiently high for viscous effects to be confined to thin boundary layers adjacent to blade and annulus surfaces. Progress may be made by calculating free stream and boundary layer regions separately and then matching the two. The initial decoupling, however, is rarely complete and may break down in particular flow regimes, such as that of separation.

When theoretical predictions proceed from potential flow calculations to consideration of real and viscous flows, three-dimensional influences begin to affect applications of the cascade model. The flows are not completely two-dimensional and consideration of the governing Navier-Stokes equations reveals significant terms arising from curvature and rotational effects which lead to substantial modifications to the two-dimensional flow model [7.2].

Nevertheless, the complexities of modelling the viscous flow through even a single blade row are such that the desirable fully three-dimensional solutions appear, at best, expensive in computer time and, at worst, impossible. Fortunately the blade profile boundary layer may be regarded as almost two-dimensional, modified slightly by three-dimensional effects, unless separation takes place [7.3]. It is therefore perfectly reasonable to continue to attack the turbomachinery flow problem according to the schema outlined by Wu [1.9] depicted in Fig. 1.3. We shall therefore be treating profile boundary layers which are assumed to be macroscopically two-dimensional.

In this chapter an attempt is made to show the real value of considering viscous flow for cascades. This value lies not only in enabling the profile drag to be calculated and areas of separated flow to be avoided, but also in providing a condition for determining the cascade outlet angle and thus giving a unique pressure distribution.

The boundary layer concept for laminar and turbulent flows and its theoretical and experimental application to cascades will be discussed. Prediction of transition from a laminar layer to a turbulent layer is of some importance for turbomachinery blading; regrettably this topic must still be approached in an empirical manner. For boundary layers of thickness comparable with that of the blading the real flow departs increasingly from the inviscid flow; this is especially so when separated flow regions and wakes are present.

Finally the author argues that viscous flow concepts in any event have a fundamental role in determining the flow around a blade and the loading that it will sustain. These viscous approximations to the real flow are outlined in the final part of this chapter.

7.1 Boundary Layer Concepts and Applicability

Viscous flows of Newtonian fluids are described by the Navier-Stokes equations and the continuity equation. For two-dimensional incompressible flows of constant viscosity

$$\frac{\partial u}{\partial x} + \frac{\partial v}{\partial y} = 0 \tag{1.5}$$

$$\rho\frac{\partial u}{\partial t} + \rho u\frac{\partial u}{\partial x} + \rho v\frac{\partial u}{\partial y} = \rho g - \frac{\partial p}{\partial x} + \mu\left(\frac{\partial^2 u}{\partial x^2} + \frac{\partial^2 u}{\partial y^2}\right) \tag{7.1}$$

$$\rho\frac{\partial v}{\partial t} + \rho u\frac{\partial v}{\partial x} + \rho v\frac{\partial v}{\partial y} = \rho g - \frac{\partial p}{\partial y} + \mu\left(\frac{\partial^2 v}{\partial x^2} + \frac{\partial^2 v}{\partial y^2}\right). \tag{7.2}$$

If steady flow is postulated and the viscosity set equal to zero the first order Euler equation (1.10) for inviscid flows is readily retrieved.

The presence of the viscosity term in equations (7.1) and (7.2) renders those equations second order and relatively intractable. Some explicit solutions exist for laminar flows and numerical calculation schemes are appearing for low Reynolds' Number turbulent flows. However complex boundary conditions and Reynolds' Numbers extending significantly into the turbulent regime have so far inhibited solutions.

The major accomplishment of Prandtl's boundary layer theory was to make simplifications which turned the elliptic equations (7.1) and (7.2) into one parabolic equation which lent itself to various solution strategies by marching rather than the more difficult field solution.

Prandtl assumed $v \ll u$ and $\partial/\partial x \ll \partial/\partial y$. This resulted in the elimination of the $\partial^2 u/\partial x^2$ term.

If the boundary layer is sufficiently thin, and the surface curvature sufficiently small, static pressure variations within the boundary layer in the y direction may be neglected. Furthermore $\partial p/\partial x$ may be obtained directly from the potential flow at the edge of the boundary layer. Upon application of the above simplifications equation (7.2) is eliminated entirely.

The remaining equation for steady flow is

$$\rho u\frac{\partial u}{\partial x} + \rho v\frac{\partial u}{\partial y} = -\frac{\partial p}{\partial x} + \mu\frac{\partial^2 u}{\partial y^2}\,. \tag{7.3}$$

and this equation is to be solved along with the continuity equation.

A further simplification is made by integrating equation (7.3) with respect to y. Taking advantage of the definitions of displacement and momentum thickness the momentum integral equation for the boundary layer is obtained:-

$$\frac{\tau_w}{\rho U} = \frac{d\theta}{dx} + \frac{(2 + H)\theta}{U} \cdot \frac{dU}{dx} . \tag{7.4}$$

Laminar and turbulent boundary layers may be measured and predicted in many different ways. The iterative approach, of calculation method development, followed by experimental verification, followed by refinement of the theory and subsequent recalculation, has generally paid handsome dividends. From time to time attempts are made to compare the results of different boundary layer theories and users are able to appraise the most promising lines of attack [7.4, 7.5]. It is not possible to survey the field in this work and for a more comprehensive account the reader is referred to the classical work of Schlichting[2.25]; a more recent text is that of White[7.6].

It cannot be said that either laminar or turbulent layers predominate in turbomachinery blading. Certainly high Reynold's Numbers and free-stream turbulence levels tend to favour the latter and high turbulence levels are known to be generally present in turbines[3.87] and compressors[2.27]. Laminar and transitional layers are also known to be present[7.7] over much of the blading in cases of practical interest and to increase in extent with increasing Mach Number[7.8]. Klein[7.9] has argued that turbulent boundary layers on blades hardly ever reach the fully developed state for which the calculations are performed.

Two examples illustrate the consequences of this pluralistic behaviour. Firstly, the fan rotor blading of turbofan engines is often observed to have a laminar boundary layer over up to 70% of the chord although this may deteriorate to a transitional or turbulent state as the surface degrades. The accurate prediction of the development of this layer is most important for the setting of blade incidence and for determining the throat area and hence the mass flow. Secondly, in the turbine blading of the same engine it is most important to ascertain whether a laminar or a turbulent layer is present at a given location since this determines directly the heat transfer coefficient and hence the blade temperature.

The two applications described above are additional to the most common application of boundary layer theory, which is to determine the drag of the blading. The accuracy of loss prediction in a cascade relates directly to the ability to predict the momentum thickness of the boundary layer in the trailing edge plane, whether this be laminar or turbulent.

7.1.1 Experimental techniques. Measurement of boundary layers and wakes usually means traversing the flow, either with a pitot tube or a hot-wire probe. For the small blades generally present in turbomachines, as represented by aerofoils in cascade this is obviously difficult. Very few investigators have worked with cascades having a chord greater than 150mm and there are accordingly not many published accounts of blade boundary layer measurements on cascade blading or in rotating machines. This situation is now improving but boundary layer measurements on blades are still quite rare.

The compactness of a pitot tube is limited by response time considerations. The boundary layers on blading of 50mm chord will typically have an absolute thickness of 2mm and such blading is therefore already too small for high quality aerodynamic measurements. It is desirable to use a Preston tube for skin friction evaluation but again physical size and curvature effects normally rule this out on blading having a smaller chord than about 150mm. Nevertheless Oldfield et al[7.8] have demonstrated the use of a flattened pitot tube for transition detection, having an external width of only 0.3mm, on cascade blades of 80mm chord.

Hot wires can be smaller but again it is difficult to conceive of good aerodynamic information being obtained on blading of less than 150mm chord. Fortunately a few measurements of boundary layers on blading exist and some comparisons of theory with measurements from a turbomachinery environment are possible[7.10, 7.11].

The application of heated thin film gauges to a ceramic blade has proved to be a valuable technique for identifying transition regions[7.8]. The use of this technique with suitable wide-bandwidth instrumentation should prove valuable for investigations of boundary layer state and of heat transfer rates.

Measurement of the extent of the transition region is important for some blading. The only certain means of identification of the state of the boundary layer is by determination of the intermittency factor which is defined as the fraction of the elapsed time for which the flow at a given point is observed to be turbulent. The intermittency varies from zero for a laminar layer, through intermediate values for transitional layers, to unity for a turbulent layer. A continuous readout of intermittency during traversing is desirable; this requires some form of intermittency meter. Although a simple device based on threshold sensing[7.12] is adequate for most experimental work it may be desirable for maximum precision for the meter to incorporate a differentiating circuit.

It might be thought that optical techniques would offer opportunities for evaluation of turbomachinery boundary layers. A promising start was made by Faulders[7.13]. Sensitive interferometry techniques were used but the results were really confined to observation of transition location and did not permit detailed comparison between theory and experiment.

Similarly attempts to use laser anemometry for this purpose have been unsuccessful because, although laser anemometry is promising for free stream flows, it has been much less successful for boundary layer flows due to spurious effects from wall interference.

For the present therefore it appears that the most fruitful approach to measurement of blade boundary layers will be the use of miniature hot wire and heated film techniques, on the blading of stationary cascades as well as in axial compressors and turbines. This route requires much painstaking and tedious effort but is quite essential if confidence is to be gained in the use of boundary layer calculations in a turbomachinery environment.

7.1.2 Calculation methods. Boundary layer calculation methods for use on cascade blading may be tested by comparison with experimental data from a wide range of sources. In addition to data from cascades the whole body of aerodynamic experimentation is available. Similarly theories developed in very different contexts can often be used with confidence in turbomachinery applications.

One interesting approach to loss prediction is somewhat specific to compressor blading. Lieblein[2.20] applying the momentum integral equation (7.4) to cascade blades, was able to demonstrate that the wake momentum thickness could be expressed as a primary function of the ratio of maximum surface velocity to outlet velocity (diffusion ratio).

Several simplifying assumptions were required to attain this relationship but the analysis of cascade data at the time and in many subsequent uses of the 'Diffusion Factor' approach, demonstrate a very good loss correlation at the minimum loss incidence. Use of this diffusion factor approach in general produces a much better loss prediction for compressor blading than consideration of lift coefficient or other such parameters. The approach is also more reliable than the use of boundary layer theory for compressor blade losses - for the present.

Nevertheless for turbine blading and for new types of compressor blading the application of boundary layer theory is becoming almost essential.

The reader is counselled against the argument that laminar flow does not persist for any significant distance along the aerofoil surface. The simplifying assumption that the boundary layer is turbulent from the leading edge, has been made many times in the past; it is often wrong and is not necessarily conservative. It is much better to assume that the boundary layer commences as a laminar one and to apply a suitable transition criterion (see section 7.2).

The laminar boundary layer is in any event readily calculated with acceptable accuracy by methods such as that of Thwaites[7.14] for which a program listing is given in ref. 7.15.

If equation (7.3) is considered at the wall

$$\mu\left(\frac{\partial^2 u}{\partial y^2}\right)_w = \frac{dp}{dx} - \rho U\frac{dU}{dx} .$$

It is useful to define two parameters l and m such that

$$l = \frac{\theta}{U}\left(\frac{\partial u}{\partial y}\right)_w . \qquad (7.5)$$

and

$$m = \frac{\theta^2}{U}\left(\frac{\partial^2 u}{\partial y^2}\right)_w \qquad (7.6)$$

Accordingly

$$m = -\frac{\theta^2}{\nu}\cdot\frac{dU}{dx} = -\lambda_\theta \qquad (7.7)$$

m is seen to be that parameter of the velocity profile which represents the ratio of pressure gradient forces to viscous forces whilst l is a measure of the skin friction.

Now if equation (7.4) is multiplied by $(U\theta/\nu)$,

$$\frac{\tau_w\theta}{\mu U} = \frac{U\theta}{\nu}\cdot\frac{d\theta}{dx} + \frac{\theta^2}{\nu}\cdot\frac{dU}{dx}(2 + H) ,$$

and rearranging

$$\frac{U}{\nu}\cdot\frac{d(\theta^2)}{dx} = 2\left[l + m(2 + H)\right] = L(m) . \qquad (7.8)$$

Thwaites evaluated $L(m)$ from various exact solutions and produced the straight line fit

$$L(m) = 0.45 + 6m .$$

Equation (7.8) may then be integrated to give

$$\theta^2 = \theta_0^2 + \frac{0.45\nu}{U^6}\int_0^x U^5 dx \qquad (7.9)$$

Thwaites tabulated values of H(m) and L(m). As l(m) is directly proportional to the gradient of velocity profile at the wall, separation occurs when l(m) becomes zero for which the corresponding value of m is -0.082. This value of m at separation has been corrected to -0.09 experimentally by Curle and Skan[7.16].

The selection of a turbulent boundary layer theory is more difficult because of the wide range of approaches. The two broad classes of calculations for the turbulent layer are integral methods, relying upon averaging across the boundary layer, and differential methods, which establish field solutions of the partial differential equations. Differential methods are still relatively new and are being developed quickly.

An early and simple integral method was that of Truckenbrodt[7.17] using the energy integral equation. The method is based upon use of empirical relationships between the energy dissipation and R_θ and between c_f, H and R_θ, both given originally by Rotta [7.18], and also upon a relationship between H' (energy thickness/momentum thickness) and H. Truckenbrodt showed by introducing the expression for energy dissipation into the energy integral equation and assuming a mean value of H', that the local momentum thickness could be obtained as a simple integral of velocity.

High free stream turbulence levels are a particularly strong influence on turbulent boundary layer development. Evans and Horlock[7.19] have developed an integral method which incorporates free stream turbulence effects.

A class of integral methods which has received extensive recent development is that of entrainment methods, originally due to Head[3.63]. It is premised that a turbulent boundary layer grows by a process of entrainment of the inviscid free stream fluid into the turbulent region. The entrainment equation is obtained by integrating the continuity equation across the whole boundary layer from y=0 to y=δ. Head therefore assumed that the normalised entrainment velocity v_E is a function of shape factor H"

$$\frac{v_E}{U} = \frac{1}{U}\cdot\frac{d}{dx}\left[U(\delta - \delta^*)\right] = F(H'') \qquad (7.10)$$

where

$$H'' = (\delta - \delta^*)/\theta . \qquad (7.11)$$

Equation (7.10) may therefore be rewritten as

$$UF = \frac{d}{dx}(U\theta H'') \qquad (7.12)$$

Head also assumed that H" is related to H by

$$H'' = G(H) .$$

Expressions for F and G were readily obtained by experiment.

These expressions were solved with equations (7.12), (7.13), the momentum integral equation (7.4) and the Ludwieg and Tillman skin friction law[7.20].

The method has been improved by Head and Patel[7.21] and a program listing has been given by Cebeci and Bradshaw[7.15]. Lakshminarayana and Govindan[7.22] have successfully applied an entrainment method to various cascades and rotor blades.

A successful development of Head's method is the lag-entrainment method of Green et al[7.23]. In this 'state of the art' integral method an additional equation is given for the streamwise rate of change of entrainment coefficient F. The new physics is that a turbulent energy balance procedure first introduced by Bradshaw et al[7.24] has been adopted for an integral method.

Upon comparing five integral methods in 1965 Smith[7.4] found that if computations of skin friction and the displacement and momentum thicknesses were required the Truckenbrodt method performed most reliably. Smith in particular drew attention to the conflicting estimates of turbulent separation by the methods he surveyed. The Truckenbrodt method is now considered reasonable for favourable pressure gradients but inaccurate for strongly adverse ones. The Head method, and its developments, is generally superior and at least four integral methods surveyed at the Stanford Conference[7.5] performed more consistently. A good account of the integral and differential methods at the 1968 Stanford Conference is given by White[7.6].

The 1968 Stanford Conference was something of a watershed. Held in competition form the limits and capabilities of 20 integral methods were tested. Furthermore 11 of the fledgling differential methods were given an opportunity to demonstrate their capabilities.

In differential turbulent boundary layer methods an attempt is made to integrate the governing partial differential equations using finite difference techniques. The power of modern computers makes such an approach feasible and by the time of the Stanford Conference three such methods were considered to be good under a wide variety of test conditions.

Of these the method of Bradshaw and Ferriss[7.24] is probably the most noteworthy. Whilst treatment of the continuity and momentum relations, equations (1.5) and (7.3), is straightforward the difference between all differential methods is in the treatment of shear stress. Bradshaw and Ferriss relate dissipation length and diffusion terms empirically to shear stress (from flat plate data). They are then able to derive a partial differential equation for computing shear stress from the turbulent energy relation. Surprisingly with this as a third equation Bradshaw and Ferriss are able to apply the stable and precise method of characteristics to the solution of the resulting set of three partial differential equations. This results in a finite difference scheme which is particularly suited to flow situations having strong pressure gradients and also very amenable to modification for additional effects such as those of three-dimensionality.

7.2 Transition

Any attention paid to boundary layer calculations will be wasted if transition from laminar to turbulent flow is not correctly predicted. Such predictions should take into account the high levels of free stream turbulence present and, eventually, the effects of ordered unsteadiness, acoustical and three-dimensional disturbance, surface roughness and surface curvature.

Transition is not an instantaneous event occuring at one location on the blade's surface but rather an ensemble of such events occurring over a region of the surface statistically described by the intermittency factor[7.25]. Useful attempts to calculate transitional boundary layers have been made by McDonald and Fish[7.26] and Forest[7.27]. These finite difference formulations are based on established approaches to turbulent layer calculation supplemented by intuitive arguments or empirical information on the transition process. Despite a neglect of the stability question for laminar layers and of much of the physics of the transition process, useful engineering predictions of transitional boundary layer development have been given.

Transition is the result of the build-up of disturbances in the boundary layer[7.28], often over long distances. Ideally it should be possible to compute the growth of these disturbances until the known occurrence of transition. However this is not possible, due to the complexity of the physics involved, the uncertainty of amplification rates in differing pressure gradients and the extraneous disturbances listed above.

In effect the transition process is a race to instability between various mechanisms[7.29]. Some of these are well understood and participate in linear amplification processes. Others are less well understood or the result of a poor test environment and appear 'out of the blue' as bypasses to dramatically assist the amplification process.

A classic investigation of the viscous instability of a laminar boundary layer was that of Tollmien [7.30]. An eigenvalue solution of the Orr-Somerfield equations was found to give a characteristic 'thumb-shaped' neutral curve, as described by Schlichting [2.25]. The equally classical experiments of Schubauer and Skramsted [7.31] confirmed that if the wind tunnel turbulence level was sufficiently low it was possible to reproduce the entire neutral curve and hence confirm the validity of the Tollmien-Schlichting model. Unfortunately this classical approach neglected much of the physics of the flow including the important three-dimensional effects and their influence on the genesis and incubation of turbulent spots [7.32]. Such difficulties have naturally received much attention of late.

White [7.6] describes current understanding of the physics of transition. Initiation is by the formation of two-dimensional unstable Tollmien-Schlichting waves which propagate in a streamwise direction at less than 40% stream velocity. These begin to vary in a spanwise direction and develop three-dimensionality. From this a concentration into peaks, valleys and hairpin vortices occurs. Large shear layers develop and the vortices break down intermittently. Turbulent spots are therefore formed in the peak regions of vorticity. It is these spots which grow and coalesce into turbulent flow.

One of the best-established procedures for transition prediction is the empirical e^n method. Smith and Gamberoni [7.33] examined a number of experimental cases and calculated by linear stability theory the amplification along a surface of a small disturbance introduced at the critical Reynolds' Number. They found that in general an amplification ratio of about e^9 was necessary for transition to occur.

Independent observations of van Ingen [7.34] working on an aerofoil gave an amplification of e^7 to e^8 for transition; however in adverse pressure gradients this could rise to e^{20}.

The e^n method is tedious to apply, requiring the evaluation of the amplification integral at each streamwise location. A more convenient approach, based on the e^9 method, is the H-R_x procedure of Wazzen et al [7.35]. Unfortunately it shares the limitations of the e^9 method in not accounting for variations in surface roughness and free stream turbulence and not giving consistent predictions in regions of adverse pressure gradient.

The effect of free stream turbulence is not a variable that lends itself to numerical prediction. For this reason most useful transition data for turbomachinery purposes are experimental in origin.

A semi-empirical approach which has given a reliable guide to the effect of turbulence variation is that of van Driest and Blumer [7.36]. A vorticity Reynolds' Number was formulated and transition presumed to occur when its maximum value reached a critical level. The critical values were a function of free stream turbulence level and were determined by correlation.

Figure 7.1 presents available data for R_θ at the start and end of transition as a function of turbulence level. The sources are given by Abu-Ghannam and Shaw [7.37] and some new data [7.38] are added. The prediction curve of Van Dreist and Blumer for transition inception is also given and is seen to be in reasonable agreement with experimental data under high turbulence conditions. The correlation curve for the end of transition is that of ref. 7.37. The scatter at the start of transition masks some systematic effects of the length scale. Hislop [7.39] performed his experiments with grids of 12.7mm, 25.4mm and 50.8mm mesh and found up to 50% variation in transition Reynolds' Number with mesh size. Such 'scale' effects have universally plagued systematic experiments on transition. Despite the scatter at low turbulence levels it does appear that a reliable design guide for zero pressure gradients is available.

As soon as there is a departure from zero pressure gradient, either favourable or adverse, major disparities between existing projections emerge. The most convenient format for presenting the effects of pressure gradient on transition inception has been found to be that of Crabtree [7.40]. A correlation of the available low turbulence level data was proposed in the form of R_θ against the pressure gradient parameter, λ_θ, defined by equation (7.7). It is thus a purely local criterion making no reference to the history of the flow or its stability. Walker [7.41] has argued that transition is dependent on the shape of the pressure distribution and Abu-Ghannam and Shaw have considered history effects which should ideally be evaluated.

To use correlations of the Crabtree type one plots the locus of the developing laminar layer (either measured or calculated) in the form of R_θ as a function of λ_θ. Assuming laminar separation ($\lambda_\theta = -0.09$) is not encountered first then transition is predicted to commence where the locus crosses the Crabtree curve. For low free-stream turbulence levels in a 'clean' environment this criterion works well. The approach is also well suited to incorporation of turbulence level as a variable.

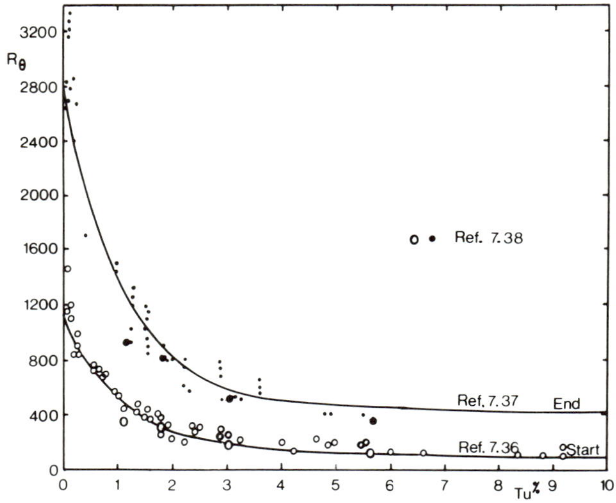

Fig. 7.1. Effect of turbulence on momentum thickness Reynolds' Number at start and end of transition for zero pressure gradients.

Since in turbomachines the level of random turbulence is usually above 4% [2.27] and often substantially higher [2.34], it is important that this variable be covered. Whilst many reliable transition data exist for zero pressure gradients this is not the case for favourable or adverse pressure gradients.

Separate plots of $R_{\theta t}$ as a function of λ_θ for three different turbulence levels are given in Fig. 7.2. The considerable divergence between the transition inception predictions of Hall [7.42], Seyb [7.43] and Abu-Ghannam and Shaw [7.37] is obvious. At zero pressure gradient the projections of Hall and Abu-Ghannam are compatible with most of the test data in Fig. 7.1 whereas the Seyb correlation appears to represent a later stage in the transition process.

Under the favourable pressure gradient conditions of most turbine blading discrepancies between various projections are considerable. Hall tentatively extrapolated the zero pressure gradient data into the favourable pressure gradient region using the stability limit curves of Pretsch[7.44] as a guide. This is now seen to give too rapid an increase of $R_{\theta t}$ with increasing favourable pressure gradient. Van Driest and Blumer were able to extend their theory to incorporate pressure gradient effects for the special case of Falkner-Skan profiles. Their predictions were substantiated experimentally by Blair[7.45] and by Abu-Ghannam and Shaw[7.37] who derived the trends given in Fig. 7.2.

In the adverse pressure gradient region Hislop originally suggested that transition inception curves of constant $R_{\theta t}$ should be drawn for a range of turbulence levels. This procedure was employed in the correlation of Hall and Gibbings[7.28]. Hall originally projected $R_{\theta t}$ curves into this region following the Pretsch limit and the curves presented in Fig. 7.2 are taken from ref. 7.42. Seyb[7.43] based his correlation on limited data from nozzle tests projected into the adverse pressure gradient region at constant $R_{\theta t}$, the justification was the heuristic argument that at high turbulence levels transition is more dependent on the inability of damping mechanisms to cope with introduced disturbances than upon the amplification of small perturbations. The curves given by Abu-Ghannam and Shaw[7.37] represent experiments at low turbulence levels. At high levels the trends are extrapolated into the adverse pressure gradient region.

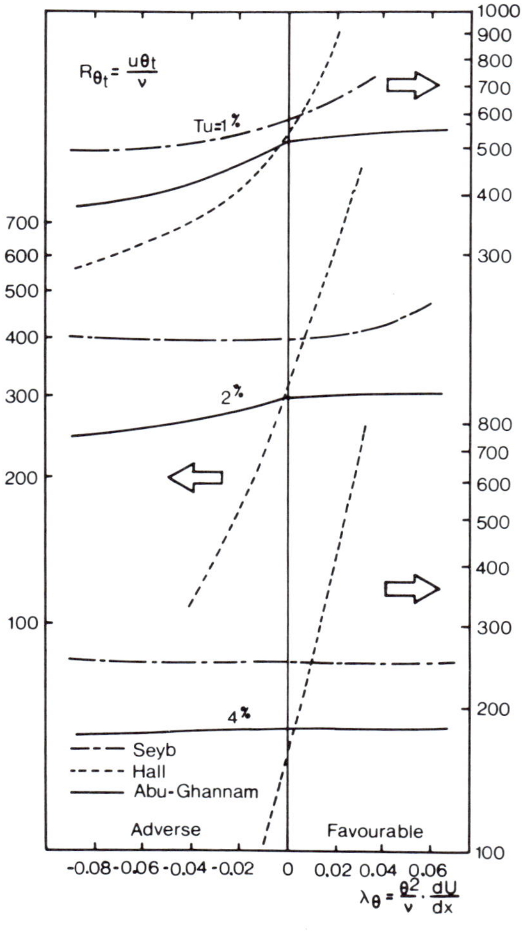

Fig. 7.2. Transition inception Reynolds' Number as a function of turbulence level and pressure gradient parameter.

Figure 7.2 shows clearly that the various projections differ drastically. Experimental data for verification of trends are scarce. For turbulence levels of 1% and 2% the projections of Abu-Ghannam and Shaw are considered to be most reliable. Spot checks, using data taken from flat plate and cascade experiments, have also shown that for low free-stream turbulence levels below 2.5% the constant $R_{\theta t}$ proposal is not unreasonable. This is also consistent with most of the data of van Ingen[7.34] and implies an amplification up to the order of e^{20}. Van Ingen's data indicate that as the pressure gradient is reduced from an adverse one towards zero a different mode is encountered in which transition occurs much sooner (amplification of the order e^7 to e^{10}).

However the constant $R_{\theta t}$ proposal has not been confirmed under high levels of free stream turbulence. The results of Schlichting and Das[2.27] and unpublished measurements by the author and co-workers suggest that for free-stream turbulence levels above a critical level (about 2.5% for typical Reynolds' Numbers) the required amplification is reduced to considerably less than e^{20} and probably less the e^9. In the adverse pressure gradient region the free-stream turbulence causes non-linear by-pass effects resulting in transition independent of the linear amplification of Tollmien-Schlichting waves.

Production of reliable experimental transition data under conditions of high turbulence level and differing pressure gradients for confirmation purposes is a matter of some urgency. This situation is not unique to cascades and is a reflection of the confusing state in which one finds transition work as a whole.

Whilst pressure gradient and turbulence level effects have been discussed there are many others which have not been represented. Length scale, non-isotropy and ordered unsteadiness are effects upon which work is still at an early stage. Some data do exist on the influences of surface roughness [7.46], noise [7.47] and heat transfer [7.48].

In observations on compressor stator blades Walker [7.41] has considered the effects of upstream wakes in transition of the blade boundary layer. In this case the predominant influence on transition was the indirect one of modification of suction peak location. This is an effect of the ordered unsteadiness of the free-stream and approaches to the consideration of ordered unsteadiness will be presented in Chapter 9.

7.3 Separation and Wakes

Any blade surface boundary layer must eventually separate. Separation may occur as a long or short bubble on the surface, as a separation region which remains detached, or at the beginning of a wake emanating from the trailing edge.

For isolated aerofoils three distinct separation regimes have been described by Jones [7.49] and subsequently by Thwaites[7.50] and Crabtree [7.51]. These correspond to boundary layer flows well known to workers on cascades of aerofoils at lower incidence.

The three regimes are:-

(i) Laminar separation occurring with $R_\delta < 500$ and subsequently a long laminar separation bubble, which gradually elongates with increasing incidence. Horton [7.52] has produced a semi-empirical theory for prediction of these bubbles based upon the different observed bubble regions. Long bubbles are often observed on the pressure distributions of compressor blading [7.53]. When such a bubble occurs it also has a marked effect on turning and loss. Roberts [7.54] has extended Horton's theory to the cascade situation.

(ii) Laminar separation occurring at or near the suction peak with $R_\delta > 500$ and a short bubble which is immediately followed by transition to a turbulent layer. As the incidence is increased the short bubble contracts and then suddenly bursts giving the familiar leading edge stall, a separated region from the leading edge which does not reattach. This phenomenon, described by McCullough and Gault [7.55], has been found to give rise to a discontinuity of performance and a hysteresis effect in isolated aerofoil work (Fig. 6 of ref. 7.55 gives a good illustration of this). Conversely van den Berg[7.56] has argued that leading-edge stalls are more often caused by turbulent separation than by bursting of a short laminar bubble.

(iii) Early transition is often followed by a region of turbulent separation at the trailing edge, which is enlarged by any increase in incidence, giving a gradual stall. It is quite common for type (iii) to be present alongside either type (i)[2.23] or type (ii) [7.57].

For many cascade flows having significant separated flow zones the initial location of separation will be known or calculable. The separated region may then be incorporated into the flow field computation by addition of a source distribution chosen to satisfy a presumed physical condition downstream of the separation location. Constancy of static pressure in the streamwise direction is a suitable condition. Flow field calculations of this kind have been performed by Geller[7.58] and Railly[7.59].

In any event, for a blade having a trailing edge of finite thickness or curvature, separation will occur as the trailing edge is approached. We are thence dealing with a wake and it is the description of wakes and the allowance for them that ultimately determines the flow around the blade.

In an extensive experimental program Lakshminarayana and colleagues have examined the wakes of compressor rotor and stator blades. Profile variations in axial, tangential and radial directions were measured using stationary and rotor-mounted three sensor hot-wire probes [7.60].

A major effect of rotation was to redistribute turbulent energy to effect amplification of radial intensities to the same levels as axial and tangential intensities[7.61].

Velocity profiles were plotted and confirmed Gaussian wake similarity laws[7.62]. This behaviour has been corroborated by Hobbs et al[7.63] for distances greater than 30% axial chord downstream of the trailing edge. Similarity was not present in the trailing edge region where wake profiles were more closely related to those of the blade boundary layers.

Reynolds, Lakshminarayana and Ravindrath[7.64] found that the rotation and blade loading effects on mean velocity and turbulence profiles were confined to the near-wake region. In this near-wake region the velocity profile was strongly three-dimensional. The radial velocities were particularly high and the decay of all velocity defect components was very rapid in the region of the trailing edge.

Downstream variations in velocity defect in ref. 7.62 were in good agreement with those of other workers. Kool and Hirsch[7.65], using an eddy viscosity model, have successfully predicted this decay and have indicated a strong sensitivity of decay rate to streamwise pressure gradient.

Data on the effect of free stream turbulence variation on wakes have been provided by Pal and Raj[7.66] and Hah and Lakshminarayana[7.67]. The latter authors have also investigated differential methods for the calculation of turbulent wakes with some success[7.68].

Although some progress has been made with differential techniques for wakes, integral methods are still simpler and more cost-effective. The wake thickness variation is most readily obtained by the following semi-empirical method which is limited to small or zero pressure gradients. For the wake form factor Spence[7.69] has obtained the relationship

$$\left(1 - \frac{1}{H_X}\right) = \frac{(1 - (1/H_T))}{\sqrt{1 + 40(x/c)}}$$

$$\therefore\ H_X = \frac{1}{1 + \left(\dfrac{((1/H_T) - 1)}{\sqrt{1 + 40(x/c)}}\right)} \qquad (7.14)$$

Assuming the power law relationship

$$\frac{\theta_X}{\theta_T} = \left(\frac{q_T/U_I}{q_X/U_I}\right)^{\frac{H_T + 5}{2}}$$

and a further empirical relationship obtained by Spence

$$\frac{q_X}{q_T} = 1 - \frac{0.8001}{\sqrt{1 + 40(x/c)}} \ ;$$

$$\therefore\ \frac{\theta_X}{\theta_T} = \left(1 - \frac{0.8001}{\sqrt{1 + 40(x/c)}}\right)^{-\left(\frac{H_T + 5}{2}\right)} \qquad (7.15)$$

Now by definition,

$$\frac{\delta_X^*}{c} = H_X\frac{\theta_X}{c} = H_X\frac{\theta_T}{c}\left(\frac{\theta_X}{\theta_T}\right) \qquad (7.16)$$

and substituting (7.14) and (7.15) into (7.16) gives

$$\frac{\delta_X^*}{c} = \frac{\theta_T/c}{\left(1 + \dfrac{1 - H_T}{H_T\sqrt{1 + 40(x/c)}}\right)\left(1 - \dfrac{0.8001}{\sqrt{1 + 40(x/c)}}\right)^{(H_T + 5)/2}} \qquad (7.17)$$

Since the boundary layer conditions at the trailing edge can be determined we are in a position to calculate the development of the wake displacement thickness.

As a first approximation, reasonable for many compressor cascades operating near design, the static pressure change across the wake is neglected, giving a straight wake. However many blades have significant degrees of trailing edge loading, imparting a finite radius of curvature to the mean line of the wake, especially in the trailing edge region.

The expression

$$\frac{dp}{dU} \simeq \frac{\rho U^2}{R} \qquad (7.18)$$

is integrated across the wake to give

$$\Delta p_{TE} = \frac{\rho U^2}{R}\left[\delta - (\theta + \delta^*)\right] \qquad (7.19)$$

This equation, linking the pressure distribution with the curvature and the boundary layer parameters may, in principle, be used in an iterative fashion thus accounting for viscous effects.

These combined effects have been ingeniously used by Oliver[7.70] who postulates a 'two vortex sheet' representation of aerofoil and wake, the two sheets being placed at the displacement and momentum thickness contours.

Prediction of wake development between the trailing edge region and downstream infinity must take account of mixing effects. The control volume wake mixing theory of Stewart[7.71] is most useful for this purpose.

Boundary layer theory is inadequate for the accurate prediction of separating flows and there has been a need for a conceptual framework which would avoid the necessity to use the full Navier-Stokes equations.

It appears that for accurate modelling of flow details in the trailing edge region the triple-deck approach[7.72] may be of value. For trailing edge flows the triple-deck approach is predicated on the localisation of separation effects to the near-wake region.

In triple-deck theory the flow over a surface is divided into three regions; an upper deck in which the flow is inviscid and irrotational, a main deck in which the flow is inviscid but rotational, and a lower deck in which the flow is viscous and rotational. At the trailing edge a region commences in which the flow from each surface forms a triple-deck. The simplification the theory brings is the assumption of constant viscous or turbulent stresses outside the lower deck. It could permit accurate prediction of separating flows and its potential application to cascades is discussed by Verdon and Werle[7.73].

7.4 Zero Nett Vorticity Theorem

It was shown in Chapter 5 that potential field calculations for cascades of profiles with rounded trailing edges do not result in unique pressure distributions and that some consideration of the viscous flow around the profiles is needed for uniqueness to be obtained.

The purpose of sections 7.4 and 7.5 is to provide a basis for the selection of a unique flow solution for a cascade having a rounded trailing edge. It will be demonstrated that the flow is determined quite simply through the 'zero nett vorticity theorem' and the 'first viscous approximation'.

Until recently, viscous flow theory was used mainly for computing the drag coefficient of each profile; it is here proposed that consideration of viscous flow is also required for the outlet angle and lift to be determined. In order to postulate such a uniqueness condition a brief summary of the physics of flow near the aerofoil trailing edge is firstly required.

Kelvin's theorem, as given by Lamb[5.65], states that "The nett flux of vorticity by convection and diffusion into any fixed closed circuit is equal to the rate of increase of circulation in the circuit," or

$$\frac{d\Gamma}{dt} = -q\gamma \; . \tag{7.20}$$

In steady flow there is therefore no nett flux of vorticity out of a fixed closed circuit enclosing the aerofoil and cutting the wake downstream. This condition of "zero nett vorticity discharge" into the downstream wake was enunciated by Taylor [7.74]. Preston [7.75] demonstrated that the circulation, Γ , was only constant for circuits enclosing the aerofoil and its boundary layers and cutting wake region streamlines at right angles.

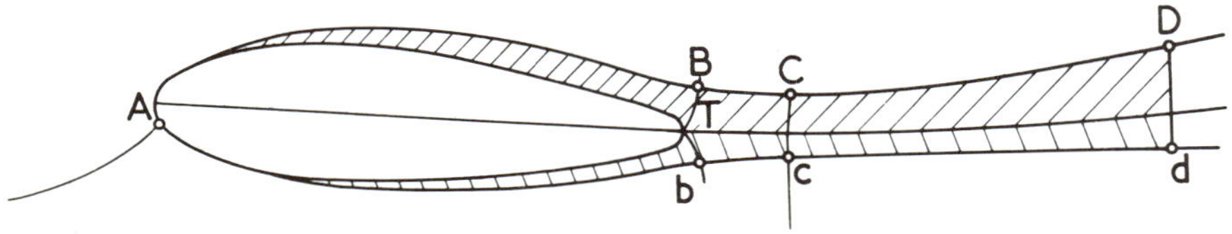

Fig. 7.3. Aerofoil with boundary layers and wake.

In Fig. 7.3, which depicts schematically the extent of the absolute thickness of boundary layer and wake on an aerofoil having a rounded trailing edge, lines BTb, Cc and Dd all cut the wake streamlines at right angles. If we assume that there is no nett vorticity discharge into the wake then the circulation in circuits BCcbB, BDdbB etc is zero. This has the important consequence, which should not be neglected, that the wake must therefore be perfectly straight from the trailing edge to a position infinitely far downstream. Preston [7.75] has demonstrated that the rate at which vorticity crosses BT is

$$\int_T^B q\gamma \, dN = -\left[p_B - p_{TS}\right]$$

and for bT

$$\int_T^b q\gamma \, dN = -\left[p_b - p_{TP}\right].$$

Now B and b are in the free stream and, by Bernoulli's theorem, $p_B = P_b$. At the point T, q=0, $\therefore P_{TS} = p_{TS}$ and $P_{TP} = p_{TP}$.

Thus if the assumption of zero nett vorticity discharge is to be satisfied

$$\int_T^B q\gamma \, dN = \int_T^b q\gamma \, dN \qquad (7.21)$$

and

$$p_{TS} = p_{TP}. \qquad (7.22)$$

Equation (7.22) can also be used for the static pressure coefficients

$$(Cp_T)_S = (Cp_T)_P . \qquad (7.23)$$

This important relationship, which states that the static pressure coefficients on the blade surface must tend to the same value as the trailing edge plane is approached from either surface, gives the basis of a condition which will be used for obtaining unique calculated pressure distributions.

7.5 First and Second Viscous Approximations

Utilising the physical understanding of the previous section we now proceed to a first approximation to the true pressure distribution. The starting point for any blade with a rounded trailing edge in cascade is a series of potential flow pressure distributions having different positions of rear stagnation point (e.g. Figs. 5.11 and 5.12). In this first approximation the aim is to select a unique pressure distribution (with corresponding outlet angle) on which to base subsequent boundary layer theory. The difficulty is that all potential flow pressure distributions are physically unrealistic in the trailing edge region, having a rear stagnation point and upstream of it an adverse pressure gradient so severe as to make separation inevitable.

The "first viscous approximation" is applied by fairing-in the pressure distributions to avoid unsustainable pressure gradients at the trailing edge. This function is achieved in a real flow by means of the displacement effect of the boundary layer near to the trailing edge. The fairing-in is thus a simplified way of allowing for the build-up in displacement thickness. However crudely the fairing-in process is applied it represents a departure from the potential flow and a first order correction for viscous effects.

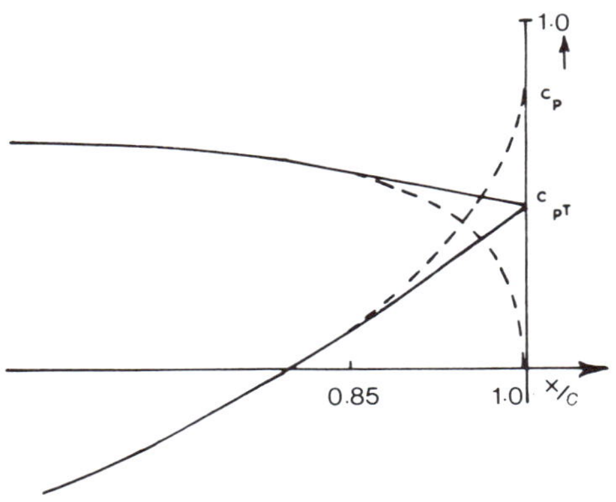

Fig. 7.4. Fairing-in procedure at trailing edge.

Following the isolated aerofoil work of Spence and Beasley [7.76], and also as a result of studies of compressor blade pressure distributions, the author originally recommended that the fairing be done by extrapolating the pressure distribution tangentially from the 85% chord location. Subsequent investigations [7.77,7.78] have shown that for most compressor blading operating near design incidence such a procedure gives acceptable accuracy. However for conditions approaching stall, and for turbine blading, application of fairing from the 85% chord location leads to inaccuracies in prediction of outlet angle. Such cases are still under investigation but it does appear that fairing from a location closer to the trailing edge is appropriate for turbine blades [7.79].

Having applied the fairing procedure, illustrated in Fig. 7.4, the family of pressure distributions, for a given inlet angle and a given range of outlet angles, is examined for the static pressure difference between the two surfaces in the trailing edge plane. According to equation (7.23) the correctly determined pressure distribution is the one for which the difference in pressure coefficients at the trailing edge is zero.

For maximum accuracy ΔCp_T can be plotted as a function of the outlet angle α_2. The α_2 for which $\Delta Cp_T = 0$ can be obtained by interpolation and the potential flow calculations re-run for the given inlet angle and this new outlet angle.

As a result of the first viscous approximation a unique pressure distribution and outlet angle are known. Following methods described in sections 7.1.2 and 7.3 it is possible to calculate the boundary layers and wake corresponding to the pressure distribution. This is not, however, the final solution since the displacement effect of the calculated boundary layer on the potential flow has not been taken into account. Preston [7.75] argued that for an isolated aerofoil this is accounted for by adding the displacement thickness around the profile and calculating the potential flow around the new profile thus derived. This "second viscous approximation" should give a revised and more accurate pressure distribution and outlet angle. Although this second computation may suffice, in principle the scheme calls for iteration until convergence is obtained.

Since for an aerofoil with an angle of attack the displacement thickness at the trailing edge is usually greater on the suction side than on the pressure side this is equivalent to a reduction of camber and incidence. Platt and Norbury [7.80] using an electrolytic tank technique, concluded that it was beneficial to include the wake effect and to round off the wake at a distance of one to two chords downstream of the trailing edge. Speidel and Scholz [7.81] simplified the rounding-off for cascades and used the concept of a 'substitute' profile slightly different from the original blade. A more powerful method using differential boundary layer calculations has been developed by Hanson et al [7.82]. An inviscid-viscous iteration scheme was developed incorporating the curvature correction term of equation (7.18) and giving well-defined displacement thickness prediction in the trailing edge region.

Successful computations of the viscous flow through blading have been made (e.g. ref. 7.83) and these have been extended to modelling of the thick trailing edges of turbine blading. However the author's view is that correct representation of the base pressure effects behind a very blunt edge depends upon the physics of the near wake region, often including the structure of a von Karman vortex street. Accordingly consideration of the base pressure problem will be deferred to Chapter 8 which deals with stalled and unsteady flows.

REFERENCES

7.1 PRANDTL, L. Uber Flüssigkeitsbewegung bei sehr kleiner Reibung. Verb. 3 int. Math - Kongr., Heidelberg 484 (1904)

7.2 RAJ, R. Formulation of a generalized mathematical model for computation of boundary layer and wake in turbomachines. Tensor N.S. 31 (1977)

7.3 HORLOCK, J.H. Boundary layer problems in axial turbomachines. in "Flow Research on Blading" L.S. Dzung ed. Elsevier (1970)

7.4 SMITH, D.J.L. Turbulent boundary layer theory and its application to blade profile design. ARC CP 868 (1965)

7.5 KLINE, S.J., MOFFATT, H.K. and MORKOVIN, M.V. Computation of turbulent boundary layers. 1968 AFOSR - IFP Stanford Conf., J. Fluid Mech. 36 pg. 481 (1969)

7.6 WHITE, F.M. Viscous Flow Theory. McGraw Hill. N.Y. (1974)

7.7 TURNER, A.B. Local heat transfer measurements on a gas turbine blade. J. Mech. Eng. Sci. 13. (1971)

7.8 OLDFIELD, M.L.G., KIOCK, R., HOLMES, A.T. and GRAHAM, C.G. Boundary layer studies on highly loaded cascades using heated thin films and a traversing probe. Trans. ASME J. Eng. Power 103, 1 pg. 237 (1981)

7.9 SCHOLZ, N. Aerodynamics of Cascades. AGARDograph-220 Trans. and revised by A. Klein (1977)

7.10 SHAALAN, M.R.A. The effect of aspect ratio in cascade testing. Ph.D. thesis, Liverpool Univ. (1967)

7.11 WALKER, G.J. The unsteady nature of boundary layer transition on an axial-flow compressor blade. ASME Paper 74-GT-135 (1974)

7.12 SHARMA, O.P., WELLS, R.A., SCHINKLER, R.H. and BAILEY, D.A. Boundary layer development on turbine airfoil suction surfaces. Trans. ASME, J. Eng. Power 104, 3 pg. 675 (1982)

7.13 FAULDERS, C.R. An interferometric study of the boundary layer on a turbine nozzle blade. Trans. ASME pg. 61 (1954)

7.14 THWAITES, B. Approximate calculation of the laminar boundary layer. Aero. Quart. 1, pg. 245 (1949)

7.15 CEBECI, T. and BRADSHAW, P. Momentum Transfer in Boundary Layers. Hemisphere, Washington (1977)

7.16 CURLE, N. and SKAN, S.W. Approximate method for predicting separation of laminar boundary layers. Aero. Quart. 8 (1957)

7.17 TRUCKENBRODT, E. A method of quadrature for calculation of the laminar and turbulent boundary layer in the case of plane and rotationally symmetrical flow. NACA TM 1379 (1955)

7.18 ROTTA, J. Schubspannungsverteilung und Energiedissipation bei turbulenten Grenschichten. Ing. Archiv. 20, 195 (1952)

7.19 EVANS, R.L. and HORLOCK, J.H. Calculation of the development of turbulent boundary layers with a turbulent free stream. Cambridge Univ. CUED/A Turbo/TR53 (1973)

7.20 LUDWIEG, H. and TILLMANN, W. Investigations of the wall shearing stress in turbulent boundary layers. NACA Rep. TM 1285 (1949)

7.21 HEAD, M.R. and PATEL, V.C. Improved entrainment method for calculating turbulent boundary-layer development. ARC R&M 3643 (1968)

7.22 LAKSHMINARAYANA, B. and GOVINDAN, T.R. Analysis of turbulent boundary layer on cascade and rotor blades of turbomachinery. AIAA J. 19 10 pg. 1333 (1981)

7.23 GREEN, J.E., WEEKS, D.J. and BROOMAN, J.W.F. Prediction of turbulent boundary layers and wakes in compressible flow by a lag-entrainment method. ARC R&M 3791 (1973)

7.24 BRADSHAW, P., FERRISS, D.H. and ATWELL, M.P. Calculation of boundary-layer development using the turbulent energy equation. J.F.M. 28 3 pg. 593 (1967)

7.25 DHAWAN, S. and NARASIMHA, R. Some properties of boundary layer flow during the transition from laminar to turbulent motion. J.F.M. 3, 418 (1958)

7.26 MCDONALD, H. and FISH, R.W. Practical calculations of transitional boundary layers. Int. J. Heat Mass Trans. 16, pg. 1729 (1973)

7.27 FOREST, A.E. Engineering predictions of transitional boundary layers. AGARD CP 224 Paper 22 (1977)

7.28 HALL, D.J. and GIBBINGS, J.C. Influence of stream turbulence and pressure gradient upon boundary layer transition. J. Mech. Eng. Sci. 14, 2 pg. 134 (1972)

7.29 MORKOVIN, M.V. Instability transition to turbulence and predictability. AGARD ograph-236 (1978)

7.30 TOLLMIEN, W. ZAMM 4, pg. 468; Trans. as NACA TM 1085

7.31 SCHUBAUER, G.B. and SKRAMSTED, H.K. Laminar boundary layer oscillations and transition on a flat plate. NACA Rep. 909 (1943)

7.32 SCHUBAUER, G.B. and KLEBANOFF, P.S. Investigation of separation of the turbulent boundary layer. NACA TN 2133 (1950)

7.33 SMITH, A.M.O. and GAMBERONI, N. Transition, pressure gradient and stability theory. Proc. 9th Int. Congr. Appl. Mech., Brussels (1956)

7.34 VAN INGEN, J.L. A suggested semi-empirical method for the calculation of the boundary layer transition region. Rep. VTH-74, Univ. Tech., Delft (1956)

7.35 WAZZAN, A.R., GAZLEY, C.Jr. and SMITH, A.M.O. $H\text{-}R_x$ method for predicting transition. AIAA J. 19 6 pg. 810 (1981)

7.36 VAN DRIEST, E.R. and BLUMER, C.B. Boundary layer transition: free stream turbulence and pressure gradient effects. AIAA J. 1 1303 (1963)

7.37 ABU-GHANNAM, B.J. and SHAW, R. Natural transition of boundary layers - the effects of turbulence, pressure gradient and flow history. J. Mech. Eng. Sci. 22 5 (1980)

7.38 GOSTELOW, J.P. and RAMACHANDRAN, R.M. Some effects of free stream turbulence on boundary layer transition. Proc. 8th Aust. Fluid Mech. Conf., Newcastle (1983)
7.39 HISLOP, G. Use of grids and gauzes to promote transition. Ph.D. thesis, Cambridge Univ. (1940)
7.40 CRABTREE, L.F. Prediction of transition in the boundary layer on an aerofoil. J. Roy. Aero. Soc. Vol. 62 (1958)
7.41 WALKER, G.J. The prediction of boundary layer development on axial flow turbomachine blades. Proc. 3rd Aust. Conf. Hyd-Fluid Mech. pg. 97 (1968)
7.42 HALL, D.J. Boundary layer transition. Ph.D. thesis, Liverpool Univ. (1968)
7.43 SEYB, N.J. Determination of cascade performance with particular reference to the prediction of the boundary layer parameters. ARC Rep. 27,214 (1965)
7.44 PRETSCH, J. Die Stabilitat einer ebenen Laminarstromung bei Druckgefalle und Druckansteig. Jb. Luftfahrtforschung, 1, pg. 58 (1941)
7.45 BLAIR, M.F. Influence of free-stream turbulence on boundary layer transition in favourable pressure gradients. Trans. ASME, J. Eng. Power 104, 4 pg. 743 (1982)
7.46 GIBBINGS, J.C. and HALL, D.J. Criterion for tolerable roughness in a laminar boundary layer. J. Aircraft 6, 2 pg. 171 (1969)
7.47 PATE, S.R. and SCHUELER, C.J. An investigation of radiated aerodynamic noise effects on boundary layer transition in supersonic and hypersonic wind tunnels. AIAA Paper 68-375 (1968)
7.48 GRANT, A.D. The effect of heat transfer on boundary layer stability in axial-flow compressors. I. Mech. E. Conf. Heat and Fluid Flow in Steam and Gas Turb. Plant. pg 252 (1973)
7.49 JONES, B.M. Stalling. J. Roy. Aero Soc. (1934)
7.50 THWAITES, B. Incompressible Aerodynamics. Oxford Univ. Press (1960)
7.51 CRABTREE, L.F. Effect of leading edge separation on thin wings in two-dimensional incompressible flow. J. Aero. Sci. vol. 24 (1957)
7.52 HORTON, H.P. A semi-empirical theory for the growth and bursting of laminar separation bubbles. ARC CP 1073 (1967)
7.53 CITAVY, J. and NORBURY, J.F. The effect of Reynolds' Number and turbulence intensity on the performance of a compressor cascade with prescribed velocity distribution. J. Mech. Eng. Sci. 19, 3 (1977)
7.54 ROBERTS, W.B. The effect of Reynolds' Number and laminar separation on axial cascade performance. Trans. ASME, J. Eng. Power 97 2 pg. 261 (1975)
7.55 MCCULLOUGH, G.B. and GAULT, D.E. Examples of three representative types of airfoil section stall at low speed. NACA TN 2502 (1951)
7.56 VAN DEN BERG, B. Role of laminar separation bubbles in airfoil leading-edge stalls. AIAA J. 19 5 pg. 553 (1981)
7.57 CARMICHAEL, A.D. Stall propagation in compressors. Ph.D. thesis, Cambridge Univ. (1958)
7.58 GELLER, W. Incompressible flow through cascades with separation. In 'Boundary Layer Effects in Turbomachines' ed. J. Surugue AGARDograph-164. pg. 171 (1972)
7.59 KADHIM, S.J. and RAILLY, J.W. An improvement in the treatment of separation in general cascades by potential flow methods. Proc. I. Mech. E. 194 17 pg.195 (1980)
7.60 RAVINDRATH, A. and LAKSHMINARAYANA, B. Rotor wake mixing effects downstream of a compressor rotor. Trans. ASME, J. Eng. Power 104, pg. 202 (1982)
7.61 LAKSHMINARAYANA, B., GOVINDAN, T.R. and REYNOLDS, B. Effects of rotation and blade incidence on properties of turbomachinery rotor wake. AIAA J, 20, 2 pg. 245 (1982)
7.62 LAKSHMINARAYANA, B. and DAVINO, R. Mean velocity and decay characteristics of the guidevane and stator blade wake of an axial flow compressor. Trans. ASME, J. Eng. Power 102, pg. 50 (1980)
7.63 HOBBS, D.E., WAGNER, J.H., DANNENHOFFER, J.F. and DRING, R.P. Experimental investigation of compressor cascade wakes. ASME Paper 82-GT-299 (1982)

7.64 REYNOLDS, B., LAKSHMINARAYANA, B. and RAVINDRANATH, A. Characteristics of the near-wake of a compressor or fan rotor blade. AIAA J. 17,9 pg. 959 (1979)

7.65 KOOL, P. and HIRSCH, CH. A prediction scheme for the decay of a turbomachine blade wake. ASME Paper 82-GT-273 (1982)

7.66 PAL, S. and RAJ, R. Characteristics of wake turbulence due to free stream turbulence environment. AIAA 80-1079 (1980)

7.67 HAH, C. and LAKSHMINARAYANA, B. Free-stream turbulence effects on the development of rotor wake. AIAA 80-1431 (1980)

7.68 HAH, C, and LAKSHMINARAYANA, B. Numerical analysis of turbulent wakes of turbomachinery rotor blades. Trans. ASME, J. Fluids Eng. 102, pg. 462 (1980)

7.69 SPENCE, D.A. Growth of the turbulent wake close behind an aerofoil at incidence. ARC CP 125 (1952)

7.70 OLIVER, A.R. A double vortex sheet model of the viscous flow near the trailing edge of a lifting aerofoil. Int. J. Heat and Fluid Flow, 3;2 (1982)

7.71 STEWART, W.L. Analysis of two-dimensional compressible- flow loss characteristics downstream of turbomachine blade rows in terms of basic boundary layer characteristics. NACA TN 3515 (1955)

7.72 BROWN, S.N. and STEWARTSON, K. Wake curvature and the Kutta condition in laminar flow. Aero. Quart. pg. 275 (1975)

7.73 VERDON, J.M. and WERLE, M.J. Development of a viscous transonic cascade analysis. UTRL Rep. R78-912864-6 (1978)

7.74 TAYLOR, G.I. Note on the connection between the lift on an aerofoil in a wind and the circulation around it. Phil. Trans. R. Soc. A. Vol. CCXXV pg.238 (1925)

7.75 PRESTON, J.H. The calculation of lift taking account of the boundary layer. ARC R&M 2725 (1949)

7.76 SPENCE, D.A. and BEASLEY, J.A. The calculation of lift slopes, allowing for boundary layer, with application to the RAE 101 and 104 aerofoils. ARC R&M 3137 (1960)

7.77 SANGER, N.L. Two-dimensional analytical and experimental performance comparison for a compressor stator section with D-Factor of 0.47. NASA TN D-7425 (1973)

7.78 MILLER, M.J. Estimation of deviation angle for axial-flow compressor blade sections using inviscid-flow solutions. NASA TN D-7549 (1974)

7.79 INOUE, M. and KUROUMARU, M. Calculation of cascade flow by means of spline fit singularity. Proc. ASME-JSME Gas Turb. Cong. Tokyo, pg. 288 (1977)

7.80 PLATT, A. and NORBURY, J.F. Measurement of the effects of viscosity on aerofoil lift using the electrolytic tank. Trans. ASME (1963)

7.81 SPEIDEL, L. and SCHOLZ, N. Untersuchungen über die Strömungsverluste in ebenen Schaufelgittern. VDI Forschungsheft 464 (1957)

7.82 HANSEN, E.C., SEROVY, G.K. and SOCKOL, P.M. Axial-flow compressor turning angle and loss by inviscid-viscous interaction blade-to-blade computation. Trans. ASME, J. Eng. Power 102, pg. 28 (1980)

7.83 RIBAUT, M. and NIESSNER, H. Influence of friction on the deviation characteristics and trailing edge pressure of a two-dimensional blade cascade. Brown Boveri Rev. 5 81 pg. 197 (1981)

CHAPTER 8

Stalled and Unsteady Flows

Unsteadiness is essential to the operation of a turbomachine. As Dean[8.1] has pointed out it is the flow fluctuations which allow work to be done on the fluid by pressure forces and this provides the basic mechanism for energy transfer.

In an even deeper sense unsteadiness is fundamental for all known flow processes. Remarkable successes have been obtained by treating apparently steady flow conditions as determinate but there should be no surprise when flow situations are encountered which do not lend themselves to treatment by steady flow aerodynamics.

Work producing machines, for example, have their performance dictated by unsteady phenomena of which observers have hitherto only perceived glimpses. The problem is particularly acute in the prediction of the flow past the turbine blading of gas and steam turbines. In the smallest turbines a sharp trailing edge is virtually impossible to make. In large ones it is often not possible for stress or blade cooling reasons. Trailing edge shapes can be square but the usual case and the one of practical interest is the rounded trailing edge. At both subsonic and supersonic speeds the static pressure in the immediate vicinity of a blunt trailing edge will usually be lower than that in the nearby free stream. This relatively low pressure is termed the base pressure. It has the effect of penalizing the profile with an additional suction or base drag. Although base drag is present for both subsonic and supersonic flows the physical cause in the two cases is different. For subsonic flows the existence of a Von Karman vortex street and its associated drag is a significant factor. In supersonic flow situations periodic vortex shedding at the trailing edge may be suppressed and a momentum deficit associated with strong expansions and compression shocks predominates.

One problem is the unpredictability of base drag; this is merely symptomatic of a lack of physical understanding of the entire base flow region. The evidence of schlieren photography now tells us that in general a Von Karman vortex street is shed from the trailing edge of a turbine blade. This street directly affects the drag, by reducing the instantaneous base pressure, and might be expected to have a substantial influence on the blade's sound, vibration and heat transfer characteristics.

Vorticity is rarely shed from a surface continuously. Periodic components in the shear layers are readily amplified until non-linear effects prescribe time delays and the shedding of discrete vortices. This process, with its inherent lags, is a fundamental and most important one in fluid dynamics. If prediction procedures are faulty it is often so by virtue of an imperfect understanding of the vortex shedding process.

The performance of work absorbing machines will be limited by stall of the blading, unless the unsteady aeroelastic phenomena impose a prior physical limit. Underlying both phenomena are somewhat related unsteady effects of the blade boundary layer and wake.

Although unsteadiness may therefore exist in flows which in no way may be considered stalled, the converse is not true. Although acceptable engineering predictions have been made, no ultimately satisfactory analysis of stalled flows can be made without consideration of the unsteady nature of the flow. If stalled flows are to be predicted successfully a deeper understanding of unsteady flows is essential.

This degree of complexity is present even for the flow around an isolated linear cascade with an undisturbed inlet flow. In a machinery environment potential flow and wake interactions with other blade rows will be present and possibly interaction with a distorted upstream or downstream flow field. These interactions have influences upon aerodynamic performance, sound and vibration levels. In the closing section of this chapter the reader is referred to analytical and experimental techniques for approaching such unsteady flow interactions.

8.1 Vortex Shedding

Steady flow at subsonic speeds past a bluff body gives rise to unsteadiness in the wake and periodic vortex shedding. The characteristic vortex pattern is one of an organized 'street' consisting of vortices alternately shed from each surface and therefore of alternating sense. This behaviour is most marked for high subcritical Reynolds' Numbers but has also been observed for laminar flows at quite low subcritical Reynolds' Numbers [8.2] and for turbulent flows [8.3].

Predictions of shedding frequency are based on the work of Strouhal [8.4]. As a result of his observations of vibrations in a circular cylinder mounted transverse to a wind the dimensionless Strouhal Number $S^* = (fd/U)$ was derived. A value of 0.21 is accurate to within about 5% for circular cylinders in the subcritical range of Reynolds' Numbers [8.5].

Although vortex streets have been computed numerically [8.6] there is, as yet, no accepted physical explanation of the Von Karman vortex street. Difficulties in interpreting the complex flow structure include the inception of shedding, the mixing rate of vorticity and the application of a stability condition.

In Von Karman's classical analysis [8.7] of the stability of two parallel lines of vortices it was shown that for an inviscid flow the arrangement has a first order instability for all arrangements except that for which $b/a = 0.281$. However more recent investigations [8.8] have shown that even for this ratio higher order instability is present. Because of this and the fact that vortex streets are known to exist in turbulent flow it has become clear that simple inviscid stability considerations do not determine the street geometry. Observed agreement with the Von Karman spacing ratio is considered fortuitous.

Kronauer [8.9] has suggested that the stability question is best resolved by stipulating that a vortex street should minimise drag by adjusting its geometrical configuration. This hypothesis has proved quite useful but lacks physical justification.

Correlations for shedding frequency based on body geometry exhibit substantial scatter. Most workers agree that the effect of the boundary layers immediately prior to separation should be accounted for. Whilst this effect is small for circular cylinders this is not so for thin profiles such as aerofoils, which also exhibit periodic shedding [8.10].

Goldberg and Florsheim [8.11] have proposed a universal Strouhal Number using the total wake momentum thickness as the characteristic length and Bauer [8.12] has used the displacement thickness in a similar way. These attempts have the shortcomings that a parameter based on momentum thickness is dependent on base pressure and one based on displacement thickness is critically dependent on axial distance in the shedding region. It also appears quite likely that multiple modes exist for vortex streets [8.13].

In view of these uncertainties it is quite understandable that observed Strouhal Numbers of vortex streets behind cascades of aerofoils should vary between 0.1 and 0.4. Because of the lack of knowledge of boundary layer parameters the physical trailing edge thickness was used in the measurements of Sieverding [8.14], Heinemann and Bütefisch [8.15] Herbert [8.16], Han and Cox [8.17] and Bryanston-Cross and Camus [8.18].

The observations of ref. 8.19 indicate that for an isentropic discharge Mach Number ranging between 1.1 and 1.35 the shedding of a periodic vortex street may be suppressed by the downstream shock wave system. These observations from turbine cascades are at least qualitatively in accord with those of Nash, Quincey and Callinan [8.20] taken on an isolated section having a blunt trailing edge. However schlieren pictures have been published [8.14, 8.15, 8.18, 8.21] showing clearly the existence of a vortex street downstream of an established shock system. It appears that above a discharge Mach Number of about 1.15 a configuration change occurs in which the vortex street no longer starts immediately downstream of the trailing edge but rather in the region of confluence between the suction surface and pressure surface streams [8.21, 8.22]. In such a case the shock system may oscillate at the shedding frequency [8.21] causing a high level of cross correlation between vortex streets shed from adjacent blades [8.14, 8.18]. It could be considered premature to include vortex street influences in flow calculations until improved physical understanding exists. Experimental verification of the Kronauer condition would be particularly useful. However if the Kronauer condition is accepted progress can be made in predicting street geometry [8.23]. Bearman [8.24], for example, has used the Kronauer condition to derive a universal Strouhal Number which is substantially independent of base pressure.

In summary, since the recent discovery of vortex streets behind turbine blading some elegant experimental work has been performed. Although the territory is still being mapped and the physics is not understood the phenomena are relatively widespread and repeatable.

In the next phase it will be necessary to take more adequate account of the physics of vortex shedding and especially its implications for blade design. New correlations will be required for the effects of vortex shedding on efficiency, acoustics, vibrations and heat transfer. For these to be consistent and comprehensive the support of numerical methods which predict vortex streets will also be required. This promises to involve further intriguing territory for exploration.

8.2 Base Pressure

Base-flow region modelling, especially for turbine blades, is rather important. In subsonic flows the losses are affected directly by thick bases. In supersonic flows the flow over the entire back face is affected. It has been concluded that "calculation methods which neglect base pressure effects are incapable of accurately calculating the flow pattern or the total pressure loss" [8.25].

8.2.1 Subsonic base flows. If the sole objective were the determination of the effect of trailing edge thickness on losses it could be argued that accumulated empirical and theoretical evidence is sufficient. Summaries of trailing edge thickness effects are given in refs. 4.4 , 8.26 , 8.27 and theoretical loss relationships are given in ref. 8.28 . Figure 8.1 taken from ref. 4.4 depicts the magnitude of the edge thickness effect.

Perry[8.26] stresses that the greatest problem in applying his own loss correlation is the choice of base pressure coefficient.

Therefore, although experimental information for loss does exist, it is dependent on improved base flow modelling for accurate systematic application. Furthermore knowledge of parameters other than blade loss is not susceptible to experimental correlation.

Nash's reviews of base flow [8.29], [8.30] still provide valuable insight into the physics of the region. Nash argued that a critical trailing edge thickness was of the order of the total momentum thickness at that location. As the physical thickness is raised above this level the static pressures in the trailing edge region fall.

In experimental work on backward-facing steps Tani[8.31] has shown that subsonic base flow depends more strongly on the mixing in the separated shear layers than on the geometry of the cavity flow. Badri Narayanan et al[8.32] have shown that a reasonably well defined similarity exists in the downstream variation of static pressure.

Unfortunately the data from backward-facing steps do not represent the full base pressure penalty for blading. Hoerner[8.33] correlated base drag for aerofoils and backward-facing steps and found that the base pressure of an aerofoil was consistently lower than that of an equivalent backward-facing step. He attributed this discrepancy to the existence of a Von Karman vortex street in the wake of the aerofoil.

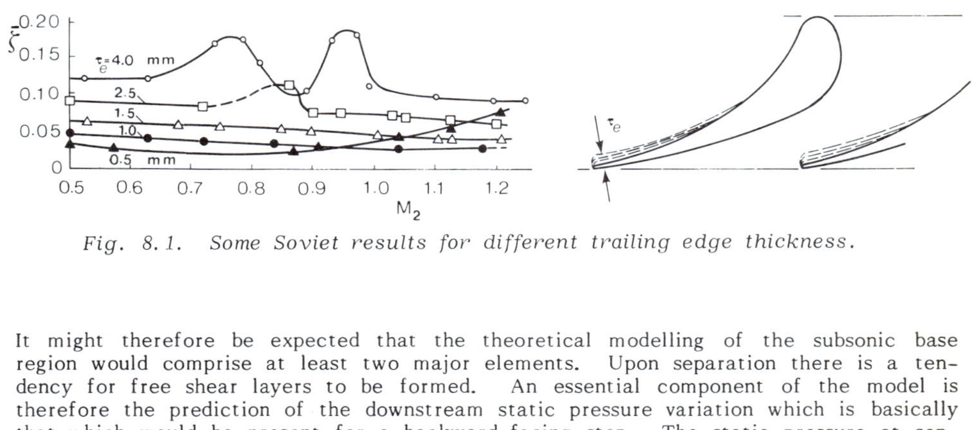

Fig. 8.1. Some Soviet results for different trailing edge thickness.

It might therefore be expected that the theoretical modelling of the subsonic base region would comprise at least two major elements. Upon separation there is a tendency for free shear layers to be formed. An essential component of the model is therefore the prediction of the downstream static pressure variation which is basically that which would be present for a backward-facing step. The static pressure at separation is closely related to the base pressure but after wake mixing has occurred it approaches the downstream free stream value. A coupling region links the separation region to the vortex street which must also be calculated if the wake is to be predicted.

All of this happens in a short distance for most base configurations and the effects are strongly interactive. It has been shown that fluid in the base region is not stagnant[8.34] but that vortex sheets fold up against the base to be ejected downstream as filaments into the vortex wake.

This is a complex process and any satisfactory theoretical description will be correspondingly complex. In the meantime correlations, relating the base pressure coefficient to the ratio of aggregate boundary layer thickness to trailing edge thickness, would be most useful. Insufficient data exist on this at present although Cox[8.35] has solved the vortex street equations to give separation velocity ratio as a function of the ratio of displacement thickness at separation to base height. The method was extended to allow for exit flow angle effects in turbine blades and included an appropriate adjustment to the potential flow solution. The base drag is highest for very thin boundary layers and becomes progressively lower as the displacement thickness approaches and exceeds the edge thickness.

8.2.2 Supersonic base flows. As the free-stream Mach Number is increased through the transonic range the characteristic sequence of events is illustrated in the schlieren photographs of Fig. 8.2. Although these were obtained from an isolated body having a slotted base they appear to represent the important features of base flows in turbine blading. The photographs are part of the work of Motallebi and Norbury[8.21].

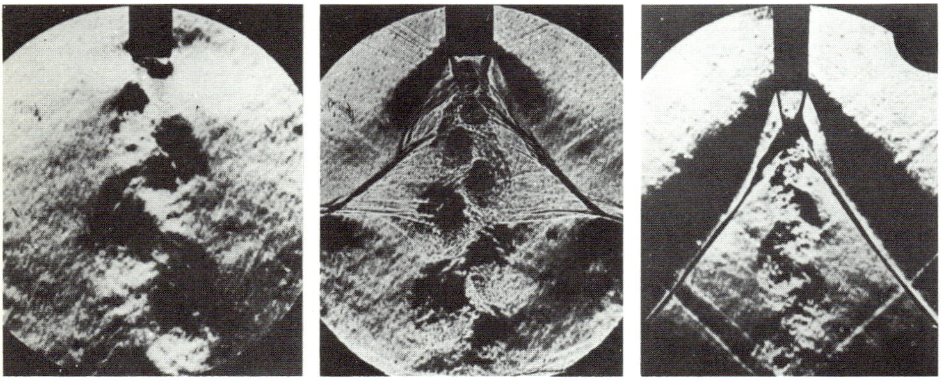

(a) $M_\infty = 0.68$ *(b)* $M_\infty = 1.06$ *(c)* $M_\infty = 1.30$

Fig. 8.2. Wake flows at three different Mach Numbers.

Periodic effects are here not suppressed upon the appearance of the shock system in the base region. Whereas under subsonic conditions (Fig. 8.2 (a)) the vortices were formed almost immediately behind the trailing edge, an expansion now occurs (in Fig. 8.2 (b)) to a base pressure corresponding to a local Mach Number of 1.7. The vortex appears to be formed between the shear layers. A lambda shock wave interacts with the shear layer at a location about one base height downstream and the whole system appears to oscillate with the shed vortices.

Expansion in Fig. 8.2 (c) is strong, to a Mach Number in the base region of nearly 1.9. The oblique trailing shocks clearly emanate from the important region of confluence of the shear layers and this is also where periodic vortex shedding commences.

The results of Sieverding et al [8.36] simulate the cascade situation more closely by arranging the imposition of an oblique shock corresponding to that from an adjacent blade. Detailed pressure surveys were made showing that, for base region Mach Numbers up to 2.25, the static pressure is reasonably constant around the entire circular trailing edge but that the isobaric mixing zone extends to little more than one half of a base height downstream. A recompression zone then occurs for a distance of two base heights downstream. Sieverding showed that for free-stream Mach Numbers above 1.2 the base Mach Number for circular and square trailing edges was identical. He considered the effect of the separation shock at the trailing edge to be important;

measurements suggested that, as a separation criterion for rounded trailing edges, this shock increased the static pressure by 15%. Such shocks are clearly visible as the abrupt termination of expansion regions in Fig. 8.2 (c).

A characteristic triangular-shaped base pressure zone exists downstream of a blunt trailing-edged turbine blade. This is terminated by the confluence region in which recompression occurs. Aerodynamic properties such as loss and turning depend on the correct modelling of the triangular base region. In addition, however, the region determines the shock waves and expansions which dominate the flow over much of the field within the blade row and downstream. A small discrepancy in base triangle representation drastically affects shock location and pressures over the whole of the back half of the turbine blade. Extreme accuracy of representation is therefore required.

Figure 8.3 represents a simplified flow model for two-dimensional supersonic base flows originated by Chapman [8.37]. The model considers the following four regions:- the flow approaching the trailing edge, the expansions emanating from the trailing edge, the free-mixing layer separating the constant static pressure wake edge from a region of semi-stagnant air and the region of confluence of the shear layers.

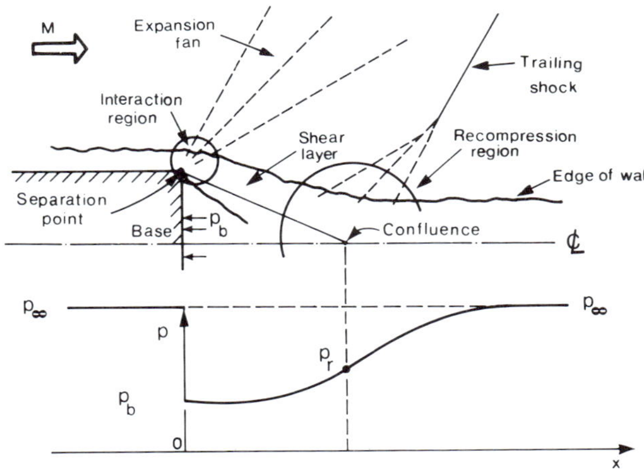

Fig. 8.3. Model of supersonic base flow.

As in the incompressible case of Chapter 5, an infinite number of solutions was found to be possible for the inviscid flow. As shown in Chapter 7, only consideration of viscosity could lead to a unique solution. The shear layers entrain air from the stagnant region and to sustain the cavity it is necessary to postulate a recompression and reversal of some entrained air back into the cavity. The unique solution thus adjusts the recompression to give just sufficient reversal for mass balance. For aerofoil sections the wake is considered fully mixed far downstream; once the recompression through the region of confluence is found the base pressure is determined.

The early theories of Chapman [8.37] and Korst [8.38] followed the above procedure. The limitations of these theories were discussed by Nash [8.30]. More sophisticated attempts to overcome the isentropic assumption and the excessive empiricism have been made by McDonald [8.39], Carriere[8.40] and Tanner[8.41] but fundamental problems remain.

The most important discharge Mach Number range for turbines is between sonic and 1.5. Some available experimental, theoretical and correlation data for base pressure in this range are plotted in Fig. 8.4. The theories of Korst [8.38] and Tanner[8.41] for the backward-facing step case are plotted with the corresponding experimental data of Tanner. The Liverpool University isolated body data of ref. 8.21 and 8.42 are plotted including the results corresponding to Fig. 8.2. Finally the correlation of Sieverding et al[8.36] is represented for converging turbine cascades over a range of wedge angle and blade curvature values.

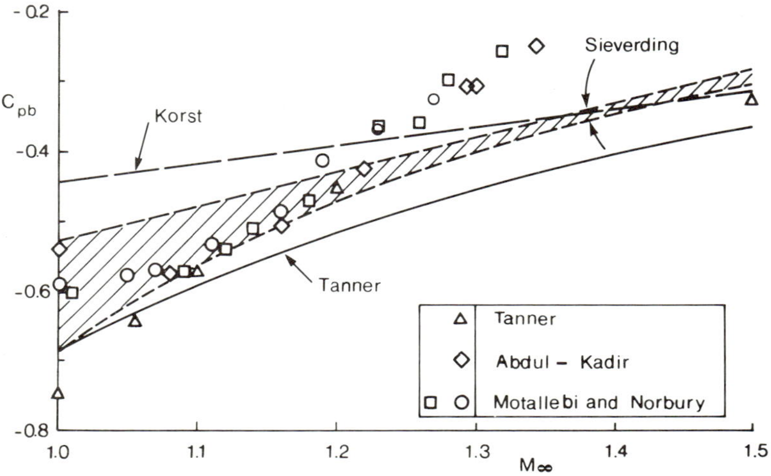

Fig. 8.4. Some supersonic base pressure data.

It appears from Fig. 8.4 that the application of backward-facing step data over the whole range is questionable. Its failure to account for a vortex street wake is in itself a serious objection. At Mach Numbers between unity and 1.2 the experimental data from isolated aerofoils and cascades are in agreement. At higher Mach Numbers there is a significant discrepancy. It is likely that genuine cascade effects are the reason for this and also possible that the isolated aerofoil data were affected by tunnel normal shock effects.

For turbine blading having thick trailing edges and supersonic outlet velocities improved modelling is necessary. This should resolve the experimental discrepancies. It appears clear that steady flow may not be assumed, that isobaric mixing is a poor assumption, and that the boundary layer shape parameter at separation is quite important [8.42, 8.43]. Although progress with viscous theory is likely further unambiguous experimental information is necessary. In this context the role of high speed cascade testing is seen as one of securing improved understanding of the physics of base flows. There is particularly seen to be a role for new techniques in flow visualisation combined with digital correlation [8.18].

8.3 Separation and Stall

In section 7.3 three distinct types of flow separation on an aerofoil were identified. These involved separation of the laminar boundary layer, with long and short bubbles, and separation of the turbulent boundary layer.

Calculation of the flow through a long laminar separation bubble remains an unsolved problem. Such a bubble will invariably have a major effect on the blade pressure distribution and an effect on the inception of a downstream turbulent boundary layer which is difficult to predict consistently.

Horton[7.52] has proposed a semi-empirical method of treating separation bubbles which involves the assumption that the static pressure is constant in the first part of the bubble and that a rapid transition then takes place, promoting entrainment, mixing and reattachment of the shear layer, with an attendant pressure rise. Work in industry has failed to confirm the predictions and Walker[8.44] has objected to the model citing measurements from an axial compressor blade which indicated extensive transition regions in the shear layer above the bubble. The constant pressure region was found to be short indicating a small laminar extent compared with transitional and turbulent regions. Walker's observations are surprising since Horton's model neglected the role of free-stream turbulence. This should promote more rapid mixing and indeed the correlation of Roberts[7.54] using, in part, the data of Evans[8.45] show such a strong free-stream turbulence effect. Data from linear cascades therefore show that Horton's model overestimates the bubble transition and reattachment lengths whilst data from the real machine show these lengths to be underestimated. The explanation could lie in the effect of the ordered unsteadiness due to blade interaction.

The short laminar separation bubble, whilst it remains short, has very little effect on the potential flow field and is therefore usually ignored in the calculation of pressure distributions. Its effect on the boundary layer is less predictable. Usually, but not always, the bubble acts somewhat as a trip wire and promotes rapid transition to turbulence. As loading increases a short bubble becomes more compact. When the length is smaller than the corresponding transition length the bubble bursts, catastrophically, into a much longer bubble or very often a completely separated region having no reattachment. In analysing this flow situation it is often assumed that the bounding streamline is a free streamline with mixing taking place downstream and no static pressure recovery. This assumption was firstly made for cascades by Cornell[8.46]. A prediction of the losses of a fully separated cascade was thereby achieved. Although a first order solution may be thus obtained, the essential physics of the flow is neglected. The inherent unsteadiness of the wake of what is effectively a bluff-body should be accounted for. In practice the shedding of discrete vortices will give rise to an additional momentum deficit as described in the previous section.

The inception of a turbulent separation region may be calculated with varying accuracy by the available turbulent layer calculation methods. Once more the bounding streamline may be assumed to be a free streamline. In a sense this is a typical mode in that for a thick trailing edge there will always be a separated region at the trailing edge and the boundary layer in this region is usually turbulent. The representation of a small turbulent separation region is therefore an extension of the correct trailing edge flow and wake modelling of the aerofoil. In this context modelling of the wake flow by a semi-infinite sting of the type used by Woods[8.47] is attractive.

In general as the blade loading rises above a certain level, flow separation becomes more than a local effect and dominates the flow field. The blade is then said to be stalled. Such situations are inherently non-linear and lead to lag and hysteresis effects. These effects are essentially time-dependent so that a stalled flow is basically an unsteady one.

A reliable definition of the presence of stall is given by the hysteresis phenomenon. If a given separated flow condition cannot be reversed by unloading the blade, i.e. if it is not possible to return to the flow regime immediately prior to the inception of major separation by simply unloading the blade, then non-linear effects have become dominant and the flow is said to be stalled. This essential role of non-linearity and hysteresis is an important feature of unsteady flows around blading.

Although most of the effects described above are detrimental to the performance of blading this is not always the case. It is quite likely that ordered unsteadiness in a free stream has a beneficial effect on boundary layer transition[8.48]. Furthermore there is a measure of controversy between engine designers concerning the effect of axial clearance between blade rows. Most designers claim that reduced clearance has a favourable effect on performance[8.49].

8.4 Stall Propagation

A constraint which often limits the pressure-rise capability of axial-flow compressors, fans and pumps is the occurrence of rotating or propagating stall. This phenomenon has caused physical damage to many compressors as a result of the periodic loading and unloading of blading which may occur at a resonant frequency. Rotating stall was first observed on the inducer vanes of a centrifugal compressor in the Whittle engines [8.50] and was subsequently described by Emmons et al [8.51].

As the loading of a diffusing blade row approaches the stall level small geometrical differences or flow disturbances tend to cause a single blade to stall first. This event is the instigator of stall for the whole blade row. However it is not usual for all blades to enter stall uniformly. Rather the stall of the initiating blade A (Fig. 8.5) creates a blockage of the passage bounded by the suction surface of A and the pressure surface of adjacent blade B. Because of this blockage 'cell', oncoming air will be deflected towards blades Z and C. Blade Z will be subjected to a more negative incidence than would otherwise occur. The incidence onto blades B and subsequently C is rendered more positive. Since the blades were already on the verge of stall, blade B will do precisely that. As a result passage BC will now become blocked. Oncoming air will be deflected to lower the incidence of blade A tending to unstall that blade and to unblock passage AB. The incidence to blade C will become more positive and that blade will then stall. The stall cell will continue to move along cascade A, B, C... in a classic propagating stall.

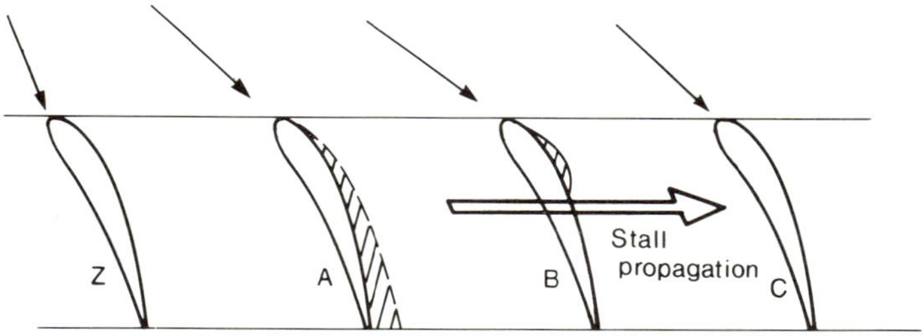

Fig. 8.5. Schematic representation of rotating stall.

Consideration of the above physical description might concentrate modelling attempts on the blockage and the inertia lag affecting transmission of the increased incidence 'signal' to the adjacent blade. Such a model was used by Marble[8.52] who appeared to achieve some reasonable predictions of propagation velocity. A contemporary model by Sears[8.53] emphasised time lags associated with the boundary layer separation process.

Early propagating stall theories, such as that of Emmons[8.51], employed linearised flow equations and assumed that blade-row characteristics might be replaced by two-dimensional cascade characteristics. A small perturbation imposed upon upstream and downstream mean flow fields would either grow or decay. This produced a stability criterion used to predict rotating stall.

Although the above linearised theories seemed to predict the cell propagation velocity satisfactorily the prediction of stall inception and of the number of stall cells remained an unsolved problem. The success of linear theory was ephemeral.

Whilst the first theoretical predictions were being made experimental work was also in progress in an attempt to improve physical understanding of propagating stall.

Stenning et al [8.54] successfully studied stall propagation in a circular cascade by taking sequences of schlieren and interferometry photographs. They noticed that during stall propagation the blades stalled from the leading edge and periodically shed nearly all bound circulation in discrete vortices. This was a discontinuous effect representing a build-up of circulation and a sudden shedding. The spacing of the blades was varied systematically and an attempt was made to plot shedding frequencies on a unique curve as a function of spacing. Although such a curve was obtained the approach would need further confirmation, especially in view of Sauer's finding [8.55] that the vortex shedding frequency for cascades decreases dramatically (e.g. from 1850 to 250 hz) as the incidence rises through the region of stall inception.

Corresponding theoretical models based upon the observation that strong propagating stalls could be accompanied by sudden shedding of leading edge vortices were proposed by Stenning and Kriebel[8.56] and by Louis[8.57] who postulated a street of point vortices, each having a strength equal to the total bound vorticity of one blade. These theories have been successfully applied although fundamental work is still required to obtain a more general version of the traditional Von Karman stability criterion.

In excellent early measurements of stall propagation in a linear cascade Carmichael [7.57] made pressure transducer measurements of fluctuations around a blade. Although Carmichael's main stalling mode was of the trailing-edge type there was evidence of a periodic separation from the leading edge. Wood[8.58] and Sovran[8.59] have also recorded observations of propagating stall in a linear cascade.

Some good recent measurements have been made by Lakhwani and Marsh[8.60], Cossar et al [8.61] and by Sexton et al[8.62] using on-rotor instrumentation. Many of the resulting pressure distributions show evidence of the importance of the bursting of short bubbles causing complete blade stall. Such dynamic stall patterns are also receiving attention as a result of inlet flow distortion problems.

Sharma and Railly[8.63] have measured instantaneous total pressure loss characteristics during stall and these exhibited a hysteresis loop indicating a strong influence of loss time lag. Sexton and O'Brien[8.64] have represented the dynamic characteristics of a compressor rotor by transfer function in a semi-actuator disc model. The transfer function included amplitude ratio and phase shift and contained mechanisms for predicting the initiation and propagation velocity of stall cells.

In the theoretical approach of Fabri and Siestrunck[8.65] few simplifying assumptions were made but, as with many other early models the stall cell was modelled as a zone of stagnant fluid. The work of Day and Cumpsty[8.66] has shown that stall cells are in reality far from stagnant. In a multistage environment the stall cells were shown to extend axially rather than helically. Mass transport across cell boundaries is vigourous with the flow angle changing instantaneously by up to 200°.

A promising recent model is that of Takata and Nagano[8.67]. In this approach the non-linear flow equations and blade characteristics are retained and a finite difference computer solution to the equations is derived. Although objections to some of the simplifications have been raised by Greitzer[8.68] the author considers that such non-linear approaches provide a good framework for subsequent inclusion of increasing physical understanding.

Cheers and Funnel[8.69] observed that 'tip stalling' rotors generally have lower stall propagation velocities than rotors stalling at other spanwise sections. The shape of the blade's nose is the predominant feature and ref. 8.69 concludes that since blade geometry can clearly influence the form of stall and its propagation speed any theory which ignores this effect is unrealistic.

Nevertheless progress has been made in the production of global correlations for stall inception[8.70] and the pressure rise capability of axial compressor stages. Ref. 8.71 and its discussion concluded that in full-span stall a total-to-static pressure rise coefficient of 0.11 per stage represented a wide range of axial compressors. There was also found to be a critical value of stall cell blockage of 30% above which part-span stall collapsed to give full-span stall. Furthermore it was clearly shown that high values of design flow coefficient led to large stall/unstall hysteresis loops.

By considering the analogy between the pressure rise capability of a compressor stage and that of a two-dimensional diffuser Koch[8.71] has produced a correlation for compressor stalling pressure rise soundly based on considerable experimental information. This approach is particulary well suited to the preliminary design assessment of pressure rise capability.

Significant improvements to the stall margin of axial flow compressors have been obtained by the use of casing treatment and impressive results for a range of configurations have been presented by Takada and Tsukada[8.72]. Greitzer[8.68] has demonstrated that use of casing treatment results in a considerable reduction in passage blockage as stall is approached. He also argues that only those compressors which have their performance limited by wall stall rather than blade stall will respond to casing treatment. Work in industry based on extensions of the casing treatment concept shows promise of completely eliminating the stall problem in fans.

Despite the impressive gains in empirical correlation of global stall parameters, and in the improvement of stall margin by casing treatment, the theoretical prediction capabilities can not be considered satisfactory and this must be a consequence of a lack of physical insight. The author considers that further improvements will come from an understanding of unsteady separation and the vortex shedding processes associated with leading edge stall since these appear to be the predominant mode of instigation of rotating stall.

8.5 Blade Flutter and Vibrations

Whereas for aircraft wings flutter is often excited by the coupling of flexural and torsional modes, for the stiffer and denser blading of turbomachinery such coupling is less frequent. Its occurrence in turbomachinery has been studied by Halliwell and by Bendiksen and Friedmann[8.73, 8.74].

Single degree of freedom flutter is more common for blading and it occurs when the aerodynamic excitation becomes sufficiently strong to overcome mechanical damping. Mechanical damping for turbomachinery blading is usually quite low although it can be increased by devices such as pin-jointed root fixings. In any event its influence is relatively unpredictable. Conservative designers tend to ignore it and consider flutter boundaries to be governed by whether aerodynamic damping is positive (stable) or negative (unstable) indicating flutter.

In high-speed fan and compressor operation flutter presents the most usual operational constraint. Figure 8.6 indicates the commonly-encountered flutter boundaries which conspire to make testing of such machines hazardous.

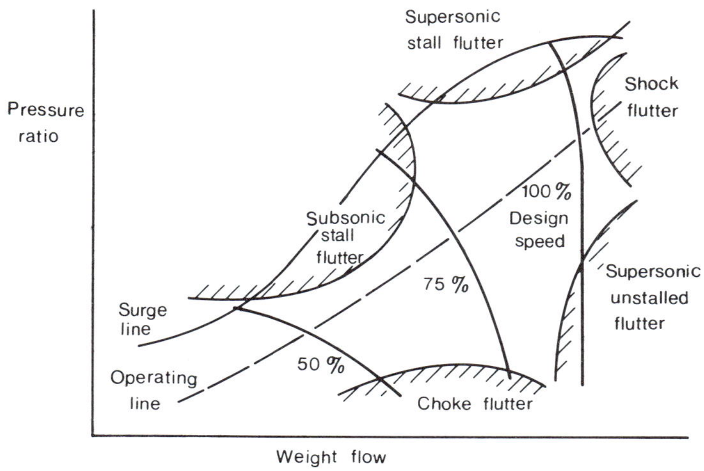

Fig. 8.6. Flutter boundaries for transonic fans.

When the blade is operating transonically and the load is reduced choke flutter may be encountered. Similarly supersonic unstalled flutter occurs in the choke region and may be predicted using supersonic small-disturbance theory[8.75]. The distinction between supersonic unstalled flutter and shock flutter depends on the blade design. However the latter condition is often caused by a strong shock wave from an adjacent blade which may excite either flexural or torsional vibration by periodically impinging on the blade. Shock flutter may impose a severe limit on overspeed capability along the operating line.

Supersonic stall flutter usually limits throttling of a transonic fan at design speed. A theory for prediction of supersonic stall flutter boundaries has been given by Adamczyk et al [8.76].

Finally, one of the most commonly encountered regions is that labelled subsonic 'stall' flutter. It had previously been considered that this region depended on flow separation and a physical coupling of stalling and aeroelastic excitation, as is the case with isolated aerofoils. However Carta and St. Hilaire [8.77] have recently argued, on the basis of systematic experimentation, that blade stall is not an essential condition for 'stall' flutter and that a more appropriate term might be 'loaded blade flutter'. The measurements of Szechenyi and Girault[8.78] are not in accordance with that argument and in ref. 8.78 it is concluded that "Flutter conditions can only occur when the flow over the blade is completely separated". Since the approaches which have been developed for flutter prediction tend to be different for unstalled flutter and 'stall' flutter for convenience the two will be considered separately.

Turbine blading is by no means exempt from flutter and therefore the somewhat different approaches which have been made to this problem will also be reviewed.

8.5.1 Unstalled flutter. The first cascade flutter analyses[8.79] were for thin aerofoils operating in counterphase. In general this case gives the greatest aerodynamic damping. Later theories[8.80] considered the interaction of blades operating in phase.

Classical linear analyses of subsonic flutter have been based on solutions for flat plates at zero mean incidence relative to a uniform stream.

In 1960 Whitehead[8.81] developed a systematic analysis and extensive tables for the forces and moments on vibrating aerofoils in cascade, subsequently extending this to include incidence effects[8.82]. Other workers have extended this approach to include torsional flutter and cambered blades. Unfortunately these solutions do not really correspond with any of the problem areas of Fig. 8.6 and are therefore limited in application. Atassi and Akai[8.83] have shown that for certain values of interblade phase angle the effects of moderate levels of thickness, camber, stagger and incidence on fluctuating lift and moment are significant.

The extension of these classical flutter theories to include subsonic compressibility effects has been covered in the review by Platzer[8.84]. Compressibility introduces acoustic resonances which, through acoustic radiation energy transfers to and from the blades, have a substantial effect on the aerodynamic damping characteristics. In effect the problems of blade vibration and blade noise generation become identical. A representative theory is that of Smith[8.85].

Verdon and Caspar[8.86] have produced a more general analysis relating the nonlinear boundary value solution for the steady flow potential to a linear variable coefficient boundary value solution for the unsteady potential. The unsteady flow is treated as a small perturbation about nonuniform mean flow and the real-blade effects of thickness, camber and incidence are covered. Carta[8.87] indicates that a prime objective of his experimental work is the verification of the theory of ref.8.86 and that the credibility of the theory at low frequency parameters and Mach Numbers is already well established. The Verdon and Caspar theory certainly appears to give an excellent prediction of the experimental results whilst still requiring further improvement in modelling the unsteady trailing edge condition.

Under the supersonic conditions of many compressor and fan blade rows the simplification to flat plate aerofoils is more realistic. Notwithstanding a supersonic relative inlet flow most transonic compressor blade rows operate with a subsonic axial velocity which introduces the complication that upstream and downstream flow fields and blade interference must be treated.

An early flat plate solution was published in 1966 by Gorelov[8.88]. Western developments followed in the 1970's [8.89], [8.90], [8.91], [8.92], [8.93] and [8.94]. Although these represent a variety of approaches there is an underlying consistency in the predictions of most theories. The possibility of torsional flutter tends to be indicated over a wide range of conditions. Less consistency was apparent when Halliwell[8.73] compared the theories of Nagashima and Whitehead[8.90] and Goldstein et al[8.94]. It was clear that neither represented observed shock locations at the tip of a transonic fan; this is hardly surprising since such shock locations can not yet be accurately predicted by steady flow theory.

Efforts have been devoted to correct representation of blade thickness and camber effects [8.95], [8.96], [8.97] which are expected to have a significant effect at low frequencies. However it is clear that better representation of shock waves and of shock-boundary-layer interactions is at least as important. Fleeter[8.98] has argued, and demonstrated, that, in view of the importance of correct physical representation, two-dimensional cascade tests are the best way for progress to be made in validating flutter theories.

8.5.2 Subsonic stall flutter. Stall flutter is thought to occur when the aerodynamic damping becomes negative as a blade stalls. If blade lift is plotted as a function of incidence the aerodynamic damping is a function of the partial derivative of lift with respect to incidence. At stall this derivative becomes negative, leading to a forced oscillation of the blades. Energy is fed into the blades from the free stream as a result of the time lags involved in separation.

Comprehensive series of tests have been summarised in refs. 8.99 and 8.100. The latter paper gives a minimum frequency parameter to be used if excessive vibration is to be avoided. Further information on this question has been obtained in ref.8.101 in which an experiment was designed to reproduce the theoretically-assumed leading edge stall.

No theory has yet been consistently successful. Limit cycle theories have been given by Sisto[8.102] and Schnittger[8.103], the lift-incidence curve is assumed to be a given non-linear shape and an arbitrary time lag may be included. The vibration amplitude for zero work is then calculated.

Thornton[8.104], using a theoretical extension of Possio's equation, incorporated experimental phase lag information and discovered a considerable loss of damping for high frequency parameters when separation was present.

For separated flows the difficulty of the problem is such that most theories treat only incompressible and inviscid flow around a cascade of flat plates operating at zero incidence. The limitations are obviously severe but as Whitehead[8.82] has shown, some measure of agreement between theory and experiment is achieved. Surveying experimental results by Tanida et al[8.105], Whitehead concluded that the available experimental information gave a progressive reduction in damping with increasing incidence and sometimes a further reduction in damping when the blades stall. These conclusions are fully in accordance with the results of Szechenyi and Girault[8.78] who, in cascade measurements at high subsonic speeds, found that flutter resulted from flow separation and was therefore related directly to incidence.

The unsteady pressures causing stall flutter of thin compressor blades were found to be mainly situated in the suction peak region of the aerofoil.

There is a disparity here with the findings of Carta and St Hilaire[8.77] who, on the basis of measurements behind a fan and on a compressor cascade, conclude that "blade stall is not an essential condition for 'stalled' flutter of airfoils in cascade". It is also concluded that the interblade phase angle is the most important parameter affecting the aeroelastic stability of cascades.

To resolve such apparent contradictions will require much more experimentation in both cascades and compressors. Substantial progress has recently been made in elegant work on the dynamic stall of isolated aerofoils[8.5]. It has become clear that, in unsteady flow, separation is not defined by the point of vanishing skin friction. In unsteady flow it is common to observe an extensive thin layer of reversed flow upstream of separation. In dynamic stall large scale vortices are formed in the suction peak region; absorbing energy from the separation shear layer they propagate downstream along the aerofoil surface.

Most of these findings are results of research on oscillating isolated aerofoils. Unsteady and dynamic stall phenomena in turbomachinery are more complex. The effect of the new physical insight from isolated aerofoils on cascade experiment and theory could assist in the resolution of the difficult problems of stall flutter.

8.5.3 Turbine blade flutter. The long and flexible last stage blades of modern steam turbines, with their substantial lacing wires, allow ample potential for flutter problems to arise. Soviet workers have been particularly active in this field. Samoylovich[8.106] considered the flutter problem for arbitrary interblade phase angles in 1962 and has since produced a reference work[8.107]. Gorelov et al[8.108] have produced an atlas of unsteady cascade data.

Researchers in the west have undertaken systematic theoretical and experimental work which has identified a wide range of flutter modes in turbine blading.

Akai and Atassi[8.109] applied their low speed theory to highly loaded turbine blades. The vibratory modes included pure bending and torsion as well as coupled modes. It was found that unstalled flutter could occur for low speed turbine cascades at typical design conditions; linear theories which neglected real thickness and loading effects were shown to incorrectly predict aerodynamic damping.

In experimental turbine cascade work at transonic speeds Rothrock et al[8.110] made comparisons between the measurements and the predictions of Smith's flat plate theory [8.85]. Again the conclusion was that aeroelastic forces were incorrectly predicted by the flat plate theory. Since the turbine blades were designed for 112° turning this was hardly surprising.

Whitehead and Grant[5.31] have applied a finite element program to vibrating turbine blades for various inter-blade phase angles and obtained very good agreement with measurements from an annular cascade rig.

Time marching programs for transonic flow through oscillating turbine cascades have been developed by Ni[8.111], Pigott and Yeh[8.112] and Takahara et al[8.113]. In the latter investigation a stalled region was calculated within the passage. Aerodynamic damping predictions were compared with values measured in a turbine cascade and gave good agreement.

Ref. 8.113 and a paper by Araki et al[8.114] were oriented towards the investigation of stall flutter in turbine blading. This is a problem in the last stage blading of some large steam turbines at part-load conditions where a strongly negative incidence may be present or where separation may be caused by shock wave impingement.

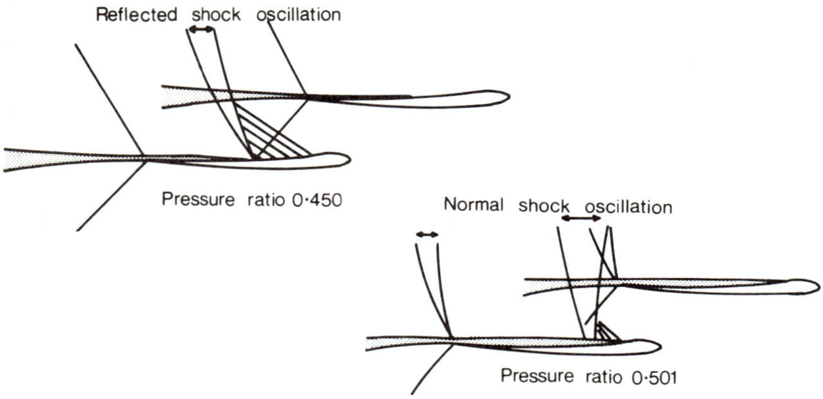

Fig. 8.7. Part-load instabilities in turbine blading.

The blading used by Araki et al[8.114], however, was of an advanced tip section design for discharge Mach Numbers up to 1.8 or higher. Negative aerodynamic damping was encountered at part-load conditions. This was unaffected by any attempt to artificially impose blade oscillations but rather coincided with severe instability problems in the passage caused by periodic oscillations of the normal and reflected shock waves at frequencies below 1 kHz. Figure 8.7 shows the conditions under which the two instability modes occurred; the unstable range corresponded to a cascade outlet Mach Number range of 0.8 to 1.33. Similar instability modes have been observed in high speed schlieren films from other turbine manufacturers[8.115]. Although the advanced converging-diverging sections have excellent efficiency at full load the part-load instability appears to be limiting their application in favour of a more stable section having very little divergence and optimised for a wider range of discharge Mach Numbers.

8.6 Blade Response to Potential Flow and Wake Interaction

There are two principal kinds of interaction between one blade row of a turbomachine and another. These are potential flow interaction and wake interaction. The co-existence of the two in rotating machinery often makes it difficult to distinguish between their effects.

Potential interaction results from the periodically changing nature of the potential flow field and its effects on neighbouring blade rows. Most blade rows experience their largest loading in the leading edge region; the suction peak will be on the blade surface but may be a small distance downstream of the leading edge. The loading across the trailing edge is comparatively small. Accordingly the potential interaction phenomenon manifests itself as the effect of the static pressure field of the downstream blade row on the flow around an upstream blade row, the two rows having a relative rotational velocity. Fluid particles in the trailing edge region of the upstream blade row will therefore experience alternate high and low static pressures at the passing frequency of the downstream blades. When the axial spacing between blade rows is increased this effect is attenuated exponentially. Potential interaction effects are therefore only of serious consequence when the axial spacing between blade rows is small.

The object of work instigated by Parker[8.116] has been consideration of potential flow interaction effects. Subsequent work has concentrated on the difficult task of separating the effects of the potential interaction of a downstream rotor row on an upstream stator row from viscous effects due to the wake of that stator row. One approach used was to energize the stator wakes by slot-blowing in order to independently control the wake interaction.

Whereas potential interaction effects do not directly involve the role of viscosity, those of wake interaction do. Wake interaction is the effect upon the flow through downstream blade rows of the wakes shed by one or more upstream blade rows and as such involves both viscous effects and the unsteady effects of the vortex wake. In contrast to potential flow interaction effects, which decay rapidly with distance from the source, wakes persist for considerable distances downstream. Wake interaction is the predominant disturbing influence in most turbomachinery flows. In any attempt to predict these disturbances it is therefore important to have reliable information on wake velocity profiles and decay rates.

Ongoing investigations of wakes are being conducted by Lakshminarayana and colleagues and some results were outlined in section 7.3. The work has resulted in correlations for velocity defect in tests on isolated aerofoils, cascades and compressor stators and rotors. These will be applied to provide information on axial clearance effects in Chapter 11.

In the simplest case of a rotor-stator pair the wake from the upstream blade row propagates in a circumferential direction as it sweeps through the downstream row. The situation is more complicated if there are more than two blade rows involved. Continuing work at The University of Tasmania[8.117] is oriented towards detailed investigation of the passage of wakes shed by upstream blade rows. Measurements at Iowa State University[8.118] were found to be compatible with the 'wake chopping' concepts of Smith[8.119] and Kerrebrock and Mikolajczak[8.120].

Gallus et al[8.121] have usefully reviewed the potential interaction theory of Lienhart[8.122] and the wake interaction theory of Henderson[8.123] in the light of their own experimental data. They conclude that in the region of close axial spacings, where wake and potential interactions are of the same order, the mere addition of potential and wake interaction components would lead to the prediction of excessive forces as a result of non-linear effects. This conclusion is in accordance with the findings of Saren and Yudin[8.124] on the basis of equivalent Soviet theories.

An insight into the nature and complexity of the nonlinear effects is given in recent measurements by Dring et al[8.125] on a turbine stage having pressure transducer intrumentation on rotor and nozzle blades. The potential field of the rotor leading edge produced strong wave motions on the suction surface of the nozzle blade. Pronounced differences in amplitude and phase across the trailing edge confirmed earlier predictions that, at values of frequency parameter representative of rotor-stator interactions, an unsteady loading would be present at the trailing edge.

In investigations on a compressor stage Fleeter et al[8.126] examined the response of the downstream stator row to the passage of rotor wakes. At representative values of frequency parameter and rotor/stator spacing a pronounced unsteady loading across the stator trailing edge was also observed in these tests. The significance of these observations was that the primary source of excitation was clearly the passage of a wake rather than a potential interaction.

Experimental studies on an oscillating aerofoil by Satyanarayana and Davis[8.127] have shown that trailing edge loading becomes significant at frequency parameters above 0.6 and grows rapidly with increasing frequency parameter. The aerofoil was uncambered and it seems likely that the effect would have been even greater for a cambered blade. More data from a wide range of blading configurations is required and especial attention will need to be paid to non-linear regimes such as that of natural vortex shedding.

The evidence is nevertheless clear that in the unsteady environment of most turbomachines both potential and wake interactions may cause periodically varying loading across the trailing edge. Calculations of blade-to-blade flow based on steady flow assumptions are deficient in this important respect. Fully unsteady calculations are required and these must have built in realistic modelling of unsteady flows at the trailing edge.

The above real effects are naturally quite complex, sometimes involving shear layers and viscosity. However considerable progress has been made by superposition of solutions for the response of blading to simple sinusoidal gusts.

The two principal gust types considered are the transverse gust of Sears[8.128] and the streamwise gust of Horlock[8.129]. These gusts are depicted schematically in Fig. 8.8.

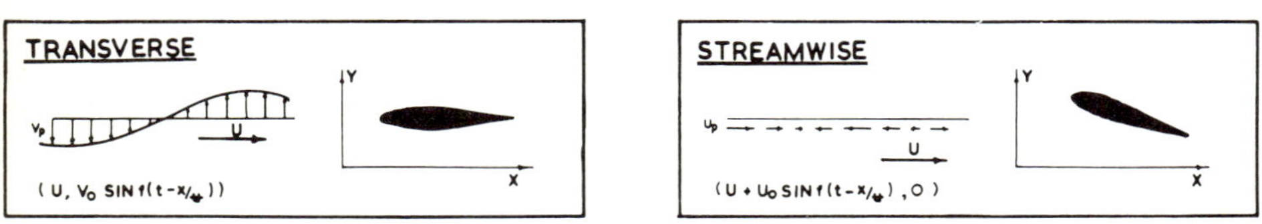

Fig. 8.8. Transverse and streamwise gusts.

Von Karman and Sears[8.130] first presented a theory for the unsteady flow around aerofoils. This was a clarification of theories originally developed by Birnbaum[8.131] and other workers. Parallel Russian work is described by Nekrasov[8.132].

Sears[8.128] showed how a vortex wake was first produced if the blade circulation is variable. The induced vorticity distributions over a thin aerofoil were presented for such wake vortices at various distances in the wake ξ' from the midpoint of the aerofoil.

Formulae for the lift and momentum were then presented for the thin aerofoil. Instead of the Joukowsky formula $L = \rho U \Gamma$, for unsteady flow it is necessary to divide the lift into three components:-

$$L = L_0 + L_1 + L_2 \, . \tag{8.1}$$

where

$$L_0 = \rho U \Gamma_0 \quad ,$$

$$L_1 = -\rho \frac{d}{dr} \int_{-1}^{1} \gamma_0(x)\, x\, dx \quad ,$$

$$L_2 = \rho U \int_{1}^{\infty} \gamma(\xi') \frac{d\xi'}{\sqrt{\xi'^2 - 1}} \quad .$$

The quasi-unsteady lift L_0 is that which would be produced if the problem were analysed as a series of steady solutions. The lift L_1, represents the lift in an ideal fluid when no circulation is produced because of the reaction of the accelerated fluid masses. $\gamma_0(x)$ is that vorticity bound to the aerofoil produced by the gust if the instantaneous velocity were maintained permanently. The lift L_2 represents the wake contribution and is the unsteady perturbation due to unsteady fluid velocity.

Sears[8.128] applied the derived expression to the case of sinusoidal transverse gusts. The result was the well-known Sears function for lift. Kemp and Sears[8.133] went on to calculate the unsteady lift on a single stator row followed by a single rotor row.

Naumann and Yeh[8.134] considered the effects of camber by partially accounting for some of the coupling between the angle of attack of the aerofoil and the unsteady flow. Goldstein and Atassi[8.135] developed a second order analysis which accounts for all of the coupling effects. These analyses are currently of limited value to turbo-machinery design in that only isolated aerofoils are considered. Henderson and Daneshyar[8.136] have considered a cascade of thin, slightly cambered aerofoils moving through a sinusoidal disturbance in an incompressible flow field. The gust response of aerofoils has also been considered by Samoylovich[8.106], Schorr and Reddy[8.137] and Hawthorne[8.138].

A conceptually simple model for flat plate cascades has been given by Rao and Jones [8.139] and was shown to be in good agreement with the results of Schorr and Reddy. Kemp and Ohashi[8.140] gave an exact solution for incompressible flow through an unstaggered flat plate exposed to generalised gusts. Whitehead's[8.81] numerical results were found to be in excellent agreement with the exact solution.

The Kemp and Sears analysis has been extended to incorporate viscous effects by Tsujimoto et al[8.141] who linearised the Navier-Stokes equations. Viscosity acted to amplify the unsteady forces but had little effect on the phase. Fujii et al[8.142] have combined the Sears and Horlock gusts to obtain good agreement between theoretical pure tone predictions and those measured from a fan rotor interacting with guide vane wakes.

One of the most general theories is that of Henderson[8.123] in which the unsteady response of a cascade is described in terms of the unsteady lift. The theory has been employed to predict a number of unsteady interaction situations[8.143].

Since the above analyses are based on small perturbation theory the coupling between steady and unsteady solutions is weak. Attempts to remove restrictions such as those of low camber and incidence and of small unsteady disturbances are being made. However in general these have yet to live up to earlier expectations[8.144].

Although somewhat beyond the scope of this work the prediction of inlet flow distortion effects is one of the most strenuous tests for unsteady theory. Non-linear effects, such as dynamic stall, abound and progress is dependent on the success of modelling loss hysteresis effects and other non-uniformities in all three coordinate directions. In 1975 Adamczyk[8.145] produced a coupled solution for the passage of a timewise periodic total pressure distortion through a highly loaded cascade.

8.7 Unsteady Flow Simulation and Measurement

There are numerous linear cascade tunnels which have been used for work in the flow through cascades of vibrating blades[8.75] and, given an intelligent approach to the establishment of inter-blade phase lags, valuable experimental information may be obtained. The more complex and expensive annular cascade has the advantage of eliminating any end effect or periodicity problems. Such a test rig has been developed and operated successfully for aerodynamic damping measurements at Mach Numbers up to 1.4 by Whitehead[5.31].

The problem of simulating an ordered fluctuating gust in a cascade wind tunnel is more difficult; the approaches developed vary considerably and three of these are briefly described.

Commerford and Carta[8.146] conducted experiments on a double circular arc aerofoil. A periodically fluctuating flow field, with a frequency parameter of 3.9, was generated by placing a transverse cylinder upstream of the test aerofoil which was then subjected to perturbations from the resulting Von Karman vortex street. The results showed that the amplitude of pressure fluctuation exceeded that predicted by Sears[8.128] and exhibited phase angles discrepant from the Sears theory.

Ostdiek[8.147] constructed an unsteady flow cascade tunnel having an oscillating inlet section which gave a sinusoidal variation in angle of attack between 6° and 14°. Tests were performed on a five-bladed cascade at low speed and for frequency parameters up to 0.11. Measured instantaneous pressure distributions were found to give good agreement with flat plate cascade theory in the leading edge region but not near the trailing edge.

The Cambridge University unsteady flow tunnel, described by Horlock[8.148] is depicted schematically in Fig. 8.9. Sinusoidal gusts are generated in a test section whose upper and lower surfaces are flexible steel sheets. A sinusoidal shape is imparted to these walls by a series of cams. Appropriate rotation of the cams causes the sine wave to propagate in the flow direction. An in-phase setting of the surfaces produces a transverse gust, a 180° phase shift gives a streamwise gust. Only gusts propagating with velocities other than that of the free stream may be simulated with this tunnel.

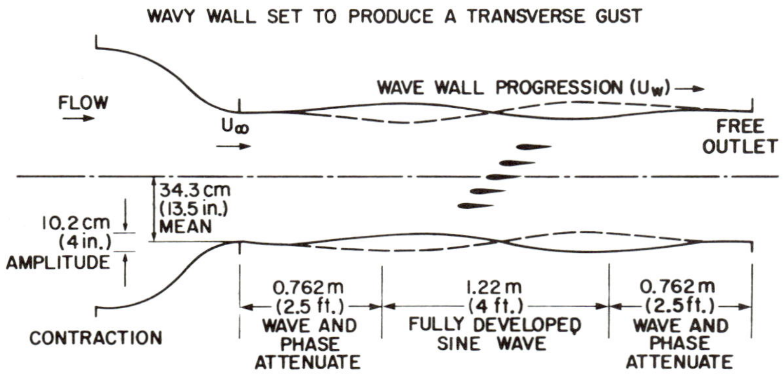

Fig. 8.9. The Cambridge University unsteady flow tunnel.

Useful measurements have been obtained on an isolated aerofoil by Holmes[8.149] and on a cascade by Satyanarayana[8.150]. Holmes found good agreement with the isolated aerofoil theory of Kemp[8.151] at zero incidence to a transverse gust. At incidence to a streamwise gust the flow separated so that thin aerofoil theory could not be applied.

Satyanarayana obtained detailed measurement of the flow past a cascade of blades. The maximum frequency parameters, based on freestream velocity and gust propagation velocity respectively, were 0.042 and 0.22. The results included instantaneous distributions of blade static pressure and of boundary layer and wake velocity profile. Agreement between measured fluctuating lift and vortex singularity theory was reasonable[8.152].

The unsteady flow measurements described are only possible because of advances in fast response instrumentation. Sensitive pressure transducers are now available with small overall dimensions of the order of 1mm having a useful frequency response of at least 60 kHz. Kerrebrock et al[8.153] have described the mounting of five of these in a sphere probe of 5mm diameter which, upon calibration, yields stagnation and static pressures, Mach Numbers and flow angles in two planes.

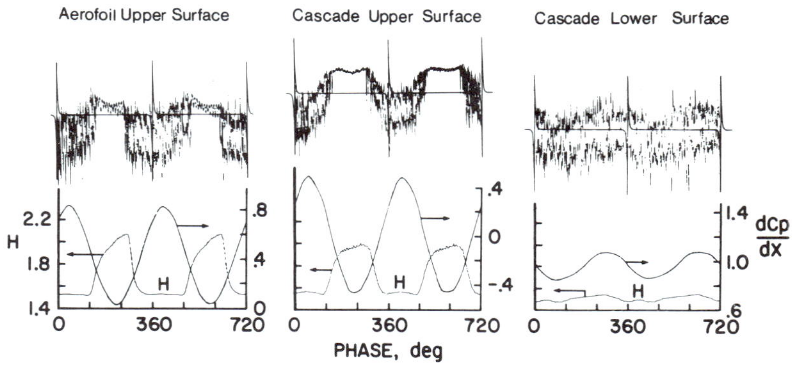

Fig. 8.10. Variation of velocity from raw data trace and of shape factor and pressure gradient from phase-lock averaging.

Hot wire techniques are well developed, present no practical frequency limitations, and are probably still the best approach to velocity profile traversing at subsonic speeds[8.154]. Hot film probes are more rugged and thus suitable for more strenuous environments. Ref. 8.155 gives an example of their use in turbine flows. Blade surface mounted thin film heat transfer gauges may be used to measure skin friction[8.156] and are used extensively in heat transfer work.

Laser anemometry can now be used successfully in an unsteady flow environment although data acquisition demands high frequency electronics and improvements are still needed in measurement time[8.157]. The combination of schlieren techniques with digital correlation has considerable potential in the study of some unsteady phenomena[8.18].

In a turbomachine rotating at constant speed a characteristic blade passing frequency is present. Most unsteady flow cascade experiments also have such a characteristic synchronising frequency. In such situations the author advocates the use of ensemble or phase-lock averaging as described in ref. 3.65. To ensure that valuable phase-unrelated information is not overlooked it is desirable to view raw data traces and phase-lock averaged data simultaneously (alongside spectrum analyses and flow visualisation information if possible). Figure 8.10 gives an example of such a superposition.

REFERENCES

8.1 DEAN, R.C.Jr. On the necessity of unsteady flow in fluid machines. Trans. ASME. J. Basic Eng. 81, 24 (1959)

8.2 TRITTON, D.J. The flow past a circular cylinder at low Reynolds' Numbers. J.F.M., 6 547 (1959)

8.3 ROSHKO, A. Experiments on the flow past a circular cylinder at very high Reynolds' Numbers. J.F.M., 10 345 (1961)

8.4 STROUHAL, V. Uber eine besondere Art der Tonerregung. Ann. Phys. Chem., N.F., 5, 10, 216 (1878)

8.5 MCCROSKEY, W.J. Some current research in unsteady fluid dynamics. 1976 Freeman Scholar Lecture. Trans. ASME, J. Fluids Eng., 99, pg. 8 (1977)

8.6 LEWIS, R.I. and PORTHOUSE, D.T.C. A generalised numerical method for bluff body and stalling aerofoil flow. ASME Paper 82-GT-70 (1982)

8.7 VON KARMAN, TH. and RUBACH, H. Uber den Mechanismus des Flussigkeits und Luftwiderstandes. Phys. Zs., 13,2 pg. 49 (1912)

8.8 WILLE, R. Von Karman vortex shedding. Adv. in App. Mech., vol.6, pg. 273, Academic Press, N.Y. (1960)

8.9 KRONAUER, R.E. Predicting eddy frequency in separated wakes. IUTAM symp. Ann Arbor, Michigan, (1964)

8.10 KRZYWOBLOCKI, M.Z. Investigation of the wing-wake frequency with application of the Strouhal Number. J. Aero. Sci. 12, 1 pg. 51 (1945)

8.11 GOLDBERG, A. and FLORSHEIM, B.H. Transition and Strouhal Number for the incompressible wake of various bodies. Phys. Fluids 9, pg. 45 (1966)

8.12 BAUER, A.B. Vortex shedding from thin flat plates parallel to the free stream. J. Aero. Sci. pg. 340 (1961)

8.13 WEIHS, D. On the existence of multiple Karman vortex street modes. J.F.M., 61, 1 pg. 199 (1973)

8.14 SIEVERDING, C.H. Unsteady flow measurements in straight cascades. Proc. Symp. Measuring Techniques in Transonic and Supersonic Cascades and Turbomachines, Lausanne, (1977)

8.15 HEINEMANN, H.J. and BUTEFISCH, K.A. Determination of the vortex shedding frequency of cascades with different trailing edge thickness. AGARD CP 227 Paper 35 (1978)

8.16 HERBERT, M.V. Base drag of single aerofoils and turbine blades at low speed. ARC 34,756 (1973)

8.17 HAN, LIT S. and COX, W.R. A visual study of turbine blade pressure-side boundary layers. Trans. ASME, J. Eng. Power 105, 1 pg. 47 (1983)

8.18 BRYANSTON-CROSS, P.J. and CAMUS, J.J. Auto and cross correlation measurements in a turbine cascade using a digital correlator. ASME Paper 82-GT-132 (1982)

8.19 LAWACZECK, O. and HEINEMANN, H.J. Von Karman vortex streets in the wakes of subsonic and transonic cascades. AGARD CP-177 Paper 28 (1976)

8.20 NASH, J.F., QUINCEY, V.G. and CALLINAN, J. Experiments on two-dimensional base flow at subsonic and transonic speeds. ARC R&M 3427 (1966)

8.21 MOTALLEBI, F. and NORBURY, J.F. The effect of base bleed on vortex shedding and base pressure in compressible flow. J.F.M., 110 pg. 273 (1981)

8.22 SIEVERDING, C.H. The influence of trailing edge ejection on the base pressure in transonic turbine cascades. ASME Paper 82-GT-50 (1982)

8.23 GOSTELOW, J.P. Trailing edge flows over turbomachine blades and the Kutta-Joukowsky condition. ASME Paper 75-GT-94 (1975)

8.24 BEARMAN, P.W. On vortex street wakes. J.F.M., 28, 4 pg. 625 (1967)

8.25 MACMARTIN, I.P. and NORBURY, J.F. The aerodynamics of a turbine cascade with supersonic discharge and trailing edge blowing. ASME Paper 74-GT-120 (1974)

8.26 PERRY, J.H. Recent developments in predicting turbine efficiency. GTCC Aero. Rep. 697 (1969)

8.27 HERBERT, M.V. A discussion on the prediction of profile loss for axial-flow turbine blades. ARC 34,757 (1973)
8.28 LIEBLEIN, S. and ROUDEBUSH, W.H. Theoretical loss relations for low-speed two-dimensional cascade flow. NACA TN 3662 (1956)
8.29 NASH, J.F. A review of research on two-dimensional base flow. ARC R&M 3323 (1967)
8.30 NASH, J.F. A discussion of two-dimensional turbulent base flows. ARC R&M 3468 (1967)
8.31 TANI, I. Experimental investigation of flow separation over a step. Grenzschichtforschung Symp., Freiburg. (H. Görtler, ed.) Springer (1958)
8.32 BADRI NARAYANAN, M.A., KHADGI, Y.N. and VISWANATH, P.R. Similarities in pressure distribution in separated flow behind backward-facing steps. Aero. Quart. pg. 305 (1974)
8.33 HOERNER, S.F. Fluid Dynamic Drag. Published by the author (1958)
8.34 CHONG, M.S., LIM, T.T. and PERRY, A.E. The vortex shedding process behind a circular cylinder. Proc. 7th Aust. Hyd. and Fluid Mech. Conf., Brisbane (1980)
8.35 COX, H.J.A. Private communication (1979)
8.36 SIEVERDING, C.H., STANISLAS, M. and SNOEK, J. The base pressure problem in transonic turbine cascades. Trans. ASME, J. Eng. Power 102;3 pg. 711 (1980)
8.37 CHAPMAN, D.R. An analysis of base pressure at supersonic velocities and comparison with experiment. NACA Rep. 1051 (1951)
8.38 KORST, H.H. A theory of base pressures in transonic and supersonic flow. Trans ASME, J. App. Mech. 23 pg. 593 (1956)
8.39 MCDONALD, H. Turbulent shear layer reattachment with special emphasis on the base pressure problem. Aero. Quart. 15 pg. 247 (1964)
8.40 CARRIERE, P. Analyse theorique du decollement et du recollement turbulents au bord de fuite d'un anbage aux vitesses supersoniques. In "Flow Research on Blading" L.S. Dzung ed. Elsevier (1970)
8.41 TANNER, M. Two different theoretical approaches to the base pressure problem in two-dimensional supersonic flow. Aero. Quart. 29, 2 pg. 114 (1978)
8.42 ABDUL-KADIR, F.F. Base pressure effects on aerofoils in transonic flow. Ph.D. thesis, Liverpool Univ. (1976)
8.43 SIEVERDING, C., DECUYPER, M., COLPIN, J. and AMANA, O. Model tests for the detailed investigation of the trailing edge flow in convergent transonic turbine cascades. ASME Paper 76-GT-30 (1976)
8.44 WALKER, G.J. Observations of separated laminar flow on axial compressor blading. ASME Paper 75-GT-63 (1975)
8.45 EVANS, B.J. Effects of free-stream turbulence on blade performance in a compressor cascade. Cambridge Univ. rep. CUED/A-Turbo/TR26 (1971)
8.46 CORNELL, W.G. The stall performance of cascades. Proc. 2nd Nat. Cong. App. Mech., ASME (1954)
8.47 WOODS, L.C. The Theory of Subsonic Plane Flow. Cambridge Univ. Press (1961)
8.48 WALKER, G.J. The unsteady nature of boundary-layer transition on an axial-flow compressor blade. ASME Paper 74-GT-135 (1974)
8.49 MIKOLAJCZAK, A.A. The practical importance of unsteady flow. AGARD CP-177 Paper 1 (1976)
8.50 CHESHIRE, L.J. The design and development of centrifugal compressors for aircraft gas turbines. War Emergency Issue No. 12. I. Mech. E. (1945)
8.51 EMMONS, H.W., PEARSON, C.E. and GRANT, H.P. Compressor surge and stall propagation. Trans. ASME 77 (1955)
8.52 MARBLE, F.E. Propagation of stall in a compressor blade row. J. Aero. Sci. 22 (1955)
8.53 SEARS, W.R. Some aspects of non-stationary airfoil theory and its practical application. J. Aero. Sci. 8;3 pg. 104 (1941)
8.54 STENNING, A.H., KRIEBEL, A.R. and MONTGOMERY, S.R. Stall propagation in axial flow compressors. NACA TN 3580 (1956)
8.55 SAUER, H. Investigations on vortex frequencies in wakes of cascade blades. Proc. 6th Conf. on Fluid Machinery, Budapest (1979)

8.56 STENNING, A.H. and KRIEBEL, A.R. Stall propagation in cascades of airfoils. Trans. ASME (1958)
8.57 LOUIS, J.F. Stall phenomena in axial flow compressors. Ph.D. thesis, Cambridge Univ. (1957)
8.58 WOOD, M.D. Stall propagation in axial flow compressors. Ph.D. thesis, Cambridge Univ. (1955)
8.59 SOVRAN, G. The measured and visualised behaviour of rotating stall in an axial-flow compressor and in a two-dimensional cascade. Trans. ASME. J. Eng. Power, 81, 1 pg. 24 (1959)
8.60 LAKHWANI, C. and MARSH, H. Rotating stall in an isolated rotor row and a single-stage compressor. I. Mech. E. Conf. Publ. 3 (1973)
8.61 COSSAR, B.F.J., MOFFATT, W.C. and PEACOCK, R.E. Compressor rotating stall in uniform and non-umiform flow. Cranfield I. T. Rep. (1978)
8.62 SEXTON, M.R., O'BRIEN, W.F. and MOSES, H.L. An on-rotor investigation of rotating stall in an axial-flow comressor. AGARD CP-177 Paper 33 (1976)
8.63 SHARMA, P.B. and RAILLY, J.W. Dynamic total head loss characteristic for an axial compressor rotor. J. Mech. Eng. Sci. 22, 6, pg. 269 (1980)
8.64 SEXTON, M.R. and O'BRIEN, W.F.Jr. A model for dynamic loss response in axial-flow compressors. ASME Paper 81-GT-154 (1981)
8.65 FABRI, J. and SIESTRUNCK, R. Rotating stall in axial flow compressors. J. Aero. Sci. (1957)
8.66 DAY, I.J. and CUMPSTY, N.A. The measurement and interpretation of flow within rotating stall cells in axial compressors. J. Mech. Eng. Sci. 20, pg. 101 (1978)
8.67 TAKATA, H. and NAGANO, S. Non-linear analysis of rotating stall. Trans. ASME, J. Eng. Power 94 pg. 279 (1972)
8.68 GREITZER, E.M. Review - Axial compressor stall phenomena. Trans. ASME, J. Fluids Eng. 102 pg. 134 (1980)
8.69 CHEERS, F. and FUNNELL, G.C. Note on the speed of stall propagation in a cascade of aerofoils. Aero. Quart. pg. 179 (1965)
8.70 DAY, I.J., GREITZER, E.M. and CUMPSTY, N.A. Prediction of compressor performance in rotating stall. Trans. ASME, J. Eng. Power 100 pg. 1 (1978)
8.71 KOCH, C.C. Stalling pressure rise capability of axial flow compressor stages. Trans. ASME, J. Eng. Power 103, 4 pg. 645 (1981)
8.72 TAKATA, H. and TSUKUDA, Y. Stall margin improvement by casing treatment - its mechanism and effectiveness. Trans. ASME, J. Eng. Power 99,1 pg. 121 (1977)
8.73 HALLIWELL, D.G. Some aerodynamic considerations in the prediction of unstalled supersonic flutter in transonic fans. "Aeroelasticity in Turbomachines" P. Suter ed. Juris-Verlag, Zurich (1981)
8.74 BENDIKSEN, O.O. and FRIEDMANN, P.P. The effect of bending-torsion coupling on fan and compressor blade flutter. Trans. ASME, J. Eng. Power 104, 3 pg. 617 (1982)
8.75 SNYDER, L.E. and COMMERFORD, G.L. Supersonic unstalled flutter in fan rotors; analytical and experimental results. Trans. ASME, J. Eng. Power 96, 4 pg. 379 (1974)
8.76 ADAMCZYK, J.J., STEVANS, W. and JUTRAS, R. Supersonic stall flutter of high-speed fans. Trans. ASME, J. Eng. Power 104, 3 pg. 675 (1982)
8.77 CARTA, F.O. and ST. HILAIRE, A.O. Effect of interblade phase angle and incidence angle on cascade pitching stability. Trans. ASME, J. Eng. Power 102, pg. 391 (1980)
8.78 SYECHENYI, E. and GIRAULT, J.PH. A study of compressor blade stall flutter in a straight cascade wind tunnel. "Aeroelasticity in Turbomachines" P. Suter ed. Juris-Verlag, Zurich (1981)
8.79 REISSNER, E. Boundary value problems in aerodynamics of lifting surfaces in non-uniform motion. Bull. Am. Math. Soc., 55, 9 pg. 825 (1949)
8.80 WOODS, L.C. On the unsteady flow through a cascade of aerofoils. Proc. R. Soc. Ser. A. No. 1172 (1955)

8.81 WHITEHEAD, D.S. Force and moment coefficients for vibrating aerofoils in cascade. ARC R&M 3254 (1960)
8.82 WHITEHEAD, D.S. Bending flutter of unstalled cascade blades at finite deflection ARC R&M 3386 (1962)
8.83 ATASSI, H. and AKAI, T.J. Aerodynamic force and moment on oscillating airfoils in cascade. ASME Paper 78-GT-181 (1978)
8.84 PLATZER, M.F. Unsteady flows in turbomachines - a review of current developments. AGARD CP-227 (1977)
8.85 SMITH, S.N. Discrete frequency sound generation in axial flow turbomachines. ARC R&M 3709 (1972)
8.86 VERDON, J.M. and CASPAR, J.R. Subsonic flow past an oscillating cascade with finite mean flow direction. AIAA J, 18, 5 pg. 540 (1980)
8.87 CARTA, F.O. Unsteady gapwise periodicity of oscillating cascaded airfoils. ASME Paper 82-GT-286 (1982)
8.88 GORELOV, D.N. Oscillations of a plate cascade in a transonic gas flow. Mekhanika Zhidkosti i Gaza. I (1) pg. 69 (1966)
8.89 KUROSAKA, M. On the unsteady supersonic cascade with a supersonic leading edge - an exact finite order theory. Trans. ASME, J. Eng. Power pg. 13 (1974)
8.90 NAGASHIMA, T. and WHITEHEAD, D.S. Aerodynamic forces and moments for vibrating supersonic cascade blades. Cambridge Univ. Rep. CUED/A-Turbo/TR59 (1974)
8.91 CARUTHERS, J.E. Theoretical analysis of unsteady supersonic flow through harmonically oscillating turbofan cascades. Ph. D. thesis, Georgia Inst. of Tech. (1976)
8.92 VERDON, J.M. Further developments in the aerodynamic analysis of unsteady supersonic cascades. Part 1: The unsteady pressure field, Part 2: Aerodynamic response predictions. Trans. ASME, J. Eng. Power 99, 4 pg. 509 (1977)
8.93 NI, R.H. A rational analysis of periodic flow perturbation in supersonic two-dimensional cascade. Trans. ASME, J. Eng. Power 101, 3 pg. 431 (1979)
8.94 GOLDSTEIN, M.E., BRAUN, W. and ADAMCZYCK, J.J. Unsteady flow in a supersonic cascade with strong in-passage shocks. J.F.M. vol.83 (1977)
8.95 CHADWICK, W.R., BELL, J.K, and PLATZER, M.F. On the analysis of supersonic flow past oscillating cascades. AGARD CP-177 Paper 29 (1976)
8.96 NAMBA, M. and MINAMI, R. Effect of mean blade loading on supersonic cascade flutter. "Aeroelasticity in Turbomachines" P. Suter ed. Juris-Verlag, Zurich (1981)
8.97 VOGELER, K. A method of characteristics solution for a finite oscillating supersonic cascade with thickness effects. "Aeroelasticity in Turbomachines" P. Suter ed. Juris-Verlag, Zurich (1981)
8.98 FLEETER, S. Experimental verification of unsteady aerodynamic models of supersonic unstalled flutter. 5th ISABE, Bangalore (1981)
8.99 CARTER, A.D.S. and KILPATRICK, D.A. Self-excited vibration of axial-flow compressor blades. Proc. I. Mech. E. 171 245 (1957)
8.100 ARMSTRONG, E.K. and STEVENSON, R.E. Some practical aspects of compressor blade vibration. J. Roy. Aero. Soc. 64 117 (1960)
8.101 YASHIMA, S. and TANAKA, H. Torsional flutter in stalled cascade. ASME Paper 77-GT-72 (1977)
8.102 SISTO, F. Stall flutter in cascades. J. Aero. Sci. 2C 598 (1953)
8.103 SCHNITTGER, J.R. Single degree of freedom flutter of compressor blades in separated flow. J. Aero. Sci. (1953)
8.104 THORNTON, B.S. Damping loss in a cascade of fluttering phased aerofoils. J. Aust. Math. Soc., 8,2 pg. 327 (1968)
8.105 TANIDA, Y., HATTA, K. and ASANUMA, T. Experimental study on flutter in cascading blades. Bull. JSME, 6,24 pg. 736 (1963)
8.106 SAMOYLOVICH, G.S. On the calculation of the unsteady flow around an array of arbitrary profiles vibrating with arbitrary phase shift. PMM 25, 1 (1962)
8.107 SAMOYLOVICH, G.S. Excitation of the fluctuations of the blades of turbomachines. Mashinestroyeniye, Moscow (1975)

8.108 GORELOV, D.N., KURZIN, V.B. and SAREN, V.E. Atlas of non-steady aerodynamic characteristics of profile cascades. Nauka, Novosibirsk (1974)
8.109 AKAI, T.J. and ATASSI, H. Stability and flutter analysis of turbine blades at low speed. "Aeroelasticity in Turbomachines" P. Suter ed. Juris-Verlag, Zurich (1981)
8.110 ROTHROCK, M.D., JAY, R.L. and RIFFEL, R.E. Time-variant aerodynamics of high-turning blade elements. Trans. ASME, J. Eng. Power 104, 2 pg. 412 (1982)
8.111 NI, R.H. Nonstationary aerodynamics of arbitrary cascades in compressible flow. Ph. D. thesis, Stevens Inst. of Tech. (1974)
8.112 PIGOTT, R. and YEH, H. Unsteady aerodynamic forces on vibrating transonic turbine cascades. ASME Paper 76-WA/GT-8 (1976)
8.113 TAKAHARA, S., ADACHI, T. and KADOYA, Y. On the numerical analysis of stall flutter in turbine cascades. "Aeroelasticity in Turbomachines" P. Suter ed. Juris-Verlag, Zurich (1981)
8.114 ARAKI, T., OKAMOTO, Y. and OHTOMO, F. Self-excited flow oscillation in the low pressure steam turbine cascade. "Aeroelasticity in Turbomachines" P. Suter ed. Juris-Verlag, Zurich (1981)
8.115 JAIKRISHNAN, K.R. Transonic steam turbine cascade measurements. Symp. Measuring Techniques in Transonic and Supersonic Cascade Flow, C.E.R.L., U.K. (1980)
8.116 PARKER, R. Pressure fluctuations due to interaction between blade rows in axial flow compressors. Proc. I. Mech. E. 183, 1 (1968-69)
8.117 LOCKHART, R.C. and WALKER, G.J. The influence of viscous interactions on the flow downstream of an axial compressor stage. 2nd ISABE, Sheffield (1974)
8.118 WAGNER, J.H., OKIISHI, T.H. and HOLBROOK, G.J. Periodically unsteady flow in an imbedded stage of a multistage axial-flow turbomachine. Trans. ASME, J. Eng. Power 101, 1 pg. 42 (1979)
8.119 SMITH, L.H.Jr. Wake dispersion in turbomachines. Trans. ASME, J. Basic Eng. 88 pg. 688 (1966)
8.120 KERREBROCK, J.L. and MIKOLAJCZAK, A.A. Intra-stator transport of rotor wakes and its effect on compressor performance. Trans. ASME, J. Eng. Power 92 pg. 359 (1970)
8.121 GALLUS, H.E., GROLLINS, H. and LAMBERTZ, J. The influence of blade number ratio and blade row spacing on axial-flow compressor stator blade dynamic load and stage sound pressure level. Trans. ASME, J. Eng. Power 104,3 pg. 633 (1982)
8.122 LIENHART, W. Berechnung der instationären Stromung durch gegeneinander bewegte Schaufelgitter und der instationären Schaufelkrafte. VDI-Berichte 193 (1973)
8.123 HENDERSON, R.E. The unsteady response of an axial flow turbomachine to an upstream disturbance. Ph. D. thesis, Cambridge Univ. (1972)
8.124 SAREN, V.E. and YUDIN, V.A. Calculation of hydrodynamic interaction of the arbitrary profile cascades in the flow of ideal liquid. Proc. 6th Conf. on Fluid Machinery, Budapest (1979)
8.125 DRING, R.P., JOSLYN, H.D., HARDIN, L.W. and WAGNER, J.H. Turbine rotor-stator interaction. Trans. ASME, J. Eng. Power 104, 4 pg. 729 (1982)
8.126 FLEETER, S., JAY, R.L. and BENNETT, W.A. Wake induced time-variant aerodynamics including rotor-stator axial spacing effects. ASME Symp. on Unsteady Aerodynamics. (1978)
8.127 SATYANARAYANA, B. and DAVIS, S. Experimental studies of unsteady trailing edge conditions. AIAA J., 16 pg. 125 (1978)
8.128 SEARS, W.R. Rotating stall in axial flow compressors. ZAMP 35,6 (1955)
8.129 HORLOCK, J.H. Fluctuating lift forces on aerofoils moving through transverse and chordwise gusts. Trans. ASME, J. Basic Eng. 90, 4 pg. 494 (1968)
8.130 VON KARMAN, TH. and SEARS, W.R. Airfoil theory for non-uniform motion. J. Aero. Sci., 5, 10 (1938)
8.131 BIRNBAUM, W. Das ebene Problem das schlagenden Flügels. ZAMM, 4, 277 (1924)

8.132 NEKRASOV, A.I. Collected works II. USAF FTD-MT-64-386 (1966)
8.133 KEMP, N.H. and SEARS, W.R. Aerodynamic interference between moving blade rows. J. Aero. Sci., 20,9 pg. 585 (1953)
8.134 NAUMANN, H. and YEH, H. Lift and pressure fluctuations of a cambered airfoil under periodic gusts and applications in turbomachinery. Trans. ASME, J. Eng. Power 95, 1 pg. 1 (1973)
8.135 GOLDSTEIN, M.E. and ATASSI, H. A complete second-order theory for the unsteady flow about an airfoil due to a periodic gust. J.F.M., 74, 4 pg. 741 (1976)
8.136 HENDERSON, R.E. and DANESHYAR, H. Theoretical analysis of fluctuating lift on the rotor of an axial turbomachine. ARC R&M 3684 (1972)
8.137 SCHORR, B. and REDDY, K.C. Inviscid flow through cascades in oscillatory and distorted flow. AIAA J, 9, 11 pg. 2043 (1971)
8.138 HAWTHORNE, W.R. Flow through moving cascades of lifting lines with fluctuating lift. J. Mech. Eng. Sci. 15, 1 pg. 1 (1973)
8.139 RAO, B.M. and JONES, W.P. Unsteady airloads on a cascade of staggered blades in subsonic flow. AGARD CP 177 (1976)
8.140 KEMP, N.H. and OHASHI, H. Forces on unstaggered airfoil cascades in unsteady in-phase motion. Trans. ASME, J. Fluids Eng. (1976)
8.141 TSUJIMOTO, Y., MURATA, S., MIYAKE, Y. and YAMAMOTO, F. An analysis of viscous effects on unsteady forces on cascade blades. Bull. JSME; 22, 165 pg. 326 (1979)
8.142 FUJII, S., NISHIWAKI, H., WATANABE, M. and KOBAYASHI, H. Some considerations on pure tone noise generation in an axial-flow fan. Proc. ASME-JSME Gas Turb. Cong., Tokyo. pg. 428 (1977)
8.143 SHEN, I-CHUNG. Unsteady pressure distributions on airfoils in cascade. Penn. State Univ. TM 80-45 (1980)
8.144 GIESING, J.P. Non-linear two-dimensional unsteady potential flow with lift. J. Aircraft, 5, 2 (1968)
8.145 ADAMCZYK, J.J. The passage of a distorted velocity field through a cascade of airfoils. AGARD CP-177 Paper 31 (1976)
8.146 COMMERFORD, G.L. and CARTA, F.O. Unsteady aerodynamic response of a two-dimensional airfoil at high reduced frequency. AIAA J. 12,1 pg. 43 (1974)
8.147 OSTDIEK, F.R. A cascade in unsteady flow. AGARD CP 177 Paper 26 (1976)
8.148 HORLOCK, J.H. An unsteady flow wind tunnel. Aero. Quart. 25 pg. 81 (1974)
8.149 HOLMES, D.H. Lift fluctuations on aerofoils in transverse and streamwise gusts. Ph. D. thesis, Cambridge Univ. (1972)
8.150 SATYANARAYANA, B. Some aspects of unsteady flow past airfoils and cascades. AGARD CP 177 Paper 25 (1976)
8.151 KEMP, N.H. On the lift and circulation of airfoils in some unsteady flow problems. J. Aero. Sci. 19, pg. 713 (1952)
8.152 SATYANARAYANA, B., GOSTELOW, J.P. and HENDERSON, R.E. A comparison between experimental and theoretical fluctuating lift on cascades at low frequency parameters. ASME Paper 74-GT-78 (1974)
8.153 KERREBROCK, J.L., EPSTEIN, A.H. and THOMPKINS, W.T.Jr. A miniature high frequency sphere probe. ASME Symp. Measurement Methods in Rotating Components of Turbomachinery. pg. 91 (1980)
8.154 EVANS, R.L. Turbulence and unsteadiness measurements downstream of a moving blade row. Trans. ASME, J. Eng. Power 97, 1 pg. 131 (1975)
8.155 JOSLYN, H.D., DRING, R.P. and SHARMA, O.P. Unsteady three-dimensional turbine aerodynamics. ASME Paper 82-GT-161 (1982)
8.156 JONES, T.V. Heat transfer, skin friction, total temperature and concentration measurements. In "Measurement of Unsteady Fluid Dynamic Phenomena" B.E. Richards ed. McGraw-Hill (1977)
8.157 WEYER, H.B. Optical methods of flow measurement and visualisation in rotors. ASME Symp. Measurement Methods in Rotating Components of Turbomachinery. pg. 99 (1980)

CHAPTER 9

Special Applications

Each turbomachinery requirement is different, having its own individual set of operating conditions, emphases and constraints, geometrical or otherwise. In this chapter we shall be looking at some special kinds of blading which result when one requirement or another is of prime interest.

Initially, boundary layer control will be considered. This term is used in a similar manner to its use for isolated aerofoil work but the use of tandem cascades will particularly be covered. An especially important technology for aircraft gas turbines is that of blade cooling and the various techniques will be reviewed with emphasis on the generation of information for design and on the effect of coolant air discharge upon blade row efficiency. The third selected topic is the use of blading having sweep and dihedral which may result from a particular three-dimensional flow field and, in some cases, offers considerable scope for improved efficiency.

9.1 Boundary Layer and Circulation Control

A considerable body of experimental data exists from numerous attempts to improve the performance of isolated aerofoils by methods which may loosely be termed 'boundary layer control'. At the same time a small number of turbomachinery aerodynamicists were applying similar techniques to compressor and turbine blading.

Two objectives of boundary layer control are as follows:-

(i) The attempt to reduce losses and heat transfer by delaying transition from a laminar to a turbulent layer.

(ii) The attempt to obtain high loading and wide stall margin by delaying separation.

The logical outcome of the latter course is circulation control which is defined by Williams [9.1] as the "artificial increase of circulation and lift over that which could be expected for a given incidence and camber in unseparated flow".

Many methods, including various forms of blown flap circulation control schemes, depend upon the suction or ejection of gas by auxiliary means. Turbomachines, having many blades, are not ideal candidates for auxiliary bleed or blow, especially when cycle efficiency is properly considered. However, such applications as turbine blade cooling do use auxiliary air indirectly for boundary layer control.

Fortunately, the employment of auxiliary air is not an essential feature of boundary layer control. In the tandem cascade method of delaying separation no external source of auxiliary air is required.

9.1.1 Tandem cascades. The designer of axial-flow compressors or turbines will usually attempt to minimise the number of stages by increasing the loading of each stage. One method of increasing the flow deflection without increasing the losses is to distribute the high deflection demanded of one blade row over two blade rows, one following the other. This is the simplest form of tandem cascade and is depicted in Fig. 9.1 (a). This kind of double blade row has been employed commercially, one instance being the last stage stator of the General Electric J-79 compressor.

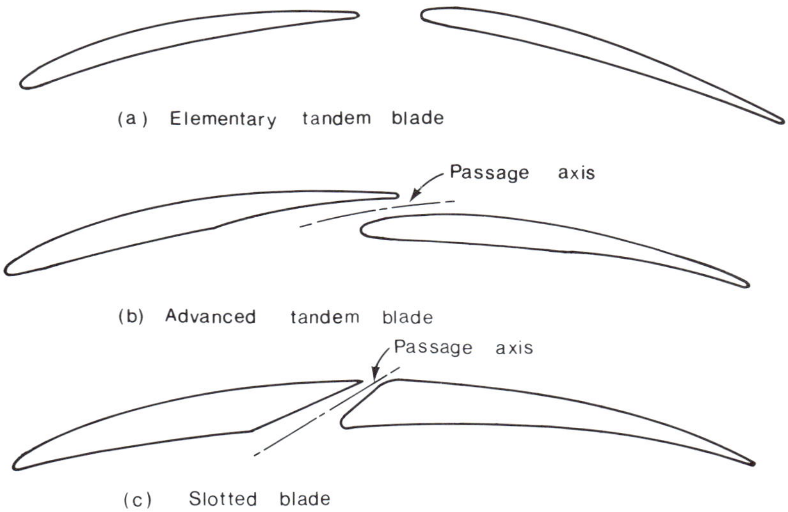

Fig. 9.1. Tandem and slotted blade types.

The purpose of tandem (or slotted) blading is usually to increase the turning or loading capability of a given blade row. This is achieved most effectively by using a nozzle-shaped slot to inject higher pressure air from the concave blade surface into the weakening boundary layer on the convex surface, upstream of the separation point. In this way separation is delayed and typically such cascades, although not possessing any advantages at the design point condition, do have a wider stall-free operating range.

Such blading dates back to German compressor work during the second world war [9.2] and has since been used in axial-flow compressors and subsonic fans. In work on a remarkable early Stuttgart rotor it was firmly established that the permissible loading for the tandem rotor was higher than that for one using a single blade row.

Sheets has presented data [9.3] on a slotted-blade blower that has a high aerodynamic loading and a high efficiency. The pressure rise achieved is higher than one would expect from experience with conventional blades although the rotor blade aspect ratio of 0.3 is somewhat unusual.

The Sheets-type blading will be referred to as tandem blading, since the forward and rear portions of each aerofoil closely resemble separate blades. In Fig. 9.1 (b & c) tandem and slotted blades are illustrated. With a properly-designed tandem arrangement the axis of the flow passage between the two aerofoils need not emerge at such a large inclination to the flow direction as when slots are cut into an existing conventional aerofoil. It is therefore expected that a properly designed tandem blade should be superior to a simply slotted blade.

Some tests [9.4] on a radial flow water pump impeller also confirm this expectation. Figure 9.2 depicts the seven configurations tested which include a basic unslotted impeller. For each configuration all vanes were treated similarly and in configurations E, F and G this involved circumferential displacement of the inner vanes with respect to the outer, thus creating a true tandem blade situation. The configurations involving simple slotting did not show any improvement in head-flow rate characteristic or efficiency but the tandem vane arrangements resulted in improvements in head rise and efficiency characteristics over the entire range of speeds and flow rates tested. The maximum improvement in flow rate, for a given head, was 7% (configuration G) and the maximum improvement in efficiency was $2\frac{1}{2}$% (configuration E).

Ref. 9.5 summarises some experimental work on tandem turbine cascades. The conclusion is that the tandem turbine blade can cope with more adverse flow conditions than an equivalent single blade whilst still producing comparable efficiency.

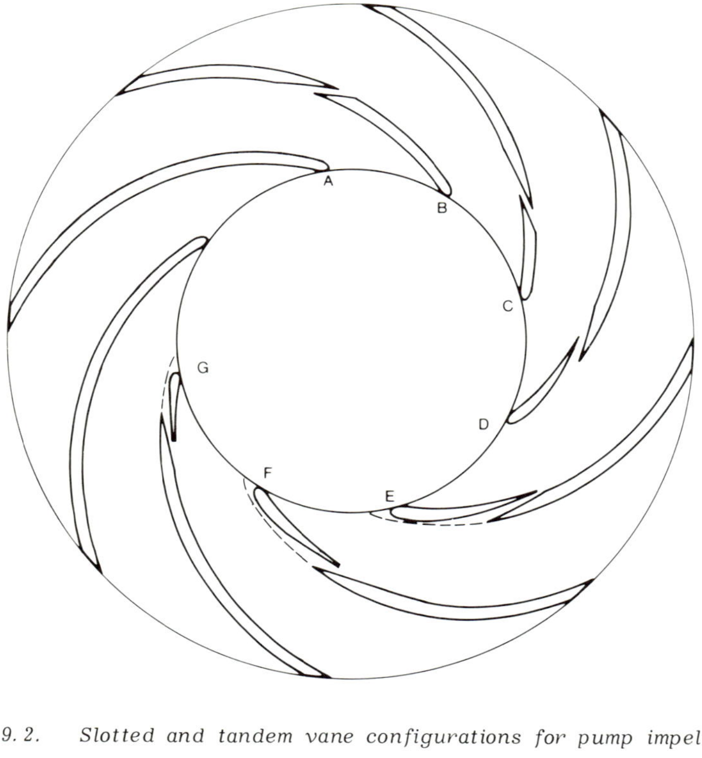

Fig. 9.2. Slotted and tandem vane configurations for pump impeller.

9.1.2 Effect of relative spacings for tandem cascade. Probably the most complete experimental survey of the effect of different circumferential and axial offsets of the second row has been given by Linnemann [9.6].

Linnemann designed a single stage blower having high deflection tandem rotor and stator, as illustrated in Fig. 9.3. The circumferential offset h^*/s could be incrementally varied, as could the axial offset a^*/c. The blades were of NACA 65-009 profile on a circular-arc camber line and the upstream and downstream blades of each tandem row had equal chord.

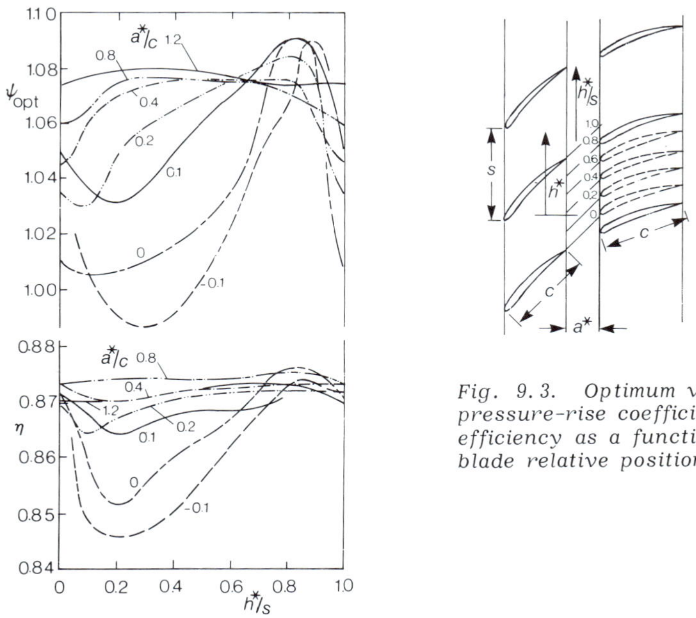

Fig. 9.3. Optimum values of pressure-rise coefficient and efficiency as a function of tandem blade relative position.

In Fig. 9.3 are plotted Linnemann's results for the pressure rise coefficient and efficiency of the rotor blading taken at the best efficiency point. The rotor results are presented for the various degrees of offset h^*/s and a^*/c; the stator results were quite similar. Linnemann noted that no substantial difference could be discerned between the performance of rotating and of fixed tandem cascades. He concluded that centrifugal effects of the boundary layer had no noticeable effect on optimum slot width or blade offset.

A comprehensive range of configurations was investigated including the elementary tandem blade of Fig. 9.1 (a) and the more sophisticated blade of Fig. 9.1 (b). It is quite clear from Fig. 9.3 that the optimum condition for circumferential spacing is with h^*/s in the region of 0.8 to 0.85. For axial spacing a^*/c should be zero or, better still, negative, giving a slight overlap of blades.

Several others have arrived at similar conclusions. Ohashi [9.7] and independently Fickert [9.8] found an optimum of h^*/s around 0.8 from wind tunnel tests of rectilinear cascades as did Marcinowski [9.9], Pal [9.10] and Railly and El-Sarha [9.11].

Railly et al[9.12] found that the effect of the wake from the upstream blade could be to constrain the practical realisation of the high suction peak on the second blade predicted by potential flow theory. Ihlenfeld[9.13] gave a criterion for optimal displacement in which the wake of the first blade grows together with the suction surface boundary layer at the trailing edge of the second blade. This implies that there should be only one peak of velocity decrement in the wake behind the cascade. Ihlenfeld also claimed that both rows should have the same loading.

Bammert and Staude[9.14] have produced calculations and test data giving optimum values of h^*/s in the region of 0.7 to 0.8. The results were incorporated into the design of a multistage axial flow compressor having a high loading, reducing considerably the number of blades required. A 100% reaction design was chosen to fully exploit the potential of tandem rotor blading. The subsequent test performance of the compressor was reasonable.

The performance of tandem blade compressors designed and tested by Linnemann, Bammert and others clearly establishes the domain of tandem blade operation as those applications for which good performance under high loading conditions is of paramount importance.

9.1.3 Theoretical work on tandem cascades. The earliest rigorous treatment was the transformation theory of Spraglin[9.15]. Spraglin's analysis included solutions to both the direct and inverse problems. The tandem cascade was mapped onto the region between two concentric circles containing suitably-located singularities. The main difficulties were in the process of transformation into a circle.

Yipp and Railly[9.16] gave further attention to the circle transformation along the lines of the Howell method[5.36]. The tandem cascade was separated into two parts, the first row being transformed to a circle whilst the second was represented by two singularities. The main virtue of the method was in the description of a new numerical treatment of the final Theodorsen transformation linking the approximate and exact circles. In cases where the second blade is small and of low loading the Yipp and Railly method should be satisfactory. However, as indicated in the previous section, most tandem cascades are likely to have rear and forward rows with approximately equal loading. The approach of ref. 9.16 would not, therefore, be valid for most tandem cascades.

Earlier theories for tandem cascades were produced by Blaho[9.17] (for thin, low-cambered blades) and by Ohashi[9.7]. Ohashi conducted extensive investigations on the mutual interference of the two rows and optimised the configuration both experimentally and theoretically. The theory was a variation of the Schlichting[3.25] method taking account of the induced velocity from the other blade row by superposition.

More recently the Martensen[3.27] method has been used successfully for tandem cascades by Pal[9.10], Wilkinson[5.75] and by the engine companies. Katsanis and McNally[9.18] have applied a finite difference technique to tandem cascades. Calculations of pressure distribution based on these methods can give good agreement with experiment provided the flow remains attached. Boundary layer calculations may then follow for prediction of blade losses; these should encompass interaction effects between the wake of the first blade and the boundary layer of the second blade.

If a simple design method for tandem cascades, suitable for incorporation into an interactive computer-aided-design procedure, is required the approach of Railly and El-Sarha[9.11] based on the Ackeret method[5.77] has much to commend it. As a zeroth order approximation two circular arc camber lines are determined. A higher approximation is made enabling first order velocities to be calculated from known vorticity and source distributions. The second approximation results are adequate for preliminary design purposes.

9.1.4 Slotted blades. We have previously observed that the performance of a properly designed tandem cascade should be superior to that of a slotted cascade. Nevertheless manufacturing considerations alone may suffice to eliminate the possibility of an optimised tandem cascade. Fickert[9.8], for example, investigated six types of compressor blades, most of which were formed from sheet metal. The slot position and extent of overlap were varied. In general the slotting of a blade resulted in increased efficiency. If the designer is constrained to put slots into an existing blade shape in order to widen the operating range there are obviously good and bad ways in which to do this.

Pratt and Whitney have undertaken substantial experimental and theoretical work on slotted blading. In a series of tests on slotted rotor and stator blading Linder and Jones[9.19] and Keenan and Bartok[9.20] attempted to substantially increase blade loading. All blading was of a basic NACA 65 series profile and the inlet flow had a high subsonic velocity. Unfortunately some aspects of the test arrangement were unsatisfactory (e.g. periodicity was poor because not all stator vanes were slotted; for mechanical reasons the slot did not extend over the entire span). However, the tests did reveal the importance of having the slot discharge into an unseparated boundary layer. The optimum slot position was found to be the 55% chord location. It was found that if careful attention was given to the slot shape then the Coanda effect could be utilized to some advantage. Mikolajczak et al[9.21] have compared theoretically derived pressure distributions (using a method similar to that of ref. 5.60) with those from the annular cascade and from their own rectilinear slotted cascade tests. The outcome was a measure of agreement when stall was absent but careful attention must be paid to viscous effects within the slot which can constrain slot effectiveness if the throat is small.

Use of tandem or slotted cascades to give variable camber and hence a wide flow deflection range, has been proposed by both major American engine companies. In two of these schemes the forward portion is flexible whereas in other schemes the forward section stays fixed and the rear portion rotates and may also translate. Possible uses include thrust vectoring and variable deflection inlet guide vanes. Okiishi et al[9.22] have tested such vanes in an annular cascade and demonstrated a wide operating range with some loss and turning penalties compared with an equivalent continuously cambered vane. Nevertheless the simplicity and versatility of such an arrangement offers a potentially valuable means of control over compressor performance characteristics.

9.1.5 Tandem cascades for high-speed flows. The advantages of tandem blading become somewhat clearer as higher Mach Numbers are encountered. The Pratt and Whitney slotted blade tests were conducted at high subsonic Mach Numbers. Sieverding [9.23] has also tested tandem blading in the high subsonic range using 6% thick double-circular-arc blading over a range of incidence. Application of back pressure was found to be necessary to give the highest efficiency. At an inlet Mach Number of 0.8 Sieverding realised 57.5^o of turning with a very high static pressure rise coefficient of 0.73 but also with a high loss coefficient of 0.115.

Supersonic compressor designs using an impulse rotor and a supersonic stator have been investigated in two German laboratories. The work at Aachen[9.24] has concentrated on stator inlet Mach Numbers up to two and that at DFVLR[9.25] on Mach Numbers above two. In all cases turning was from an inlet angle of between 30° and 50° back to axial, giving high blade loading.

Dettmering and Becker[9.24] describe testing on the two types of tandem cascade blading shown in Fig. 9.4. The design philosophy in the blading designated 'Type A' is that initially the flow is deflected in a supersonic potential vortex. The area distribution is such that rapid expansion could occur in the axial portion of the first row. This is prevented by insertion of the second row of blades forward into the first passage. Deceleration to subsonic velocities then takes place within the axial portion of the downstream row. In addition to preventing acceleration of the supersonic flow the second row has a stabilizing influence on the wake and boundary layer of the upstream row. It also gives the boundary layer of the decelerating portion a fresh start.

An alternative configuration in which the supersonic diffusion occurs in the first row and the subsonic turning in the second row was also tested at Aachen and designated 'Type B'. Such a configuration was also adopted by Fuchs and Starken[4.3]. In both sets of investigations a comparison was made with the performance of a single row stator cascade.

Total pressure ratio is plotted as a function of inlet Mach Number in Fig. 9.4 in which the normal shock loss curve is given as a reference. The Aachen results all have losses greater than that of a normal shock whereas losses recorded at DFVLR are below the normal shock levels. The disparity is especially marked for the single row cascades. It is therefore most likely to be a function of the test conditions rather than of any difference in tandem blade design philosophy. The aspect ratio for the Aachen testing was low and this may have adversely affected performance.

In all cases the performance of the tandem blading is superior to that of the single row. These differences become even stronger if static pressure rise and turning are considered. The Aachen 'Type A' blading performed consistently better than the 'Type B' blading. As the Mach Number rises through the supersonic range the losses understandably become very high. Nevertheless it is considered that the results demonstrate the feasibility of application of tandem blading to supersonic compressors for special applications.

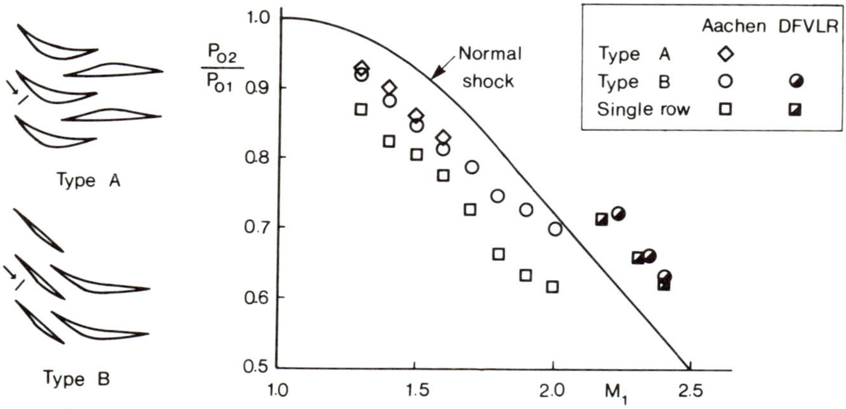

Fig. 9.4. Tandem stator configurations and total pressure ratios.

9.1.6 <u>Blown cascades and jet flaps.</u> Air of relatively high pressure is often ejected through slits or orifices in the surfaces of turbine blades for cooling purposes. Since its primary purpose is the improvement of blade cooling rather than aerodynamic performance a separate section (9.2) will be devoted to that topic.

Attempts have been made to improve the performance of highly loaded compressor blading by the judicious introduction of auxiliary air through a slit. Kiock[9.25] has given an example of the use of blowing to increase the deflection capability of a stator by up to 5°. The improvements gained would be a definite advantage to early compressor stages. High pressure air would be readily available and the question of whether the benefits would exceed the penalties becomes one of cycle considerations.

Whilst the tandem cascade concept extended the range of stall-free operation of more conventional blading, the jet flap principle gives the possibility of a dramatic increase in attainable blade lift. The jet flap consists of a stream of air blown from the trailing edge of the aerofoil, usually at an angle to the chord. Its effect is to increase circulation and to postpone stall by means of jet entrainment.

The jet flap, with its application to isolated aerofoils, appears to have been conceived independently in France, Germany and the United States in the nineteen-thirties. Its history and some basic data for jet flaps have been given by Davidson [9.26].

Use of the jet-flap principle for delaying stall in cascades of blades was suggested by Marble [9.27] and Roy [9.28]. Clark and Ordway [9.29] showed experimentally that the stalling incidence for compressor cascades could be increased by several degrees using the jet flap principle. This result was confirmed by Brocher [9.30] using an extension of Scholz's potential flow theory. Experimental work has also been performed by Paulon [9.31], Salita [9.32], Lueders and Roelke [9.5] and by Landsberg and Krasnoff [9.33]. Although the low speed cascade tests of Landsberg and Krasnoff exhibited poor periodicity and only used four flow passages, some useful indications were obtained. In particular it was shown that tangential blowing (Fig. 9.5) approximately doubled the turning effectiveness of the cascade - as compared with the more conventional normal blowing.

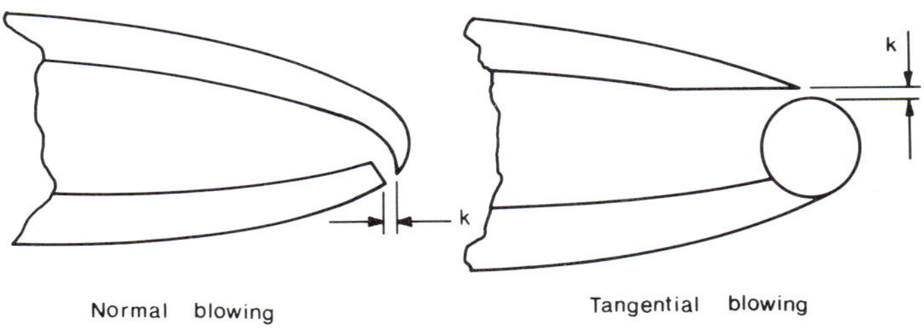

Fig. 9. 5. Jet flap aerofoil configurations.

An alternative concept, the blown flap, has been described by Stratford [9.34]. Stratford performed some experiments on this configuration in which the jet is blown from the upper surface of the aerofoil over a mechanical flap.

The jet flap is excessively sensitive to variations in jet discharge angle, effectively ruling out the use of this variable. With a blown flap control of this angle is facilitated, the jet following the hinged mechanical flap up to angles of 45° or more, depending on the momentum discharge coefficient.

The performance of an unstaggered, uncambered cascade of inlet guide vanes fitted with blown flaps has been investigated, both theoretically and experimentally, by Kruger et al [9.35]. A range of discharge coefficients and flap angles was examined and two different flaps (of length 10% and 20% of the basic chord) were tested. The tests indicated that, for a given jet coefficient, the range of flap angles for an unstalled cascade is smaller with the shorter flaps. Results with the short flap were nevertheless quite good.

Stark [9.36] has attempted to model the tests of ref. 9.35 with a singularity-based theory. Although the Stark theory ignored the presence of the flap it succeeded in representing the experimental results very closely.

9.2 Turbine Blade Cooling

Cycle studies show that turbine inlet temperature is probably the single most influential parameter for turbojet engines [9.37]. Each degree of temperature won can give over 500N of additional thrust. Turbine inlet temperature is also important for turbofans [9.38]. Means of increasing this temperature are found partially through improvements in materials technology but to a greater extent through improved aerodynamics.

Blade cooling technology has developed rapidly, being for many engines the most convenient variable for producing the inevitable 'growth version'. Figure 9.6 shows how turbine inlet temperature has risen over the years and an attempt is made to project its rise for the remainder of this century. The graph is merely indicative and factors such as rating condition, blade life and coolant flow rate should also be considered.

An eventual ceiling with existing fuels is the stoichiometric limit which for high pressure ratios is around 2500K. Combuster surfaces also need cooling and the mixed coolant reduces the corresponding turbine inlet temperature to about 2300K. Turbine inlet temperature is plotted as a sigmoid curve asymptotic to that limit. Demonstrator engines are already operating with turbine inlet temperatures over 1800K for short periods indicating the technological possibilities. Temperatures for three competing

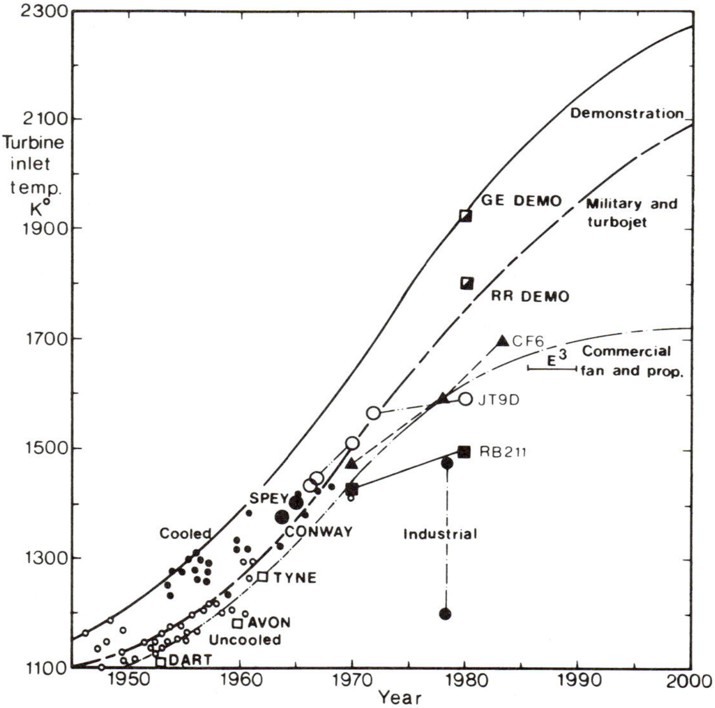

Fig. 9.6. Trends in turbine inlet temperature.

families of high by-pass engines are presented, reflecting in part the different techniques employed by the companies. The high by-pass engines may appear to be conservatively rated in comparison with some of the earlier Rolls-Royce engines, also identified. One reason for this is that whereas the optimisation process has led to high by-pass engines which operate near the maximum inlet temperature during the whole flight, in most of the earlier engines this maximum temperature condition was reached only during take-off. In a balanced thrust engine design philosophy appropriate attention would be given to both cruise and take-off conditions for the candidate aircraft.

Large coolant flows are costly in cycle efficiency and there are considerable incentives for coolant flow reduction in subsonic commercial aircraft. For this reason the trend for commercial, fan and new prop-fan engines is a levelling-off to a limit dictated by considerations of blade life and fuel economy.

These constraints are not as severe for military aircraft, which are often turbojets for supersonic application. The premium is on specific thrust rather than specific fuel consumption and turbine inlet temperatures for military engines are expected to rise at a much faster rate than those for commercial engines. Technologies for military and commercial engines have traditionally exhibited great commonality. This feature could be lost as designs become more specialised in application.

9.2.1 Blade cooling techniques. Of the available blade cooling technologies liquid cooling probably has the highest cooling potential and certainly appears to be the best candidate for very high temperature applications in stationary units. Unfortunately it is not convenient to carry large quantities of liquid coolant in long range flight and even for stationary or marine applications the associated mechanical difficulties are severe.

The development of liquid cooling technology for rotor blading has been reviewed by Bayley and Martin[9.39] with emphasis on the internal heat transfer problems. The use of steam injection for impingement and film cooling has been proposed by Jenkins and Rice[9.40]. General Electric have been working on the water cooling of stationary gas turbines for some time and the current emphasis is on a hybrid scheme for combined cycle application in which water cooling would be applied to the first stage turbine nozzle and air cooling to the remaining blade rows. The objective of water cooling in the first nozzle row is the minimisation of hot corrosion and ash deposition. Cooling is sufficient to reduce turbine inlet temperatures of the order of 1500K to values as low as 700K at the blade surface.

The principal means of cooling using high pressure air are impingement cooling, convection cooling, transpiration cooling, trailing edge bleed and discrete hole film cooling.

Most applications of impingement cooling have been on nozzle guide vanes. In impingement cooling the coolant airflow is channelled through a duct within the hollow blade. This air is then accelerated through small nozzles, emerging from the duct in a series of small jets which are directed against the internal face of the hottest regions of the blade, notably in the leading edge region. Higher local heat transfer coefficients are obtained in the stagnation point region of an impingement jet than could be obtained using normal convection cooling. The spent coolant is subsequently discharged through holes, to provide some film cooling, or through the trailing edge.

Convection cooling is the most widely used method of blade cooling. Here air, from the higher pressure compressor stages, passes radially through spanwise holes in the blade, removing heat from the blade interior and walls, and is discharged either at the blade tip or, where possible, along the trailing edge.

Turbine blades are formed in various alloys and by various manufacturing techniques. Blades for convection cooling may be either forged or cast. The holes in forged blades are formed by drilling (usually electrochemical). Investment casting technology presents a range of cooling passage options; serpentine passages are possible and ceramic cores may be used. Although suited to higher temperatures and complicated cooling geometries, cast blades have greatest difficulty in demonstrating adequate fatigue strength. Manufacturers have made great strides in improving this, the best examples being the use of blades grown from a single crystal, giving a considerable improvement in ductility, and the use of directionally-solidified eutectics.

Convection cooling generally results in turbine blades having round blading edges and thick trailing edges. Figure 9.7 from ref. 9.41 shows the cross section of a turbine blade with temperature contours superimposed; it is clear that the most difficult cooling conditions occur at the blade trailing edge.

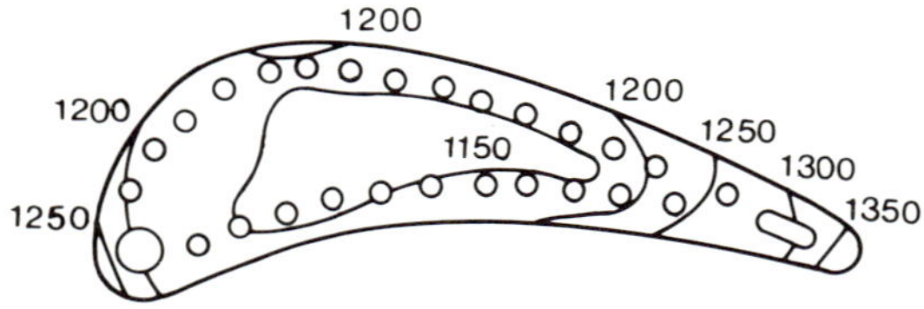

Fig. 9.7. Temperature contours on a convection-cooled blade.

At representative pressure levels, between twenty and forty atmospheres, the amount of air required for convection cooling at turbine inlet temperatures above 1600K rises dramatically. As compared with a temperature of 1400K the coolant air required for 1600K can rise by a factor of three and for 1800K by a factor of eight. Usually the passage sizes and available flows are strictly limited. Practical constraints therefore limit the sole application of convection cooling to around 1650K.

This is one contributing factor in the levelling off of turbofan blade cooling technology at about that temperature. Convection cooling is a mature technology and a breakthrough is required to a new technology if higher temperatures are to be realised.

Transpiration cooling had seemed to offer the promise of a technological breakthrough. In transpiration cooling the surface of the blade is entirely porous. Coolant air is effused through the fine pores giving the ideal situation in which the blade is insulated from the hot gases by a thin and relatively uniform blanket of coolant. The effect is twofold in that the coolant streams absorb heat upon passing through the pores in addition to the insulation effect. The main obstacles preventing implementation of transpiration cooling are the manufacturing and strength difficulties, oxidation and the clogging of the fine pores.

These difficulties are not the only problem. In some NASA tests on the aerodynamics of cooled blades the different behaviour of trailing edge slots and porous skins was investigated. The effects of coolant configuration on the aerodynamic efficiency of the blade row are shown in Fig. 9.8. The results showed that whereas with the slot the aerodynamic effects were confined to the wake, in the case of a porous skin the mainstream flow was substantially altered. The detrimental effects of discharge through pores were confirmed by turbine tests on a 750mm diameter stage[9.42].

The problems with transpiration cooling are therefore extensive. If they could be overcome there is the potential of a considerable increase in turbine inlet temperatures.

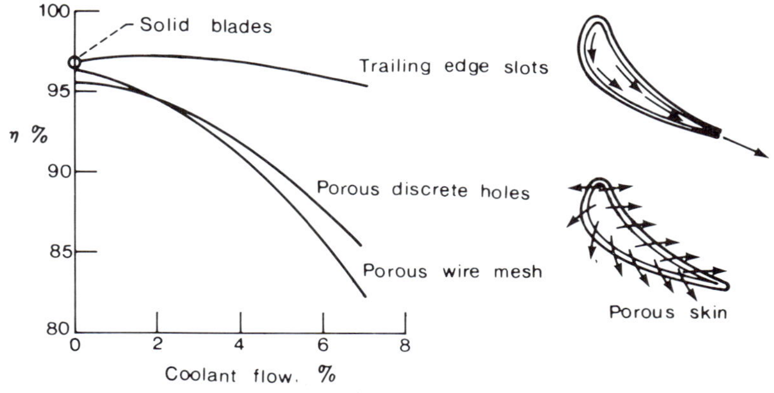

Fig. 9.8. Variation of nozzle efficiency with coolant flow for bleed through trailing edge or porous skin.

Although the porous discrete hole and wire mesh configurations of Fig. 9.8 had adverse consequences for efficiency this was not the case for base bleed through trailing edge slots. In fact it is quite usual for turbine blading having trailing edge bleed to display an efficiency improvement of between 1% and 3% at an optimum coolant flow of about 2% of primary flow [9.43]. This appears to be linked to a corresponding increase in base pressure [8.21, 8.22].

It has been noted that the trailing edges of turbine blades are particularly difficult to cool and for this reason the discharge of coolant air through a slot or row of holes in the trailing edge has been attractive to designers. Ideally this should be achieved without further thickening of the trailing edge which would tend to increase the losses (Fig. 8.1).

In tests on a turbine blade having a particularly thick trailing edge Michel and Kost [9.43] found that the mere presence of the slot, in the absence of any coolant flow, resulted in a remarkable efficiency improvement. This was attributed to the delayed formation of the vortex street and hence a weakening of the vortex strengths. Refs. 8.21 and 8.22 provide more information on the detailed wake flow structures which correspond to different bleed flow rates and the corresponding base pressures.

Detailed information is still required on heat transfer rates in the immediate vicinity of the trailing edge if trailing edge coolant discharge is to be most effectively used. It appears that high heat transfer rates are associated with periodic vortex shedding from the trailing edge and that bleed flows which optimise aerodynamic performance, by moving vortex street formation downstream to the region of confluence, should also result in a reduction in heat transfer in the trailing edge region. However this conjecture needs careful verification.

Whereas convection cooling represents a mature technology offering diminishing returns and transpiration cooling an aspirant technology which has failed to fulfil expectations, discrete hole film cooling is attracting justifiable attention.

In film cooling the coolant air is forced through an array of small holes arranged over the blade's surface. An early but common form of film cooling deployed only one row of holes on the pressure surface upstream of the trailing edge. The best prospect for a further increase in turbine inlet temperature is seen to be a combination of thermal barrier coatings with discrete hole film cooling. Laser drilling can produce 0.25mm holes of any desired non-circular shape.

Although film cooling has been employed for some time in operational engines the ejection of cooling air usually results in a reduction in turbine efficiency and any coolant air used additionally penalises the cycle efficiency. To minimise the coolant quantity and optimise its location it is necessary to predict the distribution of heat transfer coefficient over the surface. The present high level of research in this area reflects an inadequate fundamental grasp of film cooling processes.

Figure 9.7 was drawn for a convection-cooled blade but the nature of the blade temperature distribution is similar for most turbine blades. The heat transfer over turbine blade surfaces is high in the leading edge region, low in the regions having a laminar boundary layer (although it will be argued in the next section that this may not be relied upon for concave pressure surfaces) and high in the turbulent boundary layer regions. This distribution gives rise to non-uniformity of metal temperatures which leads to thermal stress problems. A principal objective of blade cooling is to minimise thermal stresses by the attainment of a uniform metal temperature. The degree to which this can be arranged by blade cooling technology directly determines the permissible temperature difference between the gas and the blade and hence the turbine inlet temperature.

9.2.2 Design for film cooling. Efficient use of cooling techniques presupposes an accurate knowledge of the temperature distribution within the blade. If the blade is solid, or has thick walls, solutions to the Laplace equation for the two-dimensional temperature distribution are helpful. It is also necessary to know the distribution of heat transfer coefficients over the blade surface. The main difficulty here is accurate prediction of the state of the boundary layers. The heat transfer coefficient is very different for laminar and turbulent boundary layers and it is therefore important to locate the transition region.

Designers generally find it more difficult to predict heat transfer coefficients over pressure surfaces than over suction surfaces. Even in the upstream regions measured heat transfer through the laminar layer is usually higher than predicted. Brown and Martin[9.44] have concluded that the current inability to predict heat transfer over the concave pressure surfaces of turbine blades is partly due to the interaction in the laminar layer between the Görtler vorticity and the mainstream turbulence but equally to inadequate understanding of the effect of these factors on the transition region. Litchfield and Norton[9.45], in theoretical and experimental studies on a transonic nozzle guide vane, have shown that a principal obstacle to accurate prediction of heat transfer coefficients is inadequacy of transition criteria especially on the concave pressure surface.

Detailed design data on heat transfer over turbine blades are sparse. Progress is best made by a combination of theoretical studies, using inviscid calculations and boundary layer theory, and cascade measurements in a high speed tunnel capable of correctly simulating temperature ratios. An example of this approach is given by Consigny and Richards[9.46] who varied Mach Number, Reynolds' Number, inlet flow angle and free-stream turbulence level.

In view of the high turbulence levels reported for turbines in Chapter 2 it seems important that these should be represented in heat transfer measurements on cascades. In their investigations of heat transfer to turbine blading Bayley and Priddy[9.47] have used a rotating squirrel cage to generate turbulence levels up to 32%.

There appears to be consensus in the literature on certain effects of free-stream turbulence. Heat transfer is increased at the leading edge stagnation point, in the laminar layers, and in the transition region, by increasing free-stream turbulence. Also high levels of heat transfer, of the order of leading edge levels, are observed in the trailing edge region. There is no such agreement on the effects of turbulence levels on the turbulent pressure surface boundary layers. Consigny and Richards [9.46] observed no change in heat transfer with changes in turbulence level whereas Krishnamoorthy[9.48] noted a strong change. The turbulence level ranges covered by the two sets of tests were quite different. Consigny and Richards varied the turbulence level over the range 0.8% to 5.2% by moving a grid in the streamwise direction. Krishnamoorthy used a novel blowing device to obtain turbulence levels between 1% and 21.3%. Krishnamoorthy's pressure surface turbulent boundary layers exhibited little variation in heat transfer for turbulence levels below 5% but a strong variation above that level.

The above uncertainties in prediction of heat transfer on concave pressure surfaces would exist regardless of the cooling mode adopted. In film cooling they are compounded by further effects which have been observed when cooling jets are introduced on convex, flat or concave surfaces.

In turbine cascade investigations of film-cooling effectiveness Ito et al[9.49] have examined the role of blade surface curvature. On the convex suction surface a near-tangential jet is pushed towards the blade by static pressure forces. For low coolant flow this results in a higher effectiveness than would pertain for a flat surface. However for high coolant flows the inertia of the jet carries it away from the blade resulting in lower effectiveness. On a concave pressure surface the effects of curvature are the reverse of those for the suction surface.

Results obtained subsequently by Dring et al [9.50] on a turbine rotor were entirely consistent with those of ref. 9.49 and with earlier work at Cambridge University. In addition Dring et al found that the pressure surface cooling jet migrated radially outwards whereas that on the suction surface did not.

The combined results of the original poor predictions of heat transfer coefficient, due mainly to the effects of Görtler vorticity and of high free-stream turbulence levels, and the lower effectiveness and possible radial migrations of cooling jets on concave surfaces, help to explain why overheating on the pressure surface of turbine blades has presented a continuing challenge in engine development.

A helpful review of film cooling research has been given by Hay[9.51]. Most progress is coming through improved experimental facilities, which are of three kinds:-

1. Low speed and low temperature facilities for fundamental work on parameters such as hole geometry and wall curvature [9.49]. Most results are presented in the form of contours or averaged plots of adiabatic wall effectiveness.
2. Work conducted on intermittent facilities such as those at M.I.T., Oxford [9.45] and VKIFD [9.46] which, in addition to Reynolds' Number and Mach Number simulation, also permit temperature ratios to be modelled.
3. Work conducted on turbine rigs. This includes industrial work under representative conditions of speed and inlet temperature as well as the low speed turbine tests of Dring et al [9.50].

Two parameters are in use for presenting film cooling information. For the studies of type 1 above and the rig testing of ref.9.50 the "adiabatic film cooling effectiveness" is used

$$\eta_{ad} = (T_g - T_{aw})/(T_g - T_c) \quad . \tag{9.1}$$

In practice η_{ad} may vary from zero (no effective film cooling) to unity (blade cooled to temperature of coolant).

An alternative parameter, which is particularly suited to use in short duration facilities such as isentropic light piston tunnels, is the "isothermal effectiveness".

$$\eta_{iso} = 1 - (q_f/q_0) \quad . \qquad (9.2)$$

Heat transfer film coefficients are defined as

$$h_f \equiv q_f/(T_{aw} - T_w)$$

and

$$h_0 \equiv q_0/(T_g - T_w)$$

so that

$$\eta_{iso} = 1 - \frac{h_f}{h_0}\left(\frac{T_{aw} - T_w}{T_g - T_w}\right) . \qquad (9.3)$$

Combining (9.1) and (9.3) the adiabatic film cooling effectiveness and the isothermal effectiveness are related by

$$\eta_{iso} = 1 - \frac{h_f}{h_0}\left(1 - \eta_{ad} \cdot \frac{T_g - T_c}{T_g - T_w}\right) . \qquad (9.4)$$

A typical 'footprint' for the variation of η_{iso} downstream of a film cooling hole is given in Fig. 9.9 taken from ref.9.52. For the example shown the product of coolant density and velocity is optimum, at 40% of the free stream value, and the free stream Mach Number is 0.5. It is clear from Fig. 9.9 that the region of effective cooling is very limited in the lateral direction and not extensive in the streamwise direction. At a sufficient distance downstream of the hole (greater than twenty hole diameters) the heat transfer coefficient is hardly affected by the presence of the film. Under these conditions $h_f \simeq h_0$ and equation (9.4) is simplified. Close to the hole the heat transfer coefficient h_f may be raised by as much as 50% by the presence of the film.

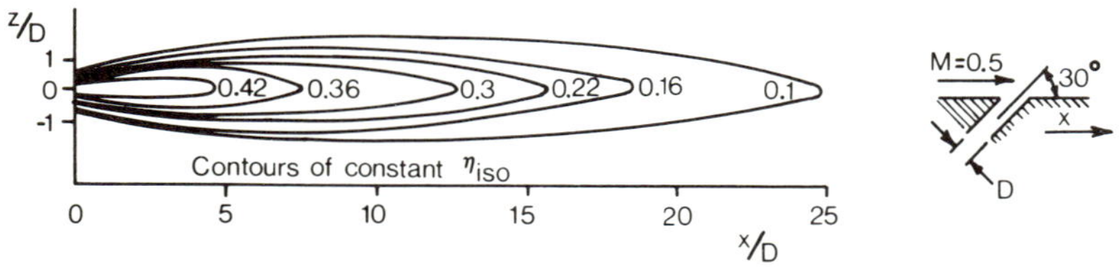

Fig. 9.9. Isothermal effectiveness contours downstream of a cooling hole.

For design prediction of the temperature distribution η_{ad} alone is insufficient and must be supplemented by a local knowledge of h_f/h_0. These two effects are combined in the non-unique parameter η_{iso} which therefore masks their independent variation.

When more than one row of cooling holes is present the designer needs to predict the effect of the upstream row on the film cooling performance of the downstream row. The superposition theory of Sellers[9.53] postulates linear addition of temperatures. The film temperature following an upstream row becomes the free stream temperature for the downstream row. Loftus and Jones[9.54] and other workers have found that such linear superposition works well for conditions representative of gas turbine engines.

9.3 Blading Having Sweep and Dihedral

Rigorous definitions, and a theoretical treatment, of sweep and dihedral in the blading of a turbomachine have been given by Smith and Yeh [1.7]. Blades are said to have sweep when the flow direction is not perpendicular to the spanwise direction and dihedral when the blade surface is not perpendicular to the end wall.

Figure 9.10 based on ref. 9.55 illustrates these definitions, showing their relationship with the equivalent definition for an aircraft wing and practical application for Francis and axial turbines.

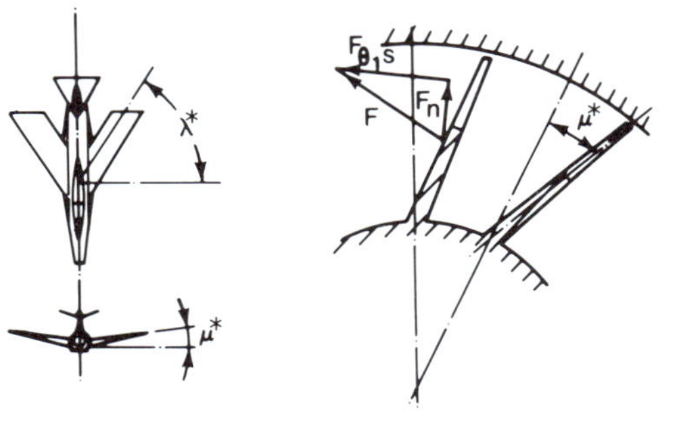

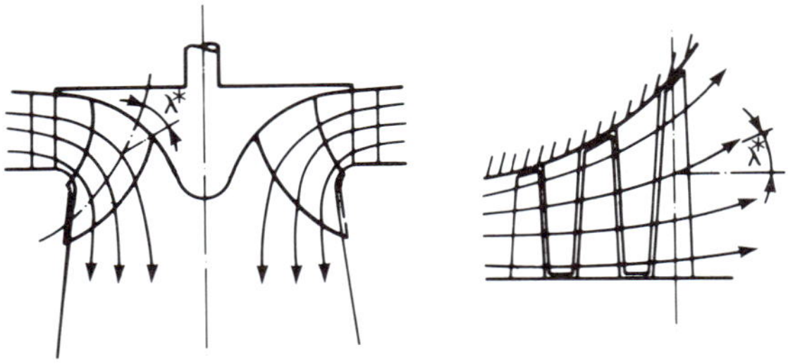

Fig. 9.10. Sweep and dihedral of lifting surfaces.

The entrance regions of axial compressors and nozzle diaphragms of steam turbines often have large slope with cone half angles up to 45°. Thus sweep may be present even when the blade is itself radial. Since the blades would generally be staggered, dihedral would also be present at the wall. In practice the dihedral definition is generalised to include as 'walls' the axisymmetric stream surfaces of the meridional flow, thus enabling the term to be used in the free stream as well as at the walls.

9.3.1 Analysis and design procedures. An axisymmetric procedure, probably of the streamline curvature [1.6] or matrix-through-flow [1.10] types, will usually form the basis of the machine design. Since properties, such as stagnation enthalpy in a stator, are constant along an axisymmetric stream surface but have different values on different surfaces, it is in the designer's interest to consider the flow on such an axisymmetric stream surface when following the flow through a blade row. In determining losses, deviation angles or generally in shaping the blade the designer will therefore wish to apply cascade aerodynamics to the blade section which is cut by the axisymmetric stream surface.

For conventional blading this appears to be relatively straightforward but when sweep and dihedral are present the question of how to view this section arises. There are various possibilities; the sections could be viewed along a line normal to themselves, or a radial line, or some other.

However, Smith and Yeh have shown that the only correct way to view this section is by looking along the blade axis. In this way a truly two-dimensional motion is described whilst preserving the convenience of the axisymmetric stream surface.

The cascade data are used in conjunction with the blade geometry and flow angles of this spanwise projection. If the axisymmetric stream surfaces have varying thickness it is the component of this in the spanwise direction which is used in, say, an axial velocity ratio correction.

To illustrate the importance of using the correct projection Lewis and Hill[9.55] have calculated the outlet angle variations due to sweep for a series of turbine cascades of infinite aspect ratio. Four cases have been considered, ranging from a nozzle with zero inlet angle to an impulse cascade. All four cascades are designed to give equal lift coefficient for zero sweep.

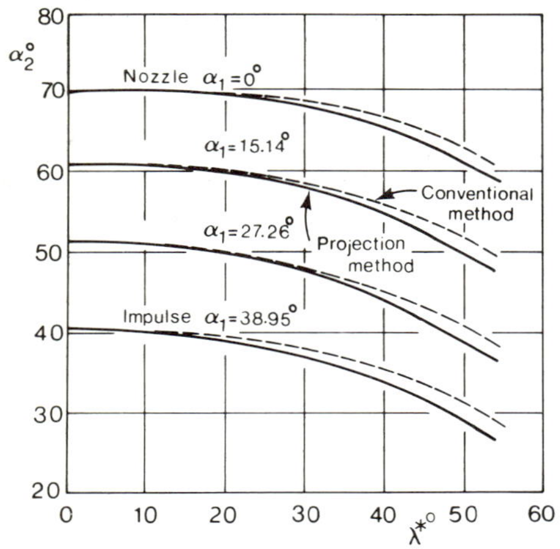

Fig. 9.11. Outlet angle variations due to sweep in turbine cascade.

The results are given in Fig. 9.11 in which "conventional method" signifies a projection of the blade section, intersected by the axisymmetric stream surface, normal to itself. "Projection method" refers to the projection of the intersected surface along the blade axis.

The conventional method is seen to predict outlet angles up to 2.5° smaller than the projection method, although admittedly this is for the large sweep angle of 50°. For a turbine nozzle blade, however, this is a serious discrepancy, giving an error of seven percent in swirl velocity for a given meridional velocity.

The bound vorticity of a swept blade may be resolved into components along, and normal to, the intersection of the axisymmetric stream surface. The normal component is directly responsible for the work done but the tangential component also has significance, causing the pressure surface streamline to move radially outwards (for swept back blade) and the suction surface streamline to move radially inwards.

For a blade of infinite span this is not significant but if the blade is bounded by end walls (hub and casing) then the constraint of these walls is such that the tangential component of vorticity causing these shifts is reduced to zero at the wall. This causes a distortion in blade pressure distributions near the wall.

The solution to this problem involves representing the blade by a lifting surface of kinked vortex lines. The wall is regarded as the plane of symmetry of these kinked lines - the so-called image technique.

The problem is posed as the determination of downwash perturbation velocity, $\underline{DW}$, over the blade surface in order for the blade pressure distributions in the wall region to be identical with those for an infinite swept cascade. These downwash velocities can obviously be used directly to modify camber and stagger. They are induced over the surface of the reference blade by its own vorticity and by the vorticities of all other blades and image blades in the cascade.

Smith and Yeh produced a thin aerofoil theory to calculate these downwash velocities, giving some numerical examples. For the examples given, and for an assumed elliptic variation of chordwise loading, it was found that a given amount of sweep had an order of magnitude greater effect than did the equivalent amount of dihedral. The sweep perturbation is strongest in the mid-chord region whereas the smaller perturbations due to dihedral are felt at the leading and trailing edges.

The downwash calculated in this way becomes infinite at the wall itself but Smith and Yeh have shown that the error involved in using an extrapolation technique, to cover the region within five percent chord of the wall, is small. Figure 9.12, taken from Smith and Yeh, shows that sweep has a powerful influence, not only on the value of downwash perturbation at the wall but also on penetration of this along the span.

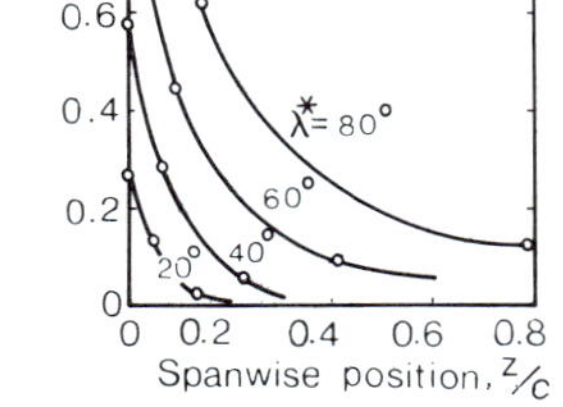

Fig. 9.12. Decay of calculated downwash perturbation due to wall effects with increasing distance from the wall.

The highly cambered nozzle blading of low pressure steam turbines often has a sweep angle of at least 45° at the casing. Whilst the image technique for end-wall influence may be used a more sophisticated calculation procedure than thin aerofoil theory is required. Graham and Lewis[9.56] have extended the Martensen method to calculate the three-dimensional flow through swept nozzle blading using alternative assumptions of constant and varying bound vorticity along the span. After checking the calculations against experimental results for 40° sweep, theoretical predictions of sweep effects using the variable spanwise loading model were made and showed a dramatic effect of sweep on pressure distribution for sweep angles greater than about 35°. As the sweep angle increases the throat of the cascade moves towards the leading edge causing a considerable increase in loading over the forward portion of the blade.

The conclusion of ref. 9.56 is that selection of turbine blade profiles by the traditional practice of normal projection to the meridional streamlines leads to suction peaks well upstream of the location traditionally considered to be the throat. This results in severe adverse pressure gradients on the latter portion of the blade surface. The real throat is the minimum gap observed by projecting the meridional streamline section along the blade axis.

To summarise, the important constituents of a procedure to correctly design for sweep and dihedral effects are threefold:-

1. An accurate axisymmetric solution, as described in Chapter 1. Many of the important body force and blockage effects of the swept blading will be accounted for at this stage.
2. Application of cascade data in the appropriate blade projection i.e. the view along the loading axis of the cross section generated by the intersection of blade and axisymmetric stream surface.
3. Application of a suitable correction for wall effects giving modified blade angles in the wall regions.

9.3.2 <u>Use to reduce effective Mach Number.</u> Swept rotor blading has been applied to high-speed fans and compressors to reduce the effective Mach Numbers. In principle the benefit is identical to that for an aeroplane wing - if properly designed the compressibility effects relate to the Mach Number component normal to the loading axis.

Successful tests have been performed on swept blading in subsonic and transonic fans and in stationary cascade tunnels. Tests in industry on a swept-back fan designed for the high transonic range resulted in encouraging levels of efficiency and stall margin. Although the transonic fan work is proprietory, useful information may be gleaned from low speed cascade and rotor tests (ref. 9.57, 9.58, 9.59) and from the high subsonic cascade investigations of Stark[9.60].

The work of Godwin[9.58] clearly shows the stall margin advantage of swept blading. The mechanism appears to be that the centrifuging of the blade boundary layer and the effects of backward sweep combine to ensure that any separated regions of flow accumulate towards the trailing edge of the tip section. Thus there is no tendency for separation to spread to other regions of the blade which would cause abrupt stall. The report also showed that two-dimensional cascade data provided an adequate basis for swept-blade design.

In an interesting series of tests on forward-swept rotors Mohammed[9.59] has shown that forward sweep may also result in advantages for stall margin and efficiency. The mechanism is precisely the opposite of that for backward sweep and the good performance is attributed to the absence of any accumulation of separated fluid at the tip. Unfortunately the forward sweep was limited to 10° but the results would justify further research on forward sweep for compressor rotors.

Figure 9.13 taken from ref. 9.60 however shows the loss in turning angle and pressure rise coefficient obtained by Stark for a cascade of NACA 65-608 blades at an inlet Mach Number of 0.9 and a Reynolds' Number of $4x10^5$. The unswept blading had a 40° stagger and a space/chord ratio of unity. The tests were carried out for zero, 20° and 40° of sweep and seem to indicate that whilst the loss of loading capability for 20° sweep is small, for 40° of sweep it is considerable. Stark attributes this result to the effect of end wall shape. Blade pressure distributions in the vicinity of one end wall are very different from those near the opposite wall.

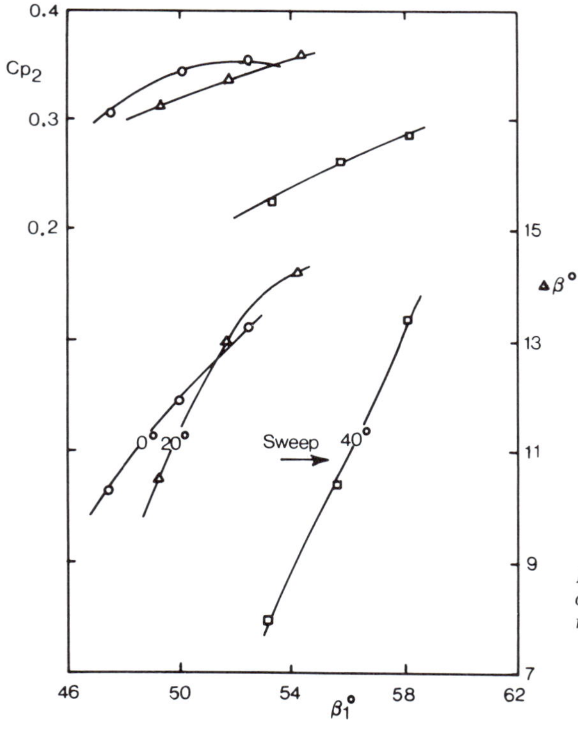

Fig. 9.13. Effect of sweep on turning angle and pressure rise coefficient.

Stark also presented pressure distributions which demonstrated clearly that the critical inlet velocity had been significantly raised by the use of sweep. Unfortunately the Stark results for swept blading were terminated at inlet angles well short of the stall value. As the inlet angle increases the pressure-rise coefficient curve for the blading having 40° sweep is continuing to rise whereas that for zero sweep has already started to fall off. In theory, an infinite swept wing produces a lower lift curve slope than the equivalent unswept wing. However, we are faced with the possibility that the effects of boundary layer movements will tend to increase the stall margin of swept blading. More information is needed to establish the validity of this effect; if it is true for a cascade then it can be expected to be more pronounced in a backward-swept rotor, where centrifugal effects will also be present, as Godwin's report has indicated, or even a forward swept rotor, as tested by Mohammed.

The role of sweep in reducing the effective Mach Number is therefore no longer in question. However, design confidence for swept rotors is dependent on a knowledge of blade loading capability and this requires further empirical evidence on stall margin effects.

9.3.3 Use for acoustic reasons. Sweep and dihedral have been used on the stators of aero-engine components to minimise noise and vibration by avoiding conditions in which the wakes of upstream rotors or stators impact upon the downstream blade rows simultaneously at all radii. In this way the amplitude of the exciting gust, at a given instant of time, measured over the entire height of the downstream blade will be reduced below the maximum which would otherwise be present at the moment of impact of the wake. Accordingly the lift of the downstream row will be responding to gradual variations of excitation rather than to an intermittent force at the upstream blade passing frequency.

The data of Sharland [9.61] indicate that by circumferential inclination of the inlet guide vanes of an axial-flow fan the noise level is considerably reduced. Figure 9.14 shows a consistent decrease in sound pressure level, which reaches a minimum for 60° of dihedral. Significant noise reductions were observed by Nemec [9.62] using inlet guide vanes inclined at around 40°, and by Benzakein [9.63] using outlet guide vanes inclined at 30°.

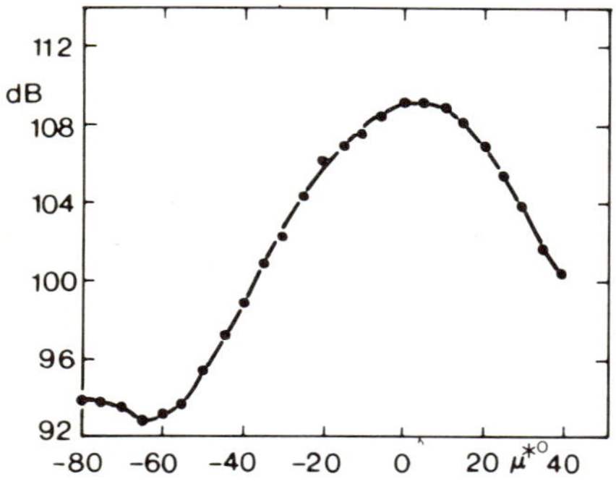

Fig. 9.14. Effect of stator circumferential inclination on fan sound pressure level at 45° to axis.

9.3.4 Sloping flowpaths and effects on meridional flow distribution. In many cases the occurrence of sweep and dihedral are 'accidental' in that they may be caused by annulus contraction or flare, or simply the need to instal blading in a region of changing radius. However, in some cases a degree of sophistication has been reached in which such effects are being used quite deliberately to improve the radial distribution of velocity and minimise separations and losses by imparting radial accelerations to the flow in a predescribed fashion.

One example of this latter course is the inner fan second stator of the CF6-6 engine shown in Fig. 9.15, where it was necessary to induce the flow from the high fan radius into the inboard location of the gas generator inlet. The bold and successful solution adopted was to use stators having suitable sweep and dihedral to impart a radially inward body force into the flow.

A somewhat similar situation arose in the development of the JT9D engine and here it was decided to boost the core compressor by stages within a downward sloping flowpath as described in ref. 9.64. The existing blading was essentially 'swept forward' with respect to the incoming flow (Fig. 9.16). The result was that separated boundary layers tended to drift towards the hub region; the ability of the blading to sustain

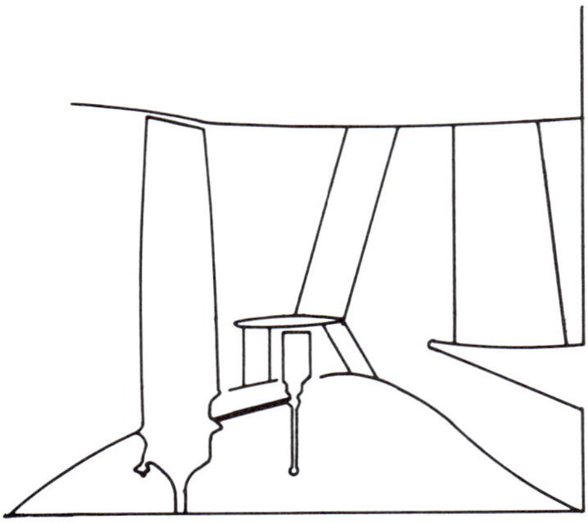

Fig. 9.15. Use of sweep and dihedral to impart downward body force.

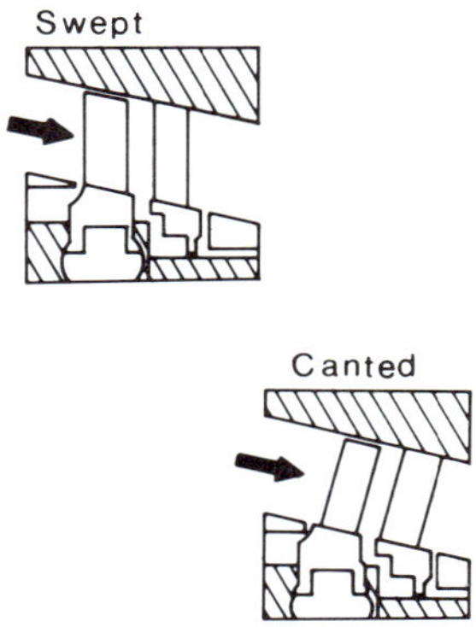

Fig. 9.16. Swept and canted blades in downward-sloping flowpath.

high loading was consequently impaired by the effect of sweep. In this situation, where transonic operation was not required, the use of swept blading had been damaging to performance and replacement blading which was 'canted', so that the loading axis was normal to the flow direction, performed well.

The moral appears to be that whereas the use of backward swept compressor blading will improve stall margin, forward swept blading, by sending the boundary layers towards the unstable hub region, may in some cases have the opposite effect.

In Soviet steam turbine practice, as reported by Deych and Troyanovsky[3.87], advantage is taken of the considerable effect of dihedral to redistribute flows radially and to reduce losses at the hub section of a stator row. Figure 9.17 shows that by inclining the blades by 13° into the flow the losses in the hub region have been greatly reduced. This has resulted from an increase in radial velocity components directed towards the hub. If the amount of inclination is excessive losses tend to increase again due to excessive skin friction.

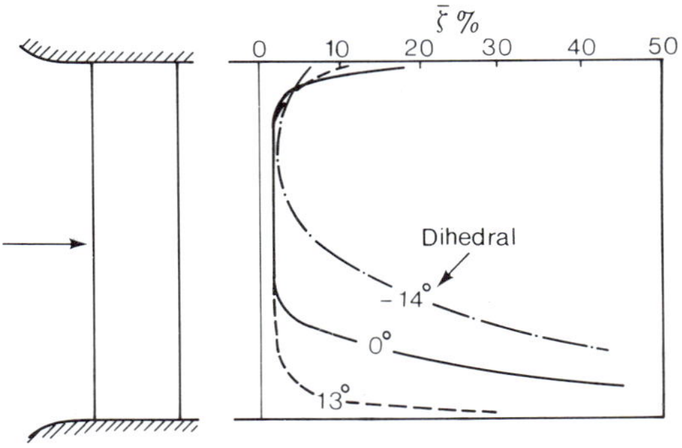

Fig. 9.17. Effect of inclination on losses behind a turbine nozzle row.

This latter point is relevant when considering sweep in compressor blading and probably contributes to the fall-off in performance observed for sweep angles greater than 30°.

Filippov and Wang[9.65] have reported that for highly-loaded nozzle blades the optimum efficiency is obtained when the dihedral varies along the span, in such a way that radial pressure gradients, especially those at the hub, are minimised. It is claimed that the resulting curvilinear "sabre-shaped" nozzle blades shown in Fig. 9.18 give greater flexibility in variation of reaction with radius and give efficiency improvements for stages having large flare. Figure 9.17 also presents the improvements in the radial loss distribution compared with conventional radially-stacked blades. At a Mach Number of 0.5 the average energy loss coefficient was reduced from 0.056 to 0.0405.

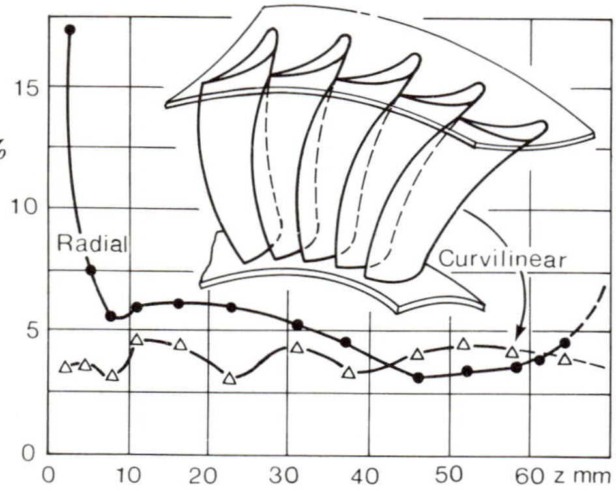

Fig. 9.18. Curvilinear nozzle blades and the associated loss improvement.

The performance of low aspect ratio turbine blading is particularly susceptible to degradation by secondary flows. Liu et al[9.66] improved the performance of a low aspect ratio nozzle row with design changes based on the fully viscous program of Dodge[3.39] and comprehensive testing. The tests confirmed the strong effect of dihedral and also the beneficial effect of curvilinear blades having 13.5° compound lean which reduced the average loss coefficient from 0.062 to 0.055. A configuration having 26° compound lean was also tested. The results indicated that increasing the suction surface spanwise concavity beyond a moderate level resulted in further loss reductions near the end walls at the expense of higher mid-span losses; the overall loss level was unchanged.

Deych et al[4.4] also made great improvements to the performance of turbine nozzle blading by profiling the end wall. The technique is applicable to low aspect ratio nozzles and energy loss coefficients obtained from different end-wall shapes are shown in Fig. 9.19. These clearly demonstrate that a moderate convergence in the throat region gives lowest losses. The improvements in energy loss coefficient ranged from 3.6% for very low aspect ratios around 0.2 with a contraction of one half of the blade discharge height, to 1.4% at aspect ratios of 0.5 with a contraction of one third of the discharge height.

The value of end-wall profiling has been confirmed in American testing[9.66]. Kopper et al[9.67] investigated the application of end-wall profiling to a high turning nozzle cascade of exit aspect ratio 0.513 and reported a 17% reduction of full passage loss relative to a planar end-wall configuration.

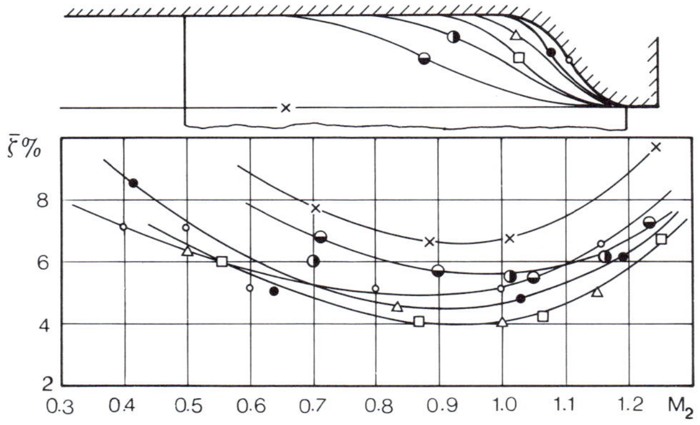

Fig. 9.19. Effect of turbine nozzle endwall profile on energy loss coefficient.

The essential features of the earlier Soviet results for both dihedral and end-wall profiling have therefore been confirmed in recent American tests. The combined effects of control over dihedral and end-wall profile are important steps toward an ultimate objective of the fully three-dimensional design of turbine nozzle passages. It seems clear that this kind of refinement will lead to significant efficiency improvements. The improved understanding of these effects also appears to be a motivating factor toward the use of longer chord, lower aspect ratio turbine blading with improved loading capability.

REFERENCES

9.1 WILLIAMS, J. In 'Boundary Layer and Flow Control; Vol. I' (G.V. Lachmann ed) Pergamon, Oxford (1961)

9.2 SCHROEDER, W.M. Axial flow compressor development at the Stuttgart Research Institute. CIOS Item No. 19, File No. XXXII-31, H.M. Stationery Office (1945)

9.3 SHEETS, H.E. The slotted-blade axial flow blower. Trans. ASME 78, 1683 (1956)

9.4 GOSTELOW, J.P. and WATSON, P.J. Investigation on staggered tandem vane pump impellers. Cambridge Univ. CUED/A-Turbo/TR 38 (1972)

9.5 LUEDERS, H.G. and ROELKA, R.J. Some experimental results of two concepts designed to increase turbine blade loading. Trans. ASME, J. Eng. Power, pg. 198 (April 1970)

9.6 LINNEMANN, H. Untersuchungen eines einstufigen Axialgebläses mit Tandemgittern. Konstruktion, Heft 4, S 128 (1964)

9.7 OHASHI, H. Theoretical and experimental investigations on tandem pump cascades with high deflection. Ing. Archiv. 27 (1959)

9.8 FICKERT, K. Investigations on the blade systems of decelerating cascades of high turning. Zeit. ver. deutsch. Ing. 93 (3D), 950 (1951)

9.9 MARCINOWSKI, H. Messungen an axialen Schaufelgittern. Das Versuchswesen der Masch. Fabr. J.M. Voith, Beitrag 24, Heidenheim (1949)

9.10 PAL, P. Untersuchungen über den Interferenzeinfluss bei Strömungen durch Tandem - Schaufelgitter. Ing. Archiv. 34, 3, 173 (1965)

9.11 RAILLY, J.W. and EL-SARHA, M.E. An investigation of the flow through tandem cascades. Proc. I. Mech. E. 180 (1966)

9.12 RAILLY, J.W., DEEB, S.D. and MOORE, M.J. Wake boundary-layer interaction in tandem cascades. Proc. I. Mech. E. 184, 3G(II) pg. 101 (1969-70)
9.13 IHLENFELD, H. Strömungsvorgänge an stark verzögernden Spaltflügelgittern. Maschinenbautechnik 14, No. 7/8 pp. 361 420 (1965)
9.14 BAMMERT, K. and STAUDE, R. Optimization for rotor blades of tandem design for axial flow compressors. Trans. ASME, J. Eng. Power 102, pg. 369 (1980)
9.15 SPRAGLIN, W.E. Flow through cascades in tandem. NACA TN 2393 (1951)
9.16 YIPP, Y.M. and RAILLY, J.W. Potential flow theory for tandem cascade by Howell's method. ARC CP 971 (1967)
9.17 BLAHO, M. Die Berechnung der Strömung für ein doppeltes Schaufelgitter. Periodica Polytechnica M4, 227-249 (1960)
9.18 KATSANIS, T. and MCNALLY, W.D. Fortran program for calculating velocities and streamlines on a blade-to-blade stream surface of a tandem blade turbomachine. NASA TN D-5044 (1969)
9.19 LINDER, C.G. and JONES, B.A. Single stage xpe.imental evaluation of slotted rotor and stator blading. Part II - Annular cascade investigation of slot location and geometry NASA CR-54545 (1966)
9.20 KEENAN, M.J. and BARTOK, J.A. Experimental evaluation of transonic stators. Data and performance report; multiple-circular-arc stator A (slotted) NASA CR-54624 (1968)
9.21 MIKOLAJCZAK, A.A., WEINGOLD, H.D. and NIKKANEN, J.P. Flow through cascades of slotted compressor blades. Trans. ASME, J. Eng. Power, pg. 57 (Jan, 1970)
9.22 OKIISHI, T.H., JUNKHAN, G.H. and SEROVY, G.K. Experimental performance in annular cascade of variable trailing-edge flap, axial-flow compressor inlet guide vanes. ASME Paper 70-GT-106 (1970)
9.23 SIEVERDING, C. Experimental data for tandem cascade in the high subsonic region. VKIFD I.N. 15 (1966)
9.24 DETTMERING, W. and BECKER, B. Steps in the development of a supersonic compressor stage. AGARD CP 34 (1968)
9.25 KIOCK, R. Improved aerodynamic performance of a highly loaded compressor blading by boundary layer control. Proc. 6th Conf. on Fluid Mach., Budapest (1979)
9.26 DAVIDSON, I.M. The jet flap. J. Roy. Aero. Soc., 25,60 (1956)
9.27 MARBLE, F.E., see Clark, E.L. An experimental study of jet-flap compressor blades. M. Aero. E. thesis, Cornell Univ. (1958)
9.28 ROY, M. Means and examples of aeronautical research in France at ONERA. J. Aero. Sci., 26 pg. 193 (1959)
9.29 CLARK, E.L. and ORDWAY, D.E. An experimental study of jet-flap compressor blades. J. Aero. Sci., 26 pg. 698 (1959)
9.30 BROCHER, E.F. The jet-flap compressor cascade. Trans. ASME, J. Basic Eng. 83, 4 pg. 401 (1961)
9.31 PAULON, J. Application due soufflage au bord de fuite au controle de la deflexion des grilles d'aubes. La Recherche Aeronautique 73 pg. 17 (1959)
9.32 SALITA, M. Lift of a finite jet-flapped cascade of airfoils at large jet angles. Penn. State Univ. Ord. Res. Lab. TM 510. 2481-06 (1967)
9.33 LANDSBERG, T.J. and KRASNOFF, E. An experimental study of rectilinear jet-flap cascades. ASME Paper 71-FE-14 (1971)
9.34 STRATFORD, B.S. Mixing and the jet flap. Aero. Quart. (May, 1956)
9.35 KRUGER, R.A., MARSH, H. and HORLOCK, J.H. The performance of a cascade fitted with blown flaps. ARC CP 526 (1961)
9.36 STARK, U. A theoretical investigation of the jet-flap compressor cascade in incompressible flow. Trans. ASME, J. Eng. Power 94, 4 pg. 249 (1972)
9.37 AINLEY, D.G. The high temperature turbo-jet engine. J. Roy. Aero. Soc. 60 pg. 549 (1956)
9.38 BAGBY, C.L. and ANDERSON, W.L. Effect of turbofan cycle variables on aircraft cruise performance. J. Aircraft, 3, 5 (1966)

9.39 BAYLEY, F.J. and MARTIN, B.W. A review of liquid cooling of high-temperature gas-turbine rotor blades. Proc. I. Mech. E. 185 18/71 pg. 219 (1971)
9.40 JENKINS, P.E. and RICE, I.G. Blading heat transfer considerations in a reheat-gas-turbine combined cycle. ASME Paper 81-GT-155 (1981)
9.41 HOOKER, S.G. The engine scene. Aeronautical Journal 74 (Jan. 1970)
9.42 MOFFITT, T.P., NOSEK, S.M. and ROELKE, R.J. Turbine aerodynamic considerations for advanced turbines. in NASA SP-259 "Aircraft Propulsion" (1970)
9.43 MICHEL, G.W. and KOST, F.H. The effect of coolant flow on the efficiency of a transonic HP turbine profile suitable for a small engine. ASME Paper 82-GT-63 (1982)
9.44 BROWN, A. and MARTIN, B.W. Flow transition phenomena and heat transfer over the pressure surfaces of gas turbine blades. Trans. ASME, J. Eng. Power 104, 2 pg. 360 (1982)
9.45 LITCHFIELD, M.R. and NORTON, R.J.G. Heat transfer measurements of a transonic nozzle guide vane. ASME Paper 82-GT-247 (1982)
9.46 CONSIGNY, H. and RICHARDS, B.E. Short duration measurements of heat transfer rate to a gas turbine rotor blade. Trans. ASME, J. Eng. Power 104, 3 pg. 542 (1982)
9.47 BAYLEY, F.J. and PRIDDY. W.J. Effects of free stream turbulence intensity and frequency on heat transfer to turbine blading. Trans. ASME, J. Eng. Power 103, pg. 60 (1981)
9.48 KRISHNAMOORTHY, V. Effect of turbulence on the heat transfer in a laminar and turbulent boundary layer over a gas turbine blade. ASME Paper 82-GT-146 (1982)
9.49 ITO, S., GOLDSTEIN, R.J. and ECKERT, E.R.G. Film cooling of a gas turbine blade. Trans. ASME, J. Eng. Power 100 pg. 476 (1978)
9.50 DRING, R.P., BLAIR, M.F. and JOSLYN, H.D. An experimental investigation of film cooling on a turbine rotor blade. Trans. ASME, J. Eng. Power 102 pg. 81 (1980)
9.51 HAY, N. Discrete hole film cooling of turbine blades. 5th ISABE, Bangalore (1981)
9.52 LOUIS, J.F., DEMERJIAN, A.M., GOULIOS, G.N., TOPPING, R.F. and WIEDHOPF, J.M. Short duration studies of turbine heat transfer and film cooling effectiveness. ASME Paper 74-GT-113 (1974)
9.53 SELLERS, J.P.Jr. Gaseous film cooling with multiple injection stations. AIAA J, 1 pg. 2154 (1963)
9.54 LOFTUS. P.J. and JONES, T.V. The effect of temperature ratios on the film cooling process. ASME Paper 82-GT-305 (1982)
9.55 LEWIS, R.I. and HILL, J.M. The influence of sweep and dihedral in turbomachinery blade rows. J. Mech. Eng. Sci. 13, 4 (1971)
9.56 GRAHAM, D. and LEWIS, R.I. A surface vorticity analysis of three-dimensional flow through strongly swept turbine cascades. J. Mech. Eng. Sci. 16, 6 pg. 425 (1974)
9.57 BEATTY, L.A., SAVAGE, M. and EMERY, J.C. Low-speed cascade tests of two 45° swept compressor blades with constant spanwise loading. NACA RM L53207 (1954)
9.58 GODWIN, W.R. Effect of sweep on performance of compressor blade sections as indicated by swept-blade rotor, unswept-blade rotor and cascade tests. NACA TN 4062 (1957)
9.59 MOHAMMED, K.P. Investigations on a single stage axial flow fan with forward swept blades. Ph. D. thesis, I.I.T., Madras (1974)
9.60 STARK, U. Strömungsuntersuchungen an gepfeilten Verdichtergittern bei kompressibler Unterschallströmung. DLR Bericht No. 67-09, Braunschweig (1967)
9.61 SHARLAND, I.J. Sources of noise in axial flow fans. J. Sound Vib. I(3) pg. 302 (1964)
9.62 NEMEC, J. Noise of axial fans and compressors: study of its radiation and reduction. J. Sound Vib. 6(2) pg. 230 (1967)
9.63 BENZAKEIN, M.J. Research on fan noise generation. J. Ac. Soc. Am. 51, 5 pg. 1427 (1972)

9.64 HARMON, H.A., MIKOLAJCAK, A.A. and MARCHANT, D. Performance of compressor blade rows in a sloping flowpath. Trans. ASME, J. Eng. Power 94, 1 pg. 61 (1972)

9.65 FILIPPOV, G.A. and WANG, ZHONG-CHI. The effect of flow twisting on the characteristics of guide rows. Teploenergetika No. 5 (1964)

9.66 LIU, H.C., BOOTH, T.C. and TALL, W.A. An application of 3-D viscous flow analysis to the design of a low-aspect-ratio-turbine. ASME Paper 79-GT-53 (1979)

9.67 KOPPER, F.C., MILANO, R. and VANCO, M. Experimental investigation of endwall profiling in a turbine vane cascade. AIAA J, 19,8 pg. 1033 (1981)

CHAPTER 10

Evaluations of Prediction Accuracy

The subject of Chapter 11 will be the application of cascade information to the design process. Before addressing that topic it is desirable to take stock of progress in cascade aerodynamics and to demonstrate the capabilities and limitations of current techniques. In this chapter the accuracy of the cascade model in predicting flow fields within a turbomachine will be assessed. This is considered an essential step in building confidence in the model's application.

The degree to which cascade theory and experience are mutually compatible for incompressible flows will be determined. It will then be shown that predictions based on cascade data can be successfully used in the design and flow analysis of real machines. Finally the difficulties posed by high speed flows will be considered and the present limitations on application of cascade information reviewed.

10.1 Incompressible Flows

The approach usually taken by researchers has been to calculate the inviscid flow field, using either the direct method approach or a design (prescribed velocity distribution) approach. Implicit in the above is selection of a blade loading criterion through application of the Kutta-Joukowsky condition to cusped blades or viscous considerations to blading having a rounded trailing edge. Measurements of blade turning, loss and pressure distribution taken from cascade tunnels have been compared with the theoretical predictions. Although both theory and experiment are subject to various sources of error the presumption has been that when good agreement is obtained between theory and experiment the answer is a correct one. The utility of that answer can only be assessed by comparison with measurements from an actual machine and such comparisons are addressed in section 10.1.3.

10.1.1 Accuracy of numerical solutions. The purpose of the exact cascade theory of section 5.2 was to provide a basis for evaluating the accuracy of numerical methods. A good example of the use of the exact solution in the refinement of finite element procedures is given in ref.5.29. In section 5.3 a number of comparisons were made for both cusped and rounded trailing edges and for varying degrees of camber. The limitations of some methods (e.g. the Schlichting method) were explored and other approaches (e.g. the Martensen method) were seen to give near-perfect accuracy.

The comparisons of section 5.3 could only facilitate conclusions on how well the numerical methods performed in solving the Laplace equation for the various blade shapes. Viscous effects were not considered and for rounded trailing edges the solution was not uniquely defined. Chapter 7 considered the effects of viscosity and it was demonstrated that these effects must be incorporated, if only by application of the simple 'fairing-in' procedure of Fig. 7.4 and the zero nett vorticity theorem, to provide a unique solution to the general case of the flow past a rounded trailing edge.

Although the comparisons with exact solutions simply evaluated the ability of methods to solve the Laplace equation for representative boundary conditions this was an essential first step; any shortcomings revealed in that examination tended to invalidate the methods as predictors of real flows. In the next step it is appropriate to consider how well some of the methods have fared when comparisons have been made between numerical solutions and cascade experiments.

10.1.2 Cascade theory and experiment. In 1951 Katzoff and Hannah[10.1], using a potential flow "interference" method, neglected the Kutta-Joukowsky condition and inserted an experimentally-determined value of lift coefficient. Good agreement between theoretical and experimental pressure distributions was obtained.

In Chapter 7 the correction of Speidel and Scholz[10.2] was mentioned. In this extensive series of comparisons the Schlichting potential theory and the Truckenbrodt boundary layer theory were used to calculate the boundary layer as far as separation and to express the displacement thickness effect in terms of an additional source-sink distribution. This resulted in a substitute profile having its trailing edge in a different position. The potential flow around the new profile was computed and results compared with measurements for pressure distribution, outlet angle and loss coefficient. The agreement was surprisingly good over a wide range of inlet angles.

Geller[7.58] has modified this approach to incorporate representation of separated regions. Neglecting displacement thickness effects Geller iterated between boundary layer and potential flow calculations to fix the separation point on each surface. Separation streamlines were then assumed to have a constant static pressure resulting from an appropriate source distribution downstream from the separation location. The agreement Geller obtained between measured and calculated pressure distributions, turning angles and drag coefficients was generally very good, even when separation was indicated well forward on the blade surface.

Objections to Geller's procedure include his disregard for displacement effects, which could clearly result in errors in the throat region, and his apparent disregard for wake loading and curvature. The NACA blade sections chosen did appear to display loading at the trailing edge and the neglect of this in the theory would explain small discrepancies between the calculations and measurements over the last 20% of chord. Nevertheless Geller's predictions were clearly very good and further work along these lines is warranted to explore the applicability of the various representations and assumptions and to apply the techniques to a wider range of blading.

For conditions around design incidence there is no longer a problem in obtaining consistent agreement between theoretical and experimental pressure distributions, turning angles and even loss coefficients. Reassuring examples for various blade shapes are given in refs. 10.3, 10.4. The author's experience is that incorporation of the first viscous approximation, described in Chapter 7, facilitates good agreement for pressure distribution, turning angle and boundary layer parameters, as illustrated in Fig. 10.1 from ref. 10.5, at incidences around design. This view is confirmed by Sanger[7.77].

However, such an approach fails to give an accurate prediction as the incidence moves significantly away from design. The results of Miller and Serovy[10.6] indicate that although very good predictions of deviation angle are given by current theoretical methods at around design incidence, for incidences 10° above or below design the discrepancies can be catastrophic.

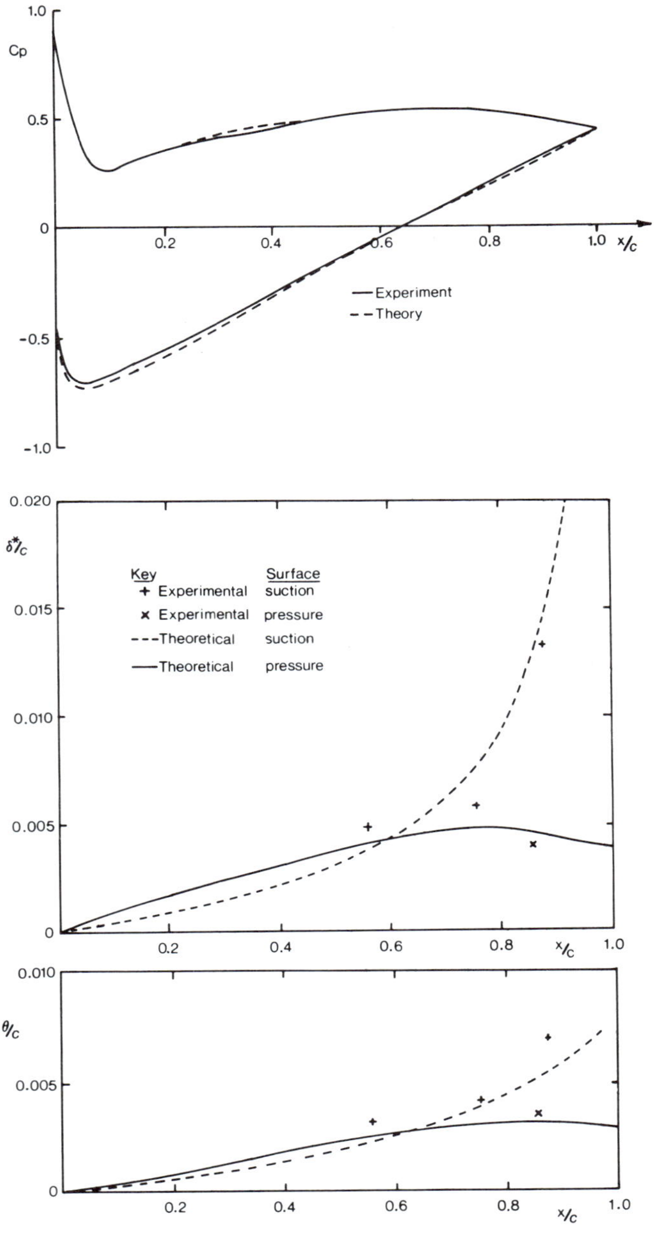

Fig. 10.1. Experimental and theoretical results for a low-speed compressor cascade.

Pressure distributions over blade surfaces are harder to calculate than deviation angles and prediction attempts may also result in serious discrepancies under off-design conditions. An indication that the strongest source of errors is flow separation is derived from an examination of two sets of data.

Series of comparisons between experimental pressure distributions and the equivalent Martensen method calculations have been given by Masek and Norbury[3.5] and Starke[10.7] for a wide range of incidence. In both sets of comparisons the agreement between theory and experiment at around the optimum incidence was reasonable. The Starke results sustained this level of agreement over the incidence range -10.25° to $+5.75^{\circ}$. The Masek and Norbury results showed severe disagreement between theory and experiment at incidences of 10.4° below design and 4.6° and 9.6° above design.

In both cases the measured pressure distributions were given for a range of axial velocity ratios. Two-dimensional conditions were readily obtained by interpolation. The Martensen method used by Starke was a modified version but the modification was confined to an appropriate incorporation of axial velocity ratio effects. This modification provides no explanation of why Starke accurately predicted pressure distributions over a wide range of incidence whereas Masek and Norbury did not; it is very clear from the latter results that the discrepancy greatly exceeded the band of axial velocity variation effects.

The explanation rather appears to reside in the different blade profiles used. Masek and Norbury were testing a 'prescribed velocity distribution' section, especially designed to minimise losses at the design point. The resulting profile had a relatively sharp wedge-shaped leading edge. As a consequence at negative incidence, 10.4° below design, a pronounced separation was present on the pressure surface at the leading edge. This led to a gross discrepancy betweeen the predicted and measured pressure distributions for the pressure surface. At an incidence 4.6° above design sufficient trailing edge separation was present to jeopardize agreement and, at 9.6° above design, the suction surface boundary layer suffered both a leading edge bubble separation and a turbulent separation over the last 40% of chord. It is clear that whilst prescribed velocity distribution blading may give low losses at the design condition its tolerance to incidence variation is severely curtailed. The resulting strong separations were the feature which invalidated flow predictions at off-design incidences.

The blading tested by Starke was a more conventional 9C7/32.5C50 section. Although strong levels of diffusion were sustained on the pressure surface at negative incidence and on the suction surface at positive incidence the only serious separation effects appear to have been confined to the last 20% of chord at $+5.75^{\circ}$ incidence. The only obvious departure of experimental results from theoretical predictions was in this region. Otherwise agreement was very good over the whole incidence range. The predictions of ref. 10.7 were facilitated by taking the experimental discharge angle as an input.

The discrepancies reported by Masek and Norbury are therefore seen to be caused by strong flow separations on a blade which was designed for optimum performance at one incidence. For conventional blades operating in the separation-free range it should be quite possible to obtain good agreement between theory and experiment. This is especially so if appropriate measures are taken to treat the trailing edge flow conditions along the lines of Chapter 7 and ref. 10.6.

Many researchers have been content to predict pressure distribution and turning angle. Consistent loss prediction has often been considered beyond existing capabilities. For conditions beyond the separation-free operating range prediction methods incorporating separated boundary layers and the resulting wakes are not well advanced. Further research towards the objective is needed. This must include improved physical understanding and prediction of separation bubbles, trailing edge flows, separated regions and loaded wakes.

An advanced prediction technique might then involve a synthesis of the approach described by Geller[7.58], the latest understanding of separation and wake behaviour and an appropriate accounting for displacement and momentum thickness effects for which Oliver's two vortex sheet model[7.70] has great merit.

10.1.3 Comparison with machine tests. Although for many flows good agreement can be obtained, between theory and cascade experiment, there is no guarantee that the result thus obtained will truly represent performance in an actual machine. Attempts have been made to establish the validity of application of cascade data by comparing either theoretical predictions or measurements from cascade tunnels with measurements from machines.

Measurements of flow angle and loss coefficient are usually obtained in a rotating machine by traversing. Simple averaging of this information behind a rotor results in errors and it is desirable to use high response hot wire or pressure transducer instrumentation for this purpose. Since the information is often contaminated by noise, appropriate signal-processing techniques[3.65] are recommended for the recovery of coherent traverse data. Pressure distributions around blading rotating at low speeds may be measured using an external manometer connected to a pressure transfer device. Good results for velocity distribution through blade passages can now be obtained using laser velocimetry techniques[10.8].

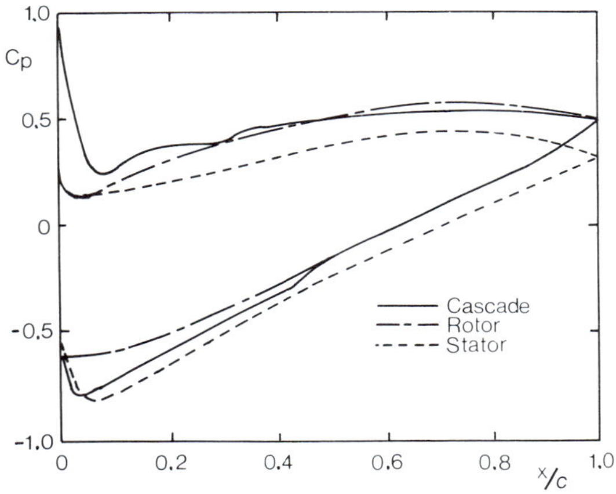

Fig. 10.2. Comparison of cascade, rotor and stator pressure distributions for low speed compressor.

In an early comparison of pressure distributions from cascade and compressor rotor experiments Westphal and Godwin[10.9] made measurements at hub, mean and tip sections. The main discrepancy in turning angle between cascade and rotor measurements was at the hub, where the compressor had a steeper slope of turning angle as a function of angle of attack. The rotor hub had a much greater stall-free operating range than the corresponding cascade section. For design conditions pressure distribution comparisons were excellent at the hub and mean sections and good at the casing. The main discrepancy resulted from a long laminar separation bubble which influenced the cascade results but not those from the rotor.

Shaw and Doyle[10.10] compared results from low speed tests on a cascade of C4 aerofoils and the corresponding section in a rotating rig. Pressure distributions were obtained at the compressor midspan section of both rotor and stator blading. By varying flow coefficient, Reynolds' Number, turbulence level and axial velocity ratio it was possible to obtain a valid comparison with the cascade data. Figure 10.2 presents results for incidences between 1.4° (cascade) and 2.2° (stator) and Reynolds' Numbers between 1.55×10^5 (stator) and 1.96×10^5 (cascade). The rotor blades exhibited more positive pressures in the suction peak region than the stator and cascade blades. The stator blading appeared to suffer from turbulent separation and as a result sustained a lower pressure rise than the cascade or rotor blading. At lower Reynolds' Numbers there was no evidence of laminar separation on either rotor or stator blading although the compressor turbulence level was high, at 4.5%.

The two-bladed rotor work of ref.3.88 showed that stationary blading needs to be subjected to high turbulence levels if the off-design performance of rotating blade rows is to be simulated. For turbulence levels above a critical value there was no significant difference between the unstalled performance of rotating and annular cascades. However, even at the highest turbulence level of 3.3%, the stationary blades stalled in a different (more abrupt) manner than the rotating blades. It was not clear whether the stalling difference was due to the flow along the span of the rotating blades, the high effective turbulence level for the rotating cascade or the difference in turbulence structure for stationary and rotating cascades.

The influences of centrifugal force and wall boundary layer skew on flow over rotor blading were shown in the work of Peacock and Overli[10.11]. These influences tended to delay corner stall at the hub. At the tip the existence of a small tip clearance (as compared with a shrouded rotor) enhanced performance by delaying corner stall. Comparisons were made in ref.10.11 of pressure distributions at rotor midspan over a 10° incidence range with corresponding cascade measurements of Rhoden[2.23]. The agreement was very good. There was reasonable agreement, over a wider incidence range, with the Martensen theory although at an incidence of +12.7^o turbulent separation on the rotor resulted in a substantial fall-off in pressures below the theoretical values. The results of ref.10.11 show that for flows which are not drastically separated it is possible to predict rotor pressure distributions from cascade theory and experiment.

A most comprehensive series of pressure distribution comparisons on compressor blading, with associated wake surveys, has been presented by Dring et al[10.12]. All

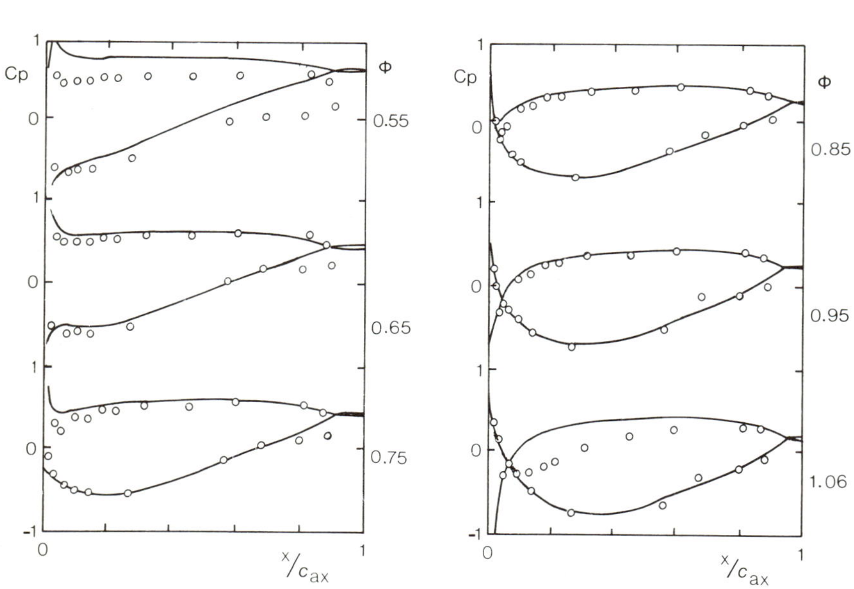

Fig. 10.3. Pressure distributions at midspan of compressor rotor.

of the measured pressure distributions were compared with corresponding calculations using the finite area method of Caspar et al[5.71]. Pressure distributions were measured on the rotor blade within 5% of hub and casing and at intermediate spanwise locations. For incidences near design the agreement was excellent over most of the span with small discrepancies at hub and casing. Midspan pressure distributions were also presented over the whole compressor characteristic. Figure 10.3 shows that whereas excellent agreement was obtained between cascade theory and rig measurements in the vicinity of the design condition the agreement outside an incidence range of -13.0^{o} to -2.5^{o} (Φ of 0.95 to 0.65) was poor. The discrepancies are reminiscent of those reported by Masek and Norbury[3.5]. At very negative incidences the steep deceleration predicted for the pressure surface leading edge region was not sustained in practice. Separation occurred on the pressure surface resulting in a marked discrepancy on that surface. The resulting blockage caused higher suction surface velocities than predicted. At high positive incidences turbulent separation occured from about the midchord location. This resulted in discrepancies on both surfaces.

In measurements of pressure distribution around turbine rotor blading Dring and Joslyn 10.13 have obtained excellent agreement between experimental data and the Caspar theory despite the presence of strong three-dimensional effects on both suction and pressure surfaces.

10.2 High Speed Flows

Just as an exact solution was needed to check the accuracy of more general numerical methods at low speeds a similar need arose for transonic flow calculations. An accurate transonic flow test case had initially seemed impossible but, as described in section 6.3.2, Hobson used hodograph theory to provide test cascades of impulse blading. Hodograph theory was subsequently used by other workers to give widely-spaced compressor blading.

In this section the comparisons between numerical techniques and the special solutions will firstly be considered. Because of the shock-free nature of the hodograph blading it is desirable to also evaluate numerical prediction methods using measurements from cascades containing shocks and such evaluations will be discussed. Finally the real value of cascade work is in predicting conditions attaining in rotating machines. Progress and remaining deficiencies in the application of cascade information to machine flows will be considered.

10.2.1 Accuracy of numerical solutions for transonic flow. In section 6.3.2 it was reported that the methods of refs. 3.86 ,6.52 and 6.53 had provided the best predictions for the Hobson cascade of Table 6.1. These comparisons are presented in Fig. 10.4. Whereas in Table 6.1 the critical Mach Number M^* was given, in Fig. 10.4 the distribution is that of local Mach Number M.

An earlier first order relaxation solution by Ives and Liutermoza[6.52] had given instabilities in the supersonic region. Although the second order solution presented has a slight tendency to predict a shock wave at the downstream part of the supersonic region this does not appear to be serious. The singularity theory of Ribaut and Vainio[6.53] gives a smooth, shock-free result which slightly underpredicts the peak Mach Number. The time-marching method of Denton[3.86] predicts the peak Mach Number but gives small errors elsewhere on the profile.

A less ambitious hodograph test case was produced by Korn. This widely-spaced compressor cascade had a peak Mach Number of 1.16 and excellent finite difference results from Ives and Liutermoza[6.52] and finite element results from Habashi et al 5.30 are presented in Fig. 10.5. Some fine predictions for a more representative supercritical compressor cascade developed by Korn using hodograph theory were given in refs. 5.30 ,6.51 and 6.52.

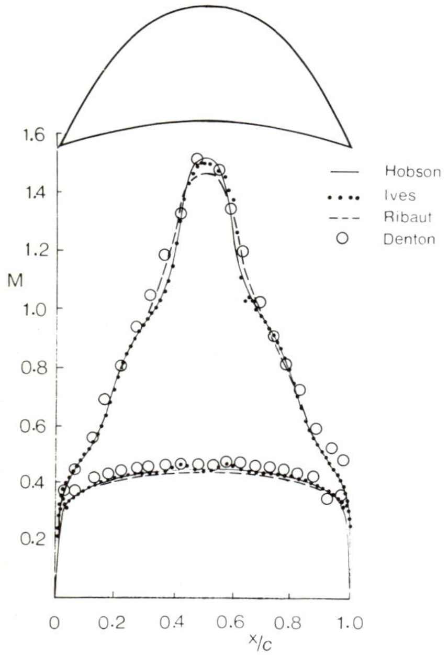

Fig. 10.4. Evaluation of numerical methods using Hobson impulse cascade.

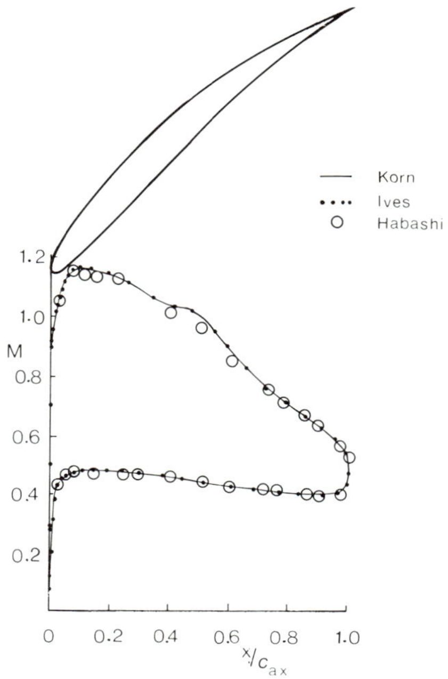

Fig. 10.5. Evaluation of numerical methods using Korn compressor cascade.

It can be concluded from the results presented in Figs. 10.4 and 10.5 that it is possible to obtain excellent numerical solutions for inviscid transonic shock-free cascade flows having local Mach Numbers up to 1.5. The methods validated in this way include singularity, time marching, finite difference and finite element solutions.

10.2.2 High speed cascade theory and experiment. In 1953 Stanitz and Sheldrake [10.14] produced a good example of agreement between measured pressure distributions and a subsonic linearised prescribed velocity distribution theory. Discrepancies increased as the downstream Mach Number was raised to 0.6 and were quite bad, due to shock formation, at higher speeds. A more recent example of the use of a design method for turbine cascades in high subsonic flows has been given by Uenishi[6.5]. The agreement between measurements and theory was reasonable.

Using the more common direct method approach Kynast[10.15] produced comparisons between a channel-flow theory[10.16] and his own measurements. For the range of subsonic Mach Numbers tested agreement was good. Lakomy[10.17] has also reported good agreement between theory and experiment for a range of cascades at high subsonic Mach Numbers.

The author has already presented subsonic comparisons for a nozzle cascade in Chapter 6. Apart from problems in blade edge definition the direct problem for unseparated subsonic flows is essentially solved.

As the Mach Number passes the critical value shock waves are formed and the prediction of experimental results becomes increasingly arduous. Where discrepancies exist it should not be assumed that the theory is necessarily in error. It was made clear in

Chapter 4 that the testing of transonic cascades is fraught with complexities which directly affect measured pressure distributions, turning and losses. The problems are most acute in the inlet flow field for compressor blading and the discharge flow field for turbine blading.

The accuracy of compressor cascade prediction is dependent on the Mach Number and shock configuration. Schmidt[10.18] designed supercritical blading for inlet Mach Numbers around 0.8 which were similar to those tested by Stephens[6.51] and had a reasonably constant Mach Number of about 1.2 over a 30% 'rooftop' on the suction surface. When side wall effects were correctly treated good agreement between theory and experiment was obtained.

As the inlet Mach Number rises to supersonic levels the quality of prediction becomes variable. The results of tests of various inlet Mach Numbers on a double circular arc cascade by DFVLR[10.19] appear to be of good accuracy. Many available time-marching programs do not predict these supersonic flows at all well although the equation-splitting relaxation procedures of Dodge[1.5] show a measure of agreement at inlet Mach Numbers up to 1.409. The procedure of Kurzrock and Novick[6.67], which solved the compressible Navier-Stokes equations with dominant viscous terms, has given excellent agreement in pressure distribution between cascade theory and experiment for an inlet Mach Number of 1.588. The inclusion of viscous effects appears to be very important to the solution of flows through supersonic compressor blading and this conclusion will be reinforced in the discussion of section 10.2.3.

For turbine blading with supersonic discharge conditions the difficulties of wave reflection and flow unsteadiness are such that any experimental data should be viewed cautiously. A common failing of high speed turbine cascade experiments is the presence of separation areas of sufficient magnitude to seriously modify the flow field. This considerably narrows the selection of experimental data suitable for comparison with numerical predictions.

Figure 6.4 presented theoretical predictions of the steam turbine blade flow of Forster [6.35]. The two streamline curvature methods used gave reasonable predictions except in the supersonic region. A method of characteristics solution was added, assuming coincidence of a straight sonic line and the geometric throat.

Test cases for high-turning turbine blading accompanied the results of McDonald's numerical method [6.61]. The agreement between theory and experiment was good for local Mach Numbers which exceeded 1.4. The method is in routine use in various establishments and performs consistently well. Farn and Whirlow[10.20] obtained excellent agreement between finite volume time-marching calculations and cascade measurements obtained at VKIFD. A gas turbine blade was run at the limit loading condition and this case, having a maximum Mach Number of about 2.3 but no passage shocks, was particularly well predicted.

Probably the most reliable test data yet published for turbine cascades containing shock waves are those of Graham and Kost [10.21]. The data for the RB blading at a discharge Mach Number of 1.15 contain a shock-induced separation but this is considered to be a genuine rather than a spurious effect. Such complex interactions are quite common on supercritical turbine blading and a useful prediction scheme should incorporate the facility for computation of local viscous effects. Another problem which has caused uncertainties for this test case, and for those from other tunnels, is the existence of periodic unsteadiness in the downstream flow field at frequencies below 1kHz. A schlieren movie of this test demonstrated the effects of the flowfield oscillations but also showed that in this case the main passage shock was fairly stable. It is not known whether the unsteady effects are identical to those of section 8.5.3 but it is concluded that the test case of ref. 10.21 is a reasonable one for evaluation of numerical predictions.

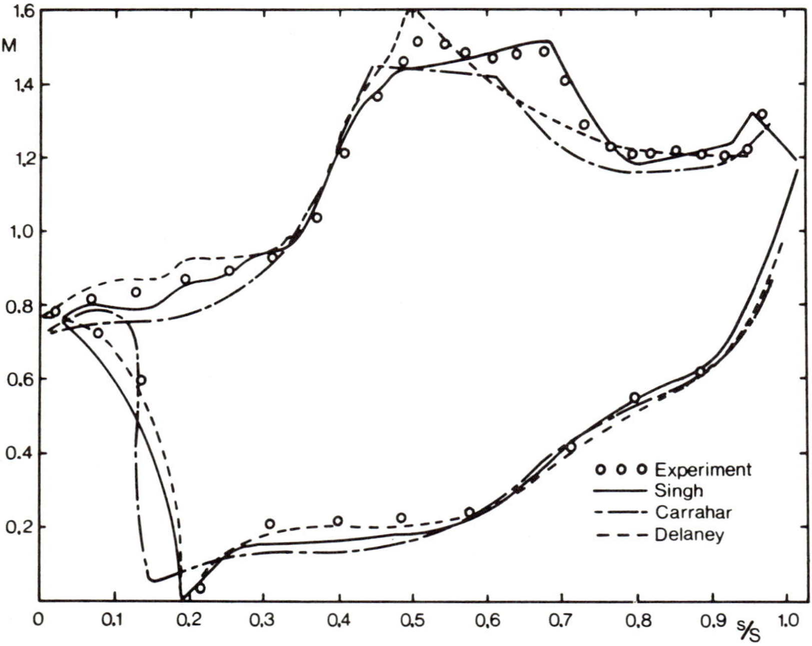

Fig. 10.6. Surface Mach Number distributions for RB profile of Graham and Kost[10.21].

Figure 10.6 presents the experimental data and corresponding predictions by three time-marching methods. The data are plotted as a function of distance around the profile rather than of axial distance in order to give good discrimination of data in the shock-boundary layer interaction region.

The first attempt to predict the results of Fig. 10.6 was made using the inviscid Denton-type time-marching program of Carrahar. The program gave a reasonable prediction but placed the shock impingement too early on the suction surface.

This defect was shared by the time-marching program of Delaney[6.70]. This was a 'hopscotch' method in which explicit and implicit finite difference formulations were used at alternate mesh points. An important feature of Delaney's approach is the use of a body-centred coordinate system for optimum representation of blade edges. Apart from the shock impingement region Delaney's method gave generally reasonable agreement with experiment.

The test data and a corresponding schlieren photograph showed clearly that a separation bubble was present in the shock impingement region. This had the effect of creating a plateau in the Mach Number distribution terminated by a rapid compression. These effects are clearly of a viscous nature and it would not be reasonable to expect an inviscid theory to predict them. The best theoretical prediction of this test case has come from a program which does incorporate viscous effects.

Singh[10.22] refined the Denton method to include an integral boundary layer calculation in each iteration. The computed boundary layers and wake were then applied to the base profile to produce a modified shape. Shock-boundary layer interaction effects were predicted using the integral method by taking very small meridional distance steps in the shock region, permitting a local thickening of the boundary layer. Figure 10.6 shows that the prediction obtained in this way was a good one, especially in the shock-boundary layer interaction region.

Time-marching methods have the advantage over potential equation solutions that they are not confined to low supersonic Mach Numbers and weak shocks. Their advantage over solutions to the full time-dependent Navier-Stokes equations is one of computing time. The latter methods are still prohibitively expensive and have inherent limitations in turbulent flows.

It appears that two features are desirable in a state-of-the-art time-marching solution. Singh has demonstrated that the capability to calculate and update boundary layer effects including those from shock wave interactions, and to incorporate these iteratively, gives a significantly improved capability. Delaney has demonstrated that the use of a body-centred grid system may give significant benefits especially in the important area of blade edge definition. The method described by Studerus[10.23] effectively incorporates both boundary layer effects and a body-centred grid.

10.2.3 Comparison with high speed rig tests. The test of agreement between cascade theory and experiment is an essential step in ensuring the validity of either procedure. It is not, in itself however, capable of testing the validity of application of cascade results, whether theoretical or experimental, to the performance of a real machine. This question can only be addressed by comparison of cascade results with those obtained from the machine. The value of annular cascades as a useful intermediate step was described in Chapter 3.

Paulon et al[3.89] have compared results of experimental research on linear, annular, fixed and rotating supersonic cascades. The agreement between plane, annular and rotating cascades was found to be good over most of the suction surface but poor on the pressure surface. The comparisons were better at low back pressure than at high back pressure due to separation difficulties in the plane cascade.

A good example of comparisons between time-marching theory and results from a corresponding annular cascade has been given by Studerus[10.23]. The annular cascade represented a high pressure turbine stator and the peak local Mach Number was around 1.4. Very good agreement between time-marching theory and the annular cascade measurements was obtained. Studerus went on to compare results from the same calculation method with measurements from the stator vanes of an air turbine. Here the peak Mach Number was 1.2 and the agreement was reasonable but not quite as close as for the annular cascade.

Available rotating rig measurements of high speed flows are mostly from transonic compressors. Instrumentation difficulties impede the collection of machine data. At high speeds blade stress concentrations make it virtually impossible to drill pressure tapping holes in rotor blades. Measurements have been made using thin surface-mounted pressure transducers and slip rings. Another technique, applicable only to the tips of unshrouded rotors, is the use of casing-mounted transducers. More recently considerable progress has been made employing laser anemometry to obtain velocity traverses at different spanwise locations.

Earlier measurements were confined to surveys of pressure distribution at the casing. Comparisons published by Prince[10.24] revealed deficiencies of time-marching calculations of a qualitative nature. Concern was expressed that measured losses across the main passage shock were less than predicted. It was also noted that the application of back pressure caused flow changes in supersonic regions on the pressure surface. Quasi-three dimensional time-marching calculations did not predict this.

It was not clear whether the discrepancies were due to:-

(i) Blunt leading edge effects causing propagation of information through a subsonic region.
(ii) Three-dimensional shock obliquity.
(iii) Annulus wall boundary layers having a strong three-dimensional effect on shock impingement on the casing.
(iv) Strong radial pressure gradients emanating from the sonic radius[10.25].

Laser anemometry traverses from a high-speed (550 m/sec) fan rotor [10.8] were relatively free from effects (i) and (ii) and also demonstrated reasonably good agreement between quasi-three-dimensional time-marching calculations and the experimental results. This suggested that effects (i) and (ii) might be important in the resolution of discrepant observations.

Fortunately some excellent traverses at different spanwise locations through a transonic compressor rotor have been made at DFVLR. These laser dual focus anemometry measurements have provided impeccable data for the evaluation of prediction methods. To complement the test data some excellent predictions have been made using time-marching methods.

A quasi-three-dimensional version of the original McDonald [6.61] theory was used in comprehensive predictions by McDonald et al[10.26]. Mach Numbers were predicted at four spanwise locations and at three different throttle settings. Three-dimensional versions of the Denton [3.86] program were used by Singh[10.27] and Sarathy [6.64]. Singh incorporated the boundary layer corrections described in the previous section and Sarathy was able to present useful data for Mach Number variation in the $r-\theta$ plane.

Each set of predictions is complementary and it is instructive to review the results together. At the peak efficiency condition the quasi-three-dimensional calculations of McDonald predict subsonic Mach Numbers over the pressure surface at 89% span from the hub whereas the measurements indicate an essentially sonic condition over most of the pressure surface in accordance with Prince's observations. The fully three-dimensional calculations of Singh and Sarathy gave predictions much closer to the measured results. The calculated $r-\theta$ plots of Sarathy, which replicate the measurements well, suggest that three-dimensional obliquity of the shock is the explanation and that when this is taken into account the shock angles and shock deceleration are correctly predicted as being smaller than the equivalent two-dimensional predictions.

Anomolous upstream propagation of disturbances in the supersonic region does seem to have occurred in the test data at the 89% span. This was not predicted by McDonald's quasi-three-dimensional calculations. Unfortunately results from the three-dimensional calculations of Singh and Sarathy which might have given insight on this phenomenon were not presented for the wide open throttle condition.

In general the McDonald predictions are impressive over a wide range of flow conditions. The Singh calculations, incorporating boundary-layer predictions, give the best accuracy and the Sarathy calculations are useful in giving information on three-dimensional effects.

Taken together the comparisons present evidence that a time-marching cascade theory can give very good predictions of transonic compressor rotor performance over a wide range of flow conditions. These predictions are enhanced by the use of a fully three-dimensional time-marching procedure and may be further refined by the incorporation of boundary layer predictions to give excellent agreement between theory and experiment.

10.3 Discussion on Validity of Theory and Experiment

Subsonic unseparated cascade flows can now be predicted with sufficient accuracy and reliability for design purposes. Comparisons of such predictions with data from rotating machines have indicated good agreement.

This conclusion may be judiciously extended to the special class of supercritical blading. Cascades can now be designed to support low supersonic local Mach Numbers without the presence of strong shocks. Calculations may be extended to the high supersonic Mach Numbers prevailing for the limit loading condition of turbine blading when there are no shocks present within the blade passage.

The presence of strong shock waves is the feature which presents most difficulty for prediction methods. Compressor cascades, for example, often have high inlet Mach Numbers; the blading supports a diffusion which is considerably greater than that on an isolated aerofoil, usually to a subsonic discharge Mach Number. The boundary layers tend to be relatively thicker than those of isolated aerofoils and shock reflections occur from adjacent blades. These are some reasons why the shocks are stronger and their structures more complex than normally encountered on an aerofoil. The robust time-marching calculations are able to cope with these difficulties whereas most methods developed for isolated aerofoils are not. It has been found that the representation of shocks in rotating machines is enhanced by use of three-dimensional time-marching and that accuracy is improved by the inclusion of boundary layer effects.

The capabilities of modern computational techniques in representing unseparated flows in two-dimensional cascades and rotating machines have therefore been demonstrated.

Flow fields for separated flows can not yet be reliably predicted. Furthermore demonstrated predictability is limited to pressure distributions and does not necessarily apply to flow angles and certainly not to loss coefficients.

For unseparated flows in turbine blading the favourable pressure gradients make the application of boundary-layer calculations to loss prediction reasonably dependable. Theoretical loss predictions only become hazardous in the presence of separation.

For compressor blading, losses are predicted most consistently by the diffusion factor approach. Two descriptions of design procedures from industry demonstrate the good agreement between cascade and compressor performance data[4.2] and the rational basis of diffusion factor-based loss correlations in boundary layer theory[10.28].

Despite the design success of the diffusion factor approach it should only be applied with considerable discretion as advanced aerofoil shapes are employed. Boundary-layer calculations should ideally be used for loss prediction but this procedure will be quite unreliable until more physical understanding is available on the phenomena of transition, separation and base flows.

This chapter has concentrated on pressure distributions and these have been shown to be quite calculable in many cases. There has been little discussion on the important question of loss prediction because techniques for this are not well advanced. The next priorities in the development of prediction techniques should be the routine, reliable and accurate prediction of losses and separated flows.

REFERENCES

10.1 KATZOFF, S. and HANNAH, N.E. Further comparisons of theoretical and experimental lift and pressure distributions on airfoils in cascade at low-subsonic speed. NACA TN 2391 (1951)

10.2 SPEIDEL, L. and SCHOLZ, N. Untersuchungen über die Strömungsverluste in ebenen Schaufelgittern. VDI Forschungsheft (1957)
10.3 ZUKOVSKI, M.I. Calculation of the flow past the turbomachine cascade. Moscow- Leningrad (1960)
10.4 NUMACHI, F. Surface pressure distribution on hydrofoil profiles in cascade. Report 2 - Accelerating cascade flow. Rep. Inst. High. Speed. Mech., Tohoku Univ., Vol 26 (1972)
10.5 GOSTELOW, J.P. The accurate prediction of cascade performance. Ph. D. thesis, Liverpool Univ. (1965)
10.6 MILLER, H.J. and SEROVY, G.K. Deviation angle estimation for axial-flow compressors using inviscid flow solutions. Trans. ASME, J. Eng. Power 97, 2 pg. 163 (1975)
10.7 STARKE, J. The effect of the axial velocity density ratio on the aerodynamic coefficients of compressor cascades. Trans. ASME, J. Eng. Power 103, 1 pg. 210 (1981)
10.8 WISLER, D.C. and MOSSEY, P.W. Practical application of LV systems to aero engine research and development. AGARD Lect. Ser. 90 (1977)
10.9 WESTPHAL, W.R. and GODWIN, W.R. Comparison of NACA 65-series compressor blade pressure distributions and performance in a rotor and in cascade. NACA TN 3806 (1957)
10.10 DOYLE, M.D.C. and SHAW, R. Pressure distributions on axial flow compressor blading and comparison with distributions on similar cascade blading. Adv. Probs. in Turbo., VKIFD (1965)
10.11 PEACOCK, R.E. and OVERLI, J. Compressor rotor blade performance in steady state operation. ARC 36,062 (1975)
10.12 DRING, R.P., JOSLYN, H.D. and HARDIN, L.W. An investigation of axial compressor rotor aerodynamics. Trans. ASME, J. Eng. Power 104, 1 pg. 84 (1982)
10.13 DRING, R.P. and JOSLYN, H.D. Measurement of turbine rotor blade flows. Trans. ASME, J. Eng. Power 103, pg. 400 (1981)
10.14 STANITZ, J.D. and SHELDRAKE, L.J. Application of a channel design method to high solidity cascades and tests of an impulse cascade with 90° of turning. NACA Rep. 1116 (1953)
10.15 KYNAST, G. Anwendung und Vergleich verschiedener Verfahren zur Berechnung der Druckverteilung an Schaufelprofilen inkompressibler Strömung. DFL-Bericht 62 (1960)
10.16 OSWATITSCH, K. and RHYMING, I. Uber den Kompressibilitätseinfluss bei ebenen Schaufelgittern starker Umlenkung. DVL-Bericht 28 (1957)
10.17 LAKOMY, C. Theoretical solution of high subsonic flow past two-dimensional cascades of airfoils. Trans. ASME, J. Eng. Power 97, 3 pg. 355 (1975)
10.18 SCHMIDT, E. Computation of supercritical compressor and turbine cascades with a design method for transonic flows. Trans. ASME, J. Eng. Power 102, 1 pg. 68 (1980)
10.19 BREUGELMANS, F.A.E. and STARKEN, H. The cascade and rotor section performance of a 9° 57' cambered DCA airfoil. Transonic Flows in Turbo., VKIFD, Brussels (1973)
10.20 FARN, C.L. and WHIRLOW, D.K. Application of time-dependent finite volume method to transonic flow in large turbines. In "Transonic Flow Problems in Turbomachinery" eds. T.C. Adamson Jr. and M.F. Platzer, Hemisphere (1976)
10.21 GRAHAM, C.G. and KOST, F.H. Shock boundary layer interaction on high turning transonic turbine cascades. ASME Paper 79-GT-37 (1979)
10.22 SINGH, U. Computation of transonic flows in cascade with shock and boundary layer interaction. VKIFD Lect. Ser. "Shock-Boundary Layer Interaction in Turbomachines" (1980)
10.23 STUDERUS, C.J. Aerodynamic effects of surface cooling-flow injection on turbine transonic flow fields. AIAA Paper 79-0048 (1979)
10.24 PRINCE, D.C.Jr. Discussion of Gostelow, J.P. Review of compressible flow theories for airfoil cascades. Trans. ASME, J. Eng. Power, pg. 77 (Jan. 1974)

10.25 EPSTEIN, A.H., KERREBROCK, J.L. and THOMPKINS, W.T.Jr. Shock structure in transonic compressor rotors. AIAA J, 17, 4 pg. 375 (1979)
10.26 MCDONALD, P.W., BOLT, C.R., DUNKER, R.J. and WEYER, H.B. A comparison between measured and computed flow fields in a transonic compressor rotor. Trans. ASME, J. Eng. Power 102, 4 pg. 883 (1980)
10.27 SINGH, U.K. A computation and comparison with measurements of transonic flow in an axial compressor stage with shock and boundary layer interaction. Trans. ASME, J. Eng. Power 104, 2 pg. 510 (1982)
10.28 KOCH, C.C. and SMITH, L.H.Jr. Loss sources and magnitudes in axial flow compressors. Trans. ASME, J. Eng. Power 98, 3 pg. 411 (1976)

CHAPTER 11

Design Application of Cascade Information

The machine designer needs a store of relationships upon which to draw when making hardware decisions. There exists no such complete compilation for cascades but in this chapter the author will attempt to reference some of the more productive sources and present data on the effects of some of the principal parameters.

These parameters are often well-defined geometric properties of a cascade, such as the thickness of a blade or its distribution along the chord. Many variables, however, are aerodynamic in nature. Some of these can be clearly defined and are outlined in a second section. Others depend upon more subtle interactions, such as tip-clearance effects and will be discussed in a third section. Finally some conclusions are drawn on the design application of cascade information and the future role of cascade aerodynamics.

11.1 The Effects of Geometric Parameters

11.1.1 Thickness/chord ratio. It has been claimed[11.1] that for low speed fans, where stress considerations permit thin blading, the use of sheet-metal blades offers an inexpensive solution without aerodynamic penalty. The losses associated with cambered plates become substantial at thicknesses around 7% [11.2]. The selection of thickness and other geometric variables for axial flow fans has been reviewed by Wallis[11.3]. The effect of thickness on drag for isolated aerofoils at low speed is shown in Fig. 11.1 taken from ref. 11.4. In general the thinnest profile which can be used for a design will have the best performance. At transonic speeds a low thickness / chord ratio of the order of 0.06 is most important.

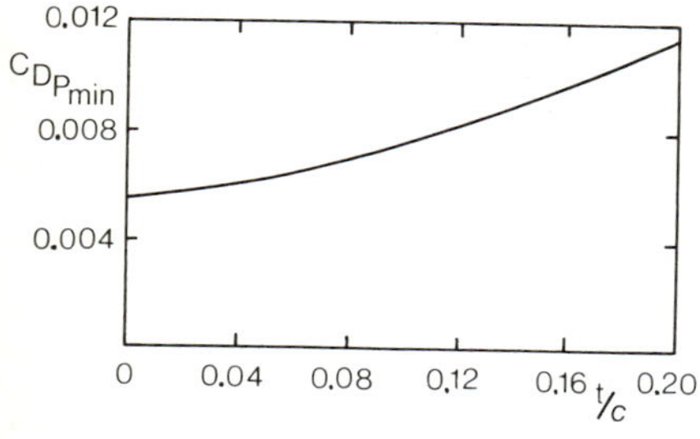

Fig. 11.1. Effect of thickness on drag for isolated aerofoils.

As a result of extensive cascade testing NACA produced corrections for minimum loss incidence angle and deviation angle as a function of various parameters including maximum thickness/chord ratio. The effects of this parameter on minimum loss incidence angle and on deviation angle are represented as proportionality coefficients K_i and K_δ respectively. These were evaluated experimentally for 65-series and double-circular-arc aerofoils and are presented in Fig. 11.2. Although these K parameters are straight multipliers on incidence and deviation angle, additional effects due to blade camber and solidity are described in ref. 5.84.

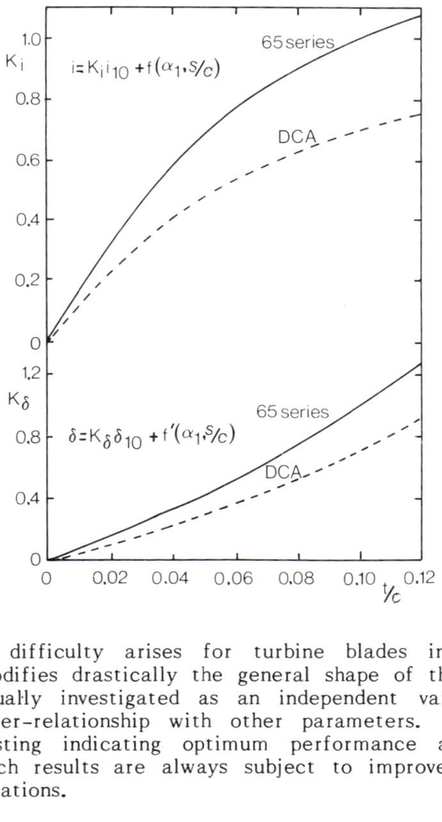

Fig. 11.2 Thickness corrections for zero-camber reference incidence and deviation angles.

A difficulty arises for turbine blades in that changing the thickness/chord ratio modifies drastically the general shape of the blade. This parameter is therefore not usually investigated as an independent variable for turbine blading because of its inter-relationship with other parameters. Results are available from turbine blade testing indicating optimum performance at thickness/chord ratios of around 0.14. Such results are always subject to improvement following detailed blade shape modifications.

11.1.2 Leading edge thickness. Because it is a difficult parameter to separate from blade shape as a whole, the leading edge thickness effect has received little attention for turbine blading. Turbine blades can stall from the leading edge, on either surface, but because the leading edge shapes (and consequent pressure gradients) of modern turbine blades are difficult to define in terms of any single parameter turbine leading edge effects appear unlikely to lend themselves to a correlation approach.

Compressor blades, on the other hand, are usually thinner. For compressor blades the leading edge thickness (generally expressed as a radius) is more readily defined and can be more critical. Transonic and supersonic compressor blading is most sensitive to leading edge thickness effects. Before that problem is considered the scanty evidence on leading edge effects at subsonic speeds will be reviewed.

The isolated aerofoil data of ref. 11.4 show that reasonable values of maximum lift coefficient are obtained with leading edge radii in the range 1% to 2% chord. As the latter value is approached, for aerofoils of 12% thickness, further rises in attainable lift coefficient are curtailed.

The results presented in Fig. 11.3 are taken from ref. 11.5 and confirm that sharp leading edges are desirable for high-speed operation. Interestingly there is some conflict between these data and isolated aerofoil data concerning operating range. Figure 11.3 indicates that sharp leading edges give a wider operating range than thin ones. The calculations of Wright [11.6] clearly indicate a wider operating range for thicker leading edges so that this point must be considered unresolved.

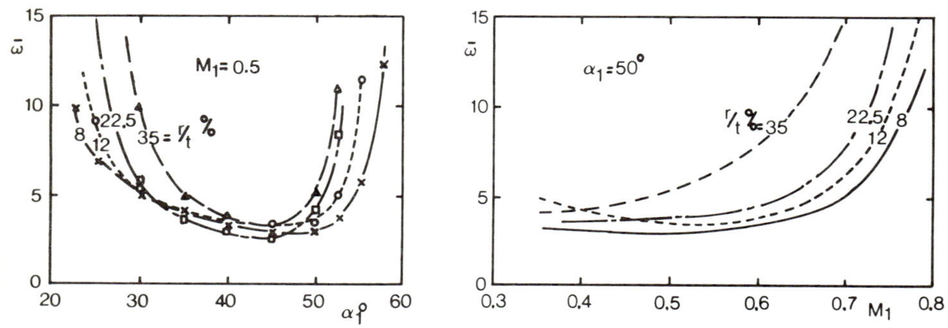

Fig. 11.3. Effect of leading-edge radius on cascade performance.

For supersonic inlet Mach Numbers the effects of leading edge thickness are particularly strong. In theory the bow shock will be detached for a non-zero leading edge radius and the prediction is confirmed by observation. Bluntness of the leading edge changes the unique incidence angle of attack. As a result of blockage the effective flow area is reduced. The axial velocity component and mass flow are reduced and the inlet flow angle is increased.

Because some proportion of the flow upstream of a blunt leading edge is effectively subjected to a normal shock compression rather than an oblique shock the total pressure losses in the leading edge region become stronger as the leading edge becomes blunter. Bow shock losses depend primarily on Mach Number and leading edge radius; other parameters are of secondary importance. Early calculations were based on the simplifying premise of simple wave flow. Klapproth [11.7] calculated the loss effect by assuming the bow shock to be of hyperbolic shape in the leading edge region. The inlet flow angle was assumed identical to that of a straight suction surface. The calculated loss was found to increase linearly with edge thickness for a given Mach Number. Subsequent results obtained by Reddy and Klapproth [11.8] using a simple wave analysis are presented in Fig. 11.4. Comparisons with the available experimental data have shown that simple wave analyses underpredict measured shock losses. The curvature of the bow shocks induces entropy gradients normal to the streamlines both upstream and within the passage. The consequent rotationality of the flow is a major contributor to losses. York and Woodard [11.9] have used a method of characteristics solution giving higher levels of loss which more closely represent the very high levels observed in experimental work. The results of York and Woodard for a cascade of double-circular-arc blades having a space/chord ratio of 0.8 are presented in Fig. 11.4. The theoretical predictions of ref. 11.9 are reported in ref. 10.28 to be in good agreement with method of characteristics predictions by Prince. An obvious and pressing requirement is for leading edge thickness effects to be predicted correctly by calculation techniques such as time-marching. For this to be possible considerable improvements in the accuracy of blade edge definition are needed for most time-marching methods.

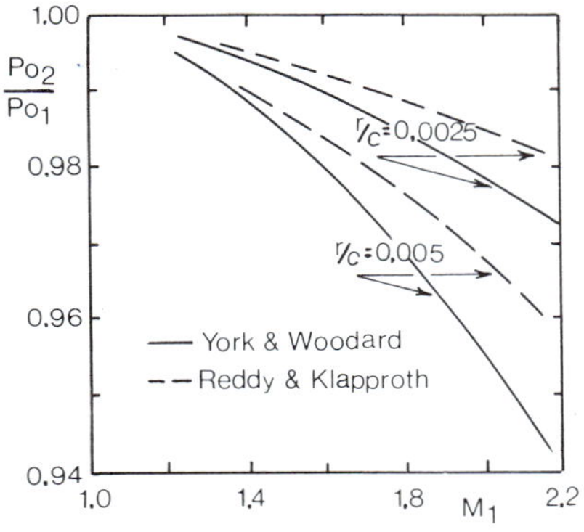

Fig. 11.4. Entrance region losses as a function of inlet Mach Number.

11.1.3 Trailing edge thickness. For compressor blading useful early measurements of the effect of trailing edge thickness were presented in ref. 11.10. As a result it may be concluded that trailing edge thicknesses up to 2% of chord should have little effect on aerodynamic performance.

Inoue and Kaneko[11.11] have performed tests on constant thickness blade sections having different reductions in trailing edge thickness. By thinning a 6% thick blade to 4% thickness at the trailing edge a significant loss reduction is obtained. Further thinning results in no improvement but rather poorer performance in deflecting the flow.

Compressor blades usually exhibit a degree of turbulent separation upstream of the trailing edge. It has been shown that this separation contributes up to one third of the total cascade loss[11.12]. Analyses of trailing edge thickness effects have been performed by Reeman and Simonis[11.13] and Stewart[7.71].

The effects of trailing edge thickness on turbine performance are more extensive than for compressors. Quite apart from vortex shedding and supersonic discharge flows there are quite pronounced trends which are best presented as correlations of experimental data. The data are presented as integrated loss coefficients; ref. 11.14 gives a good example of local pressure variation at different stations.

A considerable body of data exists relating the incremental total pressure loss resulting from trailing edge thickness variation to the value of trailing edge thickness. The correlations are most consistent when the trailing edge thickness is normalised by the throat dimension.

In a simple proportional relationship attributed to Flyugel[3.87] the incremental losses are given as a function of the edge thickness, normalised by the throat. Although stated for enthalpy losses the same proportional relationship applies approximately to total pressure losses when the Mach Number is low[1.3].

The results of many investigators (refs. 3.80, 3.87,11.15,11.16) confirm the linear relationship, only the constant of proportionality varying over a range from 0.10 to 0.27 for different cascades. In earlier work Markov[11.17] recommended a value of 0.41 but this is considerably higher than the results of other investigators.

In an important contribution Zaryankin[11.18] has shown that such a wide variation is partly explained by the fact that different investigators made measurements at different distances downstream. Zaryankin presented an approximate equation for the variation of loss increment with axial distance downstream, ξ, valid for $1.5<(\xi/\tau_e)< 6$,

$$\bar{\zeta} = (0.1 + 0.002(\xi/\tau_e))\cdot(\tau_e/\sigma) \quad . \qquad (11.1)$$

The limit $\bar{\zeta} = 0.26\,(\tau_e/\sigma)$ was recommended for values of $\xi/\tau_e > 10$ and thin trailing edges $(\tau_e/\sigma < 0.1)$.

Whilst it might be expected that the above results would be compatible with the Soviet data of ref. 3.87 the linear behaviour of the British and German data of refs. 11.15 and 11.16 is identical to that of the Soviet data. The compilations are most easily compared at low Mach Numbers since the Soviet data were presented on an enthalpy loss basis.

The proportionality constant is essentially a negative base pressure coefficient and the available data were readily correlated by the linear relation,

$$\bar{\zeta} = \bar{\zeta}_0 - C_{pb}\cdot\tau_e/(s\sin\beta_2) \quad . \qquad (11.2)$$

Nevertheless there remains a wide range of unexplained variation in C_{pb}.

In section 8.2.1 it was indicated that the ratio of boundary layer to blade thickness at the trailing edge is a principal determinant of C_{pb}. Compilations of experimental data from aerofoils having square trailing edges have been given by Hoerner[8.33] and Herbert[8.16]. The data are mutually consistent and reveal the presence of a very strong base drag ($C_{pb} \simeq -0.6$) when the boundary layer is very thin compared with the trailing edge. Hoerner used an ejector pump analogy to describe how the energetic free stream pumps away stagnant air in the base region. The boundary layer acts as an insulating sheet inhibiting the pumping effect. This analogy, which neglects periodic effects, is clearly an oversimplification.

These data fairly represent the range of C_{pb} for some blading with a square trailing edge. Cox[8.35] has provided an interpretation of the data of Bauer[8.12] for a similar situation. However as the boat-tail angle rises towards 10° or, for rounded trailing edges, as $\Sigma\theta/\tau_e$ rises above 0.1[11.19] significant base pressure recovery is obtained. The base drag of typical turbine blades is therefore weaker than that of truncated aerofoils. Trailing edge shape is an important parameter needing separate consideration. The results of Deych[3.87] indicate that lowest losses are achieved with tapered edges; rounded edges give a 50% loss increase and square edges a 100% increase. The results of ref. 11.20 suggest an even stronger effect. Further work is therefore required to account for the effects of trailing edge shape and momentum thickness on C_{pb}.

Attempts have been made to correlate base pressure coefficients in the form

$$C_{pb} = -\alpha^*(\Sigma\theta/\tau_e)^{\beta^*} \qquad (11.3)$$

where α^* and β^* are empirically-derived constants. Hoerner's data were well represented by values of α^* and β^* of 0.14 and $-\frac{1}{3}$ respectively. Herbert[8.27] has proposed that the weaker base drag of tubine blading be represented by values of 0.066 and $-\frac{1}{4}$.

Consideration of the observed range of proportionality constants, between 0.10 and 0.27, suggests that whereas the correlations of Hoerner and Cox, derived from square edged aerofoils represent too strong a base drag, Herbert's correlation tends to understate the base drag. Values of α^* and β^* in equation (11.3) of 0.1 and $-\frac{1}{4}$ would be more consistent with observed trends. This recommendation requires further investigation with attention to the outlet angle as an additional variable. There remain anomalies which might not be explained by application of this approach. For example Soviet investigators[3.87] claim that nozzle blading produces the lowest proportionality constants whereas Kacker and Okapuu[11.21] attribute the highest proportionality constants to nozzle blading. In principle, however, if the trailing edge momentum thickness can be obtained it is possible to predict C_{p_b} and hence the effects of edge thickness on loss.

11.1.4 Camber and stagger angles. Axial flow compressor development has particularly benefited from an isolation of specific geometrical parameters and the generation of appropriate data both experimentally and theoretically. Systematic descriptions of design procedures for compressors are given in refs. 1.2 and 5.84 . The setting of optimum incidence conditions for a cascade and the prediction of the resulting deviation angle are dependent upon both camber and stagger angles. The work of Carter[11.22] has provided correlations for both incidence angle and deviation angle as a function of camber and stagger.

Carter referred performance to an optimum incidence at which the cascade operated at its maximum lift/drag ratio. For selection of optimum incidence Fig. 11.5 gives as excellent design guide. Optimum incidence is presented as a function of optimum camber angle, space-chord ratio and deflection. The influences of stagger and outlet angle are assumed to be negligible. Should the designer prefer to use, rather than maximum lift/drag ratio, a criterion minimising losses and giving lower suction surface velocities, the optimum incidence values should be lowered by up to 5°, especially for blading having low values of camber.

The rule for deviation angle known as Carter's rule,

$$\delta^* = m\theta\,(s/c)^n\,, \qquad (2.1)$$

was discovered by Constant[2.14], extended by Howell[11.23] and presented in its final form by Carter[11.22].

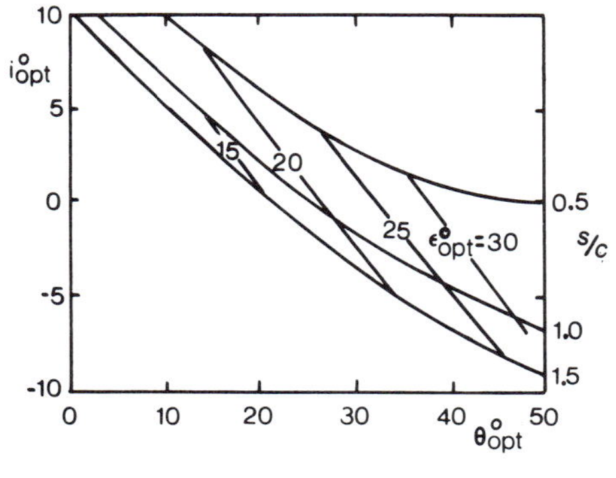

Fig. 11.5. Carter's correlation for optimum incidence.

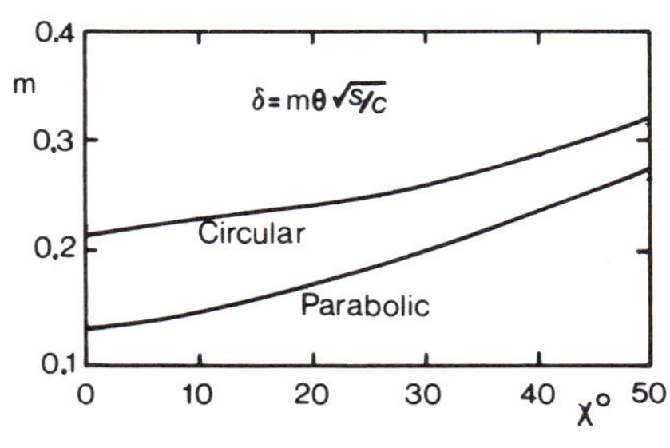

Fig. 11.6. Carter's correlation for deviation angle.

For compressor cascades n has a value of one half and for inlet guide vanes, unity. Carter contributed the curves of Fig. 11.6 which enable a suitable value of m to be extracted for camber lines having circular arcs and parabolic arcs with the maximum camber at 40% of chord.

11.1.5 Space/chord ratio. Selection of an appropriate value of space/chord ratio for compressor blading at subsonic speeds is adequately covered in the standard works on compressor design [1.2, 5.84]. The effect of space/chord ratio on incidence and deviation angles is allowed for in Carter's correlations as presented in Figs. 11.5 and 11.6. Subsequent design experience has tended to confirm these data.

Spacing effects are quite critical in transonic compressors. The spacing is determined by consideration of blade inlet angle and Mach Number which govern throat selection. Shock and diffusion losses must also be considered in establishing the optimum spacing for a given duty.

Figure 11.7, replotted from the data of Heilmann et al [11.24], presents the variation of total pressure loss coefficient in the transonic range for three different space/chord ratios. The cascade tests were on 6% thick double-circular-arc blading representative of a transonic rotor tip section. The data show clearly that at an inlet Mach Number of 1.4 performance is damaged by the use of space/chord ratios around unity typical of subsonic design practice. Transonic compressor design had been hampered for some years by the use of excessively wide tip spacings. Rotor tip space/chord ratios are now usually in the range 0.7 to 0.8.

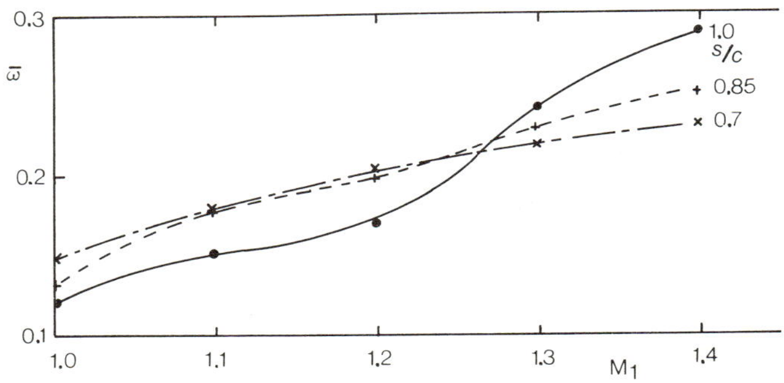

Fig. 11.7. Effect of inlet Mach Number on the loss coefficient of double-circular-arc blading at different space/chord ratios.

Blade spacing in subsonic turbines is set by the application of established correlations based on cascade data. Figure 11.8 presents a composite plot of loss coefficient as a function of outlet angle and space/chord ratio. This correlation, which was published in 1951 by Ainley and Mathieson[3.80], still provides a reliable basis for turbine design. Cohen et al[11.25] have given a helpful description of its application.

Ref. 3.80 gave correlations for secondary and tip clearance losses and, whilst these were appropriate for many turbines, Dunham and Came[3.57] found it necessary to modify the correlation to predict losses for small turbines and low aspect ratios. Kacker and Okapuu[11.21] have subsequently introduced compressibility effects and shock losses and have reduced the profile loss by 50% to allow for efficient modern blade sections. It was demonstrated that this new analysis system could predict the efficiency of a wide range of turbines to $\pm 1\frac{1}{2}$%.

Criteria for optimum space/chord ratio based on maximum lift coefficient have been given by Howell[11.26] and Zweifel[11.27]. Ref.11.28 demonstrates that significant discrepancies are present between these spacing criteria.

For turbines operating with supersonic discharge flows selection of spacing results from application of a more sophisticated design approach based on high speed cascade theory and experimental work.

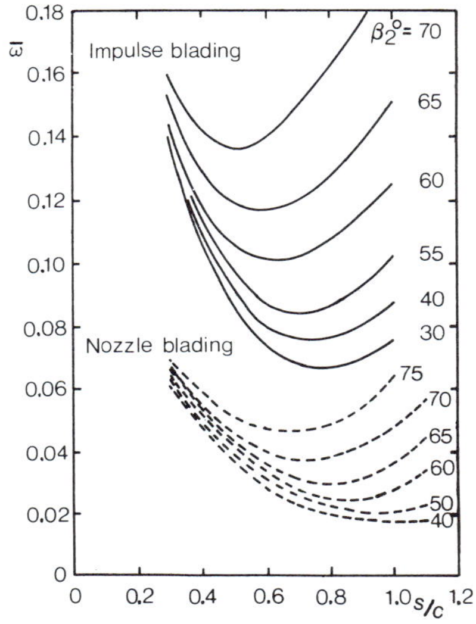

Fig. 11. 8. Turbine loss coefficient correlation of Ainley and Mathieson.

11.1.6 Shape of camber line. The most common shape used for the camber line is the simple circular arc. The use of a circular arc camber line has the advantage of leading and trailing edge blade angles which are readily-defined. This is important for accurate specification of incidence and deviation angles. The convenience of the circular arc camber line is such that other camber line shapes are often approximated by the closest corresponding circular arc. This practice is common, for example, in the use of NACA 65-series compressor blades. The originally-specified camber line for these blades, designated A_{10}, has the distinctive feature of infinite slope at the leading and trailing edges.

After weighing the merits of circular arc and NACA-style camber lines for low-speed industrial fans Wallis [11.3] decided that the convenient circular arc camber line should be modified by incorporation of a small amount of nose droop to give more favourable inlet flow conditions.

Desirable performance features of compressor blading may be readily accentuated by the use of parabolic arcs or more than one circular arc to define the camber line shape. Useful data for assessing the extent of the trade-offs is given in ref.11.5 .

Blades with the position of maximum camber forward of the midchord tend to have good low speed performance, with high stalling lift, a low critical Mach Number (or cavitation velocity) and a high choking mass flow. Blades with the position of maximum camber aft of the midchord will have a lower stalling lift coefficient at low speeds, a lower choking mass flow, but will benefit from having a higher drag critical Mach Number or cavitation velocity.

This latter advantage is so important that for fan blading operating at 427 m/sec (1400 ft/sec) there is a 2% efficiency advantage and an 8% stall margin advantage associated with the use of a camber line having a straight forward portion as compared with a circular-arc camber line. The results of Fig. 11.9 drawn from the results of ref.11.29 demonstrate this effect. Here camber ratio is defined as the ratio of camber occurring in the supersonic portion of the flow to the total camber. Stall margin is taken to be the percentage by which the ratio (weight flow/pressure ratio) at design point exceeds that quantity at the stall point.

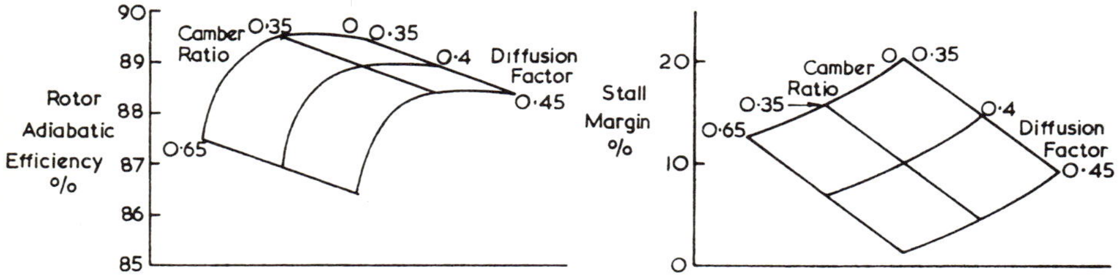

Fig. 11.9. Effect of camber ratio and diffusion factor on efficiency and stall margin for transonic compressors.

Similar considerations apply to turbine blading. Many workers, but most notably Deych and Troyanovskiy[3.87], have shown that whereas for subsonic flows convex curvature is appropriate for the rear portion, as Mach Numbers increase through transonic to supersonic the optimum shape changes from convex curvature, through zero to concave curvature corresponding to the converging-diverging nature of the passage.

11.1.7 Thickness distribution. According to Carter[11.5] variation in position of maximum thickness has much the same effect as varying the position of maximum camber. This conclusion was limited to the operation of compressor blading at subsonic speeds but again Carter was able to present data to support the statement.

Some shapes commonly employed for subsonic fan and compressor blading are given in Table 11.1. The C7 and NACA-65 section having maximum thickness at 40% of chord are most suitable for operation at high subsonic speeds.

x/c%	0010	C4	65-Series	C7	T6
0	0	0	0	0	0
1.25	1.58	1.65	1.17	1.51	1.17
2.5	2.18	2.72	1.57	2.04	1.54
5.0	2.96	3.08	2.18	2.72	1.99
7.5	3.50	3.62	2.65	3.18	2.37
10.0	3.90	4.02	3.04	3.54	2.74
15	4.46	4.55	3.67	4.05	3.40
20	4.78	4.83	4.14	4.42	3.95
30	5.00	5.00	4.76	4.86	4.72
40	4.84	4.89	5.00	5.00	5.00
50	4.40	4.57	4.81	4.86	4.67
60	3.80	4.05	4.15	4.43	3.70
70	3.05	3.37	3.16	3.73	2.51
80	2.19	2.54	1.99	2.78	1.42
90	1.21	1.60	0.81	1.65	0.85
95	0.67	1.06	0.31	1.09	0.72
100	0.11	0	0	0	0
L.E. RAD	1.10	1.20	0.69	1.20	1.20
T.E. RAD		0.60		0.60	0.60

Table 11.1. Ordinates (y/c%) for 10% thick blade sections.

For operation in the transonic range it is necessary to move the location of maximum thickness towards the rear. The midchord location is well suited to low transonic operation, as reflected in the popular double-circular-arc sections. For higher transonic speeds it is desirable to move the position of maximum thickness even further aft. Under such circumstances use of multiple circular arc sections as described in ref. 11.29 is recommended.

A typical turbine blade form, for use with a parabolic camber line (maximum camber around 40% chord) is the T6 form given in Table 11.1. Other such series of turbine blades were developed by NACA[11.30]. A method of blade profiling often used in the USSR is the construction of lemniscate curves defined by the equation

$$(x^2 + y^2)^2 = k^2(x^2 - y^2) \quad . \tag{11.6}$$

An advantage of this approach is that the point of greatest curvature may be placed at any section of the channel. Smooth changes of curvature exist at all stations.

Although many turbine designers have made use of the above approaches there is an increasing tendency for designers to perform calculations by the direct and prescribed velocity distribution methods described in Chapters 5 and 6 to arrive at an optimum blade shape for each particular section. The additional effort involved has improved turbine performance substantially and has well repaid the investment.

11.2 The Effects of Aerodynamic Parameters

11.2.1 Incidence angle. The performance of a rotating machine, especially of the work-absorbing variety, is greatly influenced by the design decisions made in the setting of incidence angle. The incidence setting is dependent on the specified operating conditions; maximisation of such parameters as efficiency, loading and operating range each calls for a different incidence setting. A typical design decision will tend to favour one of these at the expense of the others.

For compressor blading at subsonic speeds the results given by Carter, as described in section 11.1.4, provided reliable setting data. The value of incidence thus obtained should be reduced if the minimum loss condition is desired. Ref. 5.84 is also an invaluable source of blade incidence-setting data. For assessing the range of off-design operation resulting from incidence variations the data of Howell[11.23] are most useful.

Good correlations are available for assessing the effects of incidence variation on turbine cascade performance. The works of Soderberg and of Ainley are particularly relevant and are described by Horlock[1.3].

Because of the adverse pressure gradients involved and the generally sharper leading edges, compressor blading tends to be much more sensitive to changes in incidence angle than turbine blading. This is especially the case as supersonic speeds are approached when the incidence becomes uniquely specified. This phenomenon is explained by Scholz[7.9]. The importance of correctly aligning blading in high speed compressor cascade testing is emphasized by Mikolajczak et al[4.2] who quote a 20% loss variation for 1° incidence misalignment.

11.2.2 Reynolds' Number. In Fig. 2.6 the effects of flight altitude and Mach Number on blade Reynolds' Number were summarised. For aircraft engines in flight the operational blade Reynolds' Number may vary by a factor of 15. This would typically correspond to the range $10^5 < Re < 1.5 \times 10^6$ for a medium thrust fighter engine.

The most comprehensive series of experiments on Reynolds' Number effects in compressor cascades was performed by Rhoden[2.23]. The Reynolds' Number range between 3×10^4 and 5×10^5 was covered for C4 cascades having cambers of 20°,

30° and 40° over a range of stagger and incidence. Rhoden's data were substantially in agreement with Howell's predictions[11.23] for positive incidences but revealed additional performance deterioration at negative incidence. The occurrence of complete laminar separation at low Reynolds' Numbers was a major contributor to performance degradation.

Composite plots of loss coefficient for compressor cascades as a function of Reynolds' Number are given in ref. 5.84. These show that as the Reynolds' Number is reduced to a critical value between 0.5×10^5 and 2.5×10^5 the losses rise abruptly. The deviation angle has a similar tendency. The critical Reynolds' Number varies appreciably with blade shape, incidence angle, turbulence level (Fig. 2.7) and aspect ratio[11.31]. Performance variation below the critical point is also unpredictable.

A representative example of current compressor design usage[10.28] assigns an increase in momentum thickness as $Re^{-0.166}$ with a decrease in Reynolds' Number to a critical value of 2×10^5. Below that value the loss is taken to increase as $Re^{-0.5}$.

Considerations for aircraft turbines are, at first sight, similar. For axial turbines Horlock[1.3] suggests that a loss variation proportional to $Re^{-0.25}$, as given by Soderberg, is representative in the turbulent regime. The corresponding variation given by Ainley and Mathieson[3.80] is $Re^{-0.2}$.

A reasonably comprehensive discussion of Reynolds' Number effects is given in ref. 11.32. Caution should, however, be exercised in making general statements because of the wide differences in blade shape. For example the cascade tests of Bridle[11.33] indicated that reaction blading was less sensitive to Reynolds' Number variation than impulse blading. This conclusion is at variance with the corresponding Soviet data[4.4]

In Fig. 11.10(a) the results of a number of cascade tests are plotted to illustrate the wide variation in turbine blade performance with Reynolds' Number. Scholz[7.9] quoted

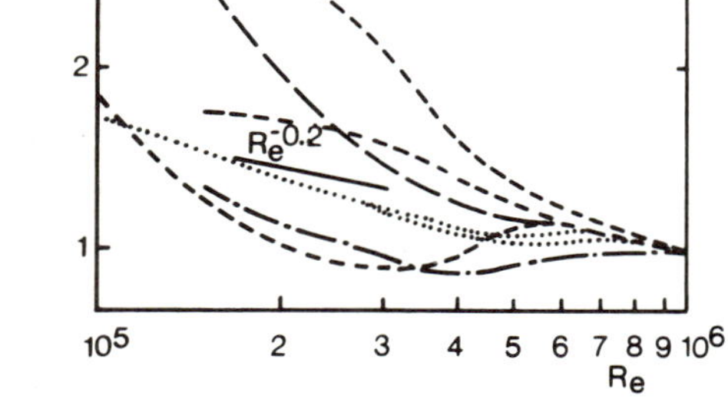

(a) Results of Dring and Heiser for $10^5 < Re < 10^6$.

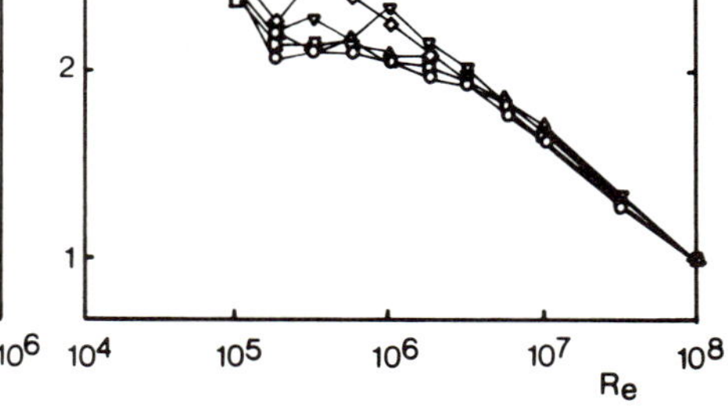

(b) Results of Denton for $10^4 < Re < 10^8$.

Fig. 11.10. Effects of Reynold's Number on normalised loss of turbine cascades.

a number of cascade tests covering a similar Reynolds' Number range and exhibiting similarly erratic behaviour. In ref. 11.34, from which Fig. 11.10(a) was drawn, Dring and Heiser comment "...no single scaling rule appears to apply. It is upon application of the analyses that are now available that the situation begins to become clear".

A wider perspective is in order and is given by consideration of steam turbine blading. For modern condensing plant the blade Reynolds' Number may fall by a factor of 1000 from inlet to discharge. Craig and Cox[11.35] were accordingly impelled to remark "...any correlation of cascade data which neglects the Reynolds' Number of test is of little value". Denton's numerical results[11.36] for a wide variety of nozzle and rotor blading cover the range $10^4 < Re < 10^8$; Fig. 11.10(b) gives typical examples of these re-plotted and normalised to the loss value at $Re = 10^8$.

Whereas Fig. 11.10(a) displayed erratic behaviour that of Fig. 11.10(b) is relatively orderly. In the turbulent regime ($Re > 10^6$) the loss varies approximately as $Re^{-0.2}$. In the laminar regime ($Re < 2 \times 10^5$) the loss varies more rapidly, approximately as $Re^{-0.4}$. In both regimes the results are closely bunched, thus enhancing predictability. In the crossover range between the laminar and turbulent regimes the behaviour is apparently as erratic as that of Fig. 11.10(a). The behaviour of turbine blading in the crossover range is unpredictable because of the inadequate state of existing information on boundary layer transition. The observations of Dring and Heiser are undoubtedly true for the Reynolds' Number range which they considered. Fortunately outside of that range the behaviour of blading is more predictable.

The Reynolds' Number dependency of both compressor and turbine blading is therefore similar in that outside the critical range, of approximately $10^5 < Re < 10^6$, fairly similar power law dependency prevails. The behaviour in the crossover range $10^5 < Re < 10^6$ tends to be less predictable for both compressors and turbines. Within this range there are important differences for the two classes of blading. For compressor blading the changes in the critical range are relatively abrupt ones associated with boundary-layer separation phenomena. Hysteresis in the loss-Reynolds' Number characteristic may be present and its magnitude may be affected by the turbulence level[7.53]. Figure 2.7 showed that the location of the abrupt loss increase is a function of turbulence level. If performance of compressor blading in this crossover range is to be predicted further progress in predicting laminar and turbulent separation will be necessary. For turbine blading separation is less common and boundary-layer transition determines performance in the critical range. As explained in Chapter 7 transition predictions which correctly account for free stream turbulence effects are not yet available. The flow in the crossover range is intermittently turbulent and the availability of reliable data for the start and end of transition should facilitate the predictions of such apparently disorganised behaviour as that of Fig. 11.10(a).

11.2.3 Free-stream turbulence level. The amplitude of velocity fluctuations within the blade rows of turbomachines is generally a significant proportion of the free-stream velocity level. The turbulence in a machine is not isotropic nor particularly random, having fundamental frequencies due to blade-passing phenomena etc.[8.154] . Nevertheless the use of modern instrumentation techniques makes it possible to discriminate between periodic and random effects and to subsequently superpose the results of appropriate cascade tests.

The discussion of Fig. 2.7 in the previous section has indicated the favourable role which free-stream turbulence can play in lowering the critical Reynolds' Number. Turbulence level and Reynolds' Number are closely linked in this way and it was natural to ask whether turbulence variations could be expressed in terms of an effective Reynolds' Number. Drawing on the work of Martlew[2.35] it was argued in section 2.6 that although closely related, the Reynolds' Number and turbulence level should be treated as separate variables. By using grids placed upstream of the cascade Martlew obtained turbulence levels of 0.18%, 1.5% and 3%. He investigated the effect of these alterations upon pressure distributions and concluded that the previously accepted practice of equating turbulence level with an effective Reynolds' Number, based upon sphere tests, represented an over-simplification of the situation. Martlew obtained large performance variations for different turbulence levels and demonstrated that cascade tests should be carried out in a reasonably turbulent flow and that the turbulence level should always be quoted.

An example of this realisation is given in ref.11.37 where it is concluded that "in order to obtain realistic performance data, cascade tests of turbine blades should be conducted with an air turbulence level compatible with that experienced in the turbine".

In measurements downstream of the blading of a crudely-designed fan Kiock [11.38] measured high turbulence levels up to 21% behind the first rotor. This level did not rise appreciably after two subsequent rows of blades. Kiock obtained the results of Fig. 2.9 in an aerodynamically-designed six-stage compressor where the turbulence was fairly constant at levels between 4.8% and 8% after the first three stages.

Kiock did observe a pronounced and consistent tendency for measured turbulence level to vary with Reynolds' Number in the critical Reynolds' Number range. At low Reynolds' Numbers the turbulence level was around 14% falling to a level around 4% at Reynolds' Numbers above 10^5.

The measurements of ref. 11.39 indicated a turbulence level varying between 2.2% and 3.4% in a single stage axial compressor. The measurements by Evans in Fig. 2.10 have turbulence levels varying between 2% at the maximum flow condition and 4.3% at stall.

Section 2.6 referred to disturbance levels (in a 500MW steam turbine) of 18% and a random turbulence level in a turbine stage of 8.8%.

The above range of free-stream turbulence levels, between 2% and 21%, may therefore be taken to be that which should be simulated in cascade experiments. It is customary to design most wind tunnels for as low a turbulence level as possible. This practice has been preserved for cascade wind tunnels and has the advantage that an initially low turbulence level can be raised in a controlled fashion whereas the converse procedure is not usually possible. Because of the disparitiy between good wind tunnel practice ($Tu < 0.3\%$) and machine operation ($2\% < Tu < 21\%$) many cascade results obtained without a suitable turbulence generating device are quite suspect.

Obviously the very high turbulence levels of many turbines are difficult to simulate in a controlled fashion. Schultz et al [4.28] cite the generation of a turbulence level of 4.2% by placing an array of horizontal rods of diameter 12.7mm spaced 31.1mm apart some 413mm upstream of the blades. It was reported that attempts to raise the turbulence level above 5% had resulted in a field which was neither homogeneous nor isotropic.

To simulate very high turbulence levels different techniques are required. Gibson in the discussion of ref.2.27 has reported the use of a moving grid. Bayley's solution 9.47 is to place a rotating squirrel cage upstream of the blading. A turbulence level of up to 32% appears to be achievable in this way. Krishnamoorthy [9.48] was able to obtain a level of 21.3% in a controlled fashion using a novel blowing device.

Although variation in turbulence level does affect the behaviour of laminar and turbulent [11.40] boundary layers its main local effects are on transition and the laminar separation bubble. As with the effect of Reynolds' Number, this behaviour is not yet predictable and thus overall correlations for the effect of turbulence level on blade performance are not particularly meaningful. For the present, attention is concentrated on the effect of turbulence on transition and on laminar separation.

Because of the considerable amount of experimental work on flat plates confidence in the zero pressure gradient data of Fig. 7.1 is high for transition inception, although data on completion of the transition process are fewer and more scattered.

The trends for favourable pressure gradients are now well established although more information is needed on transition length under these conditions.

Whilst the data of Abu-Ghannam and Shaw[7.37] are considered reliable for adverse pressure gradients and turbulence levels under 2%, reliable data are not yet available at higher turbulence levels.

The crucial question is whether there is a critical turbulence level under adverse pressure gradient conditions. Figure 11.11, taken from Schlichting and Das[2.27], indicates that such a critical level exists. The state of the suction surface boundary layer of a NACA 65-series compressor blade is plotted as a function of turbulence level. These cascade tests by Kiock indicate the existence of a laminar separation bubble at low turbulence levels. As the turbulence level is raised above a critical level of 2.5% the bubble is suppressed and transition moves immediately to the leading edge. Kiock also observed that the critical turbulence level was a function of Reynolds' Number. A reduction in Reynolds' Number from 1.6 x 10^5 to 0.9 x 10^5 increased the critical turbulence level from 2.5% to 4.5%.

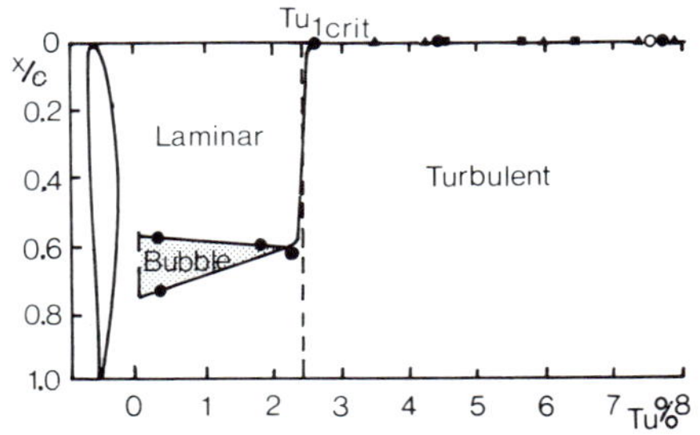

Fig. 11.11. State of suction surface boundary layer as a function of turbulence level.

Unpublished work performed by the author's students is consistent with the concept of a critical turbulence level. The turbulence level was varied up to 4.5% and extensive measurements were made under adverse pressure gradient conditions. The results began to diverge from the Seyb projections on a $R_{\theta t}$ - λ_θ plot at a turbulence level of 2.4%. By 3.2% the results were tending more towards the projections of Hall and at 4.5% this behaviour was firmly established. It was also observed that the λ_θ parameter did not adequately represent pressure gradient variations for high turbulence levels.

Evidence of the existence of a critical turbulence level is quite widespread but procedures have not yet been developed to account for this in the design process. More systematic experiments are required at high turbulence levels under adverse pressure gradients.

Some interesting data are available concerning the influence of free-stream turbulence on the length of a laminar separation bubble[7.53]. Klein[7.9] has also provided a thoughtful discussion of turbulence effects on compressor blades.

11.2.4 Surface roughness. Helpful discussions of the effects of surface roughness on blade performance have been given by Klein [7.9], Forster[11.41] and Koch and Smith[10.28].

Although it had been considered that roughness could have no effect on the skin friction loss of a laminar layer the results of Bammert and Milsch[11.42] from compressor cascades indicate that roughness can cause an increased momentum thickness for both laminar and turbulent layers. Roughness most certainly has an effect on transition and, beyond a certain threshold level, has a strong effect on skin friction in the turbulent regime.

A blade is termed 'hydraulically smooth' when the roughness peaks are contained within the laminar sublayer. Schlichting[2.25] has given as a criterion for hydraulic smoothness on flat plates a value of roughness Reynolds' Number, based on sand-grain roughness and free-stream velocity, of less than 100. A criterion for compressor blades has been given by Koch and Smith[10.28].

According to ref. 7.9 the allowable hydraulic roughness is generally lower than the critical roughness for transition. If transition is caused by roughness then the turbulent layer will also be affected by roughness.

Relationships between root-mean-square and centre-line-average measurements are discussed in ref. 10.28 and different types of surface finish are compared in terms of sand-grain roughness in ref. 11.41. The influence of the roughness of various engineering finishes on turbulent boundary layer drag is discussed in ref. 7.9.

The effect of roughness of a given size is represented by a horizontal line on the right hand side of a plot such as Fig. 11.10(b). As the Reynolds' Number is raised to a point where the roughness elements protrude through the laminar sublayer the momentum thickness and loss are no longer a function of Reynolds' Number but instead become a function of roughness size.

Blade performance is far more vulnerable to roughness at high Reynolds' Numbers than at low Reynolds' Numbers. For example at a Reynolds' Number of 10^8 the losses of blading having an equivalent sand-grain roughness of 0.01% chord are about 2.3 times those of a similar blade which is hydraulically smooth. If the same blading is operated at a Reynolds' Number of 10^6 there is no difference in performance between the rough blade and the smooth blade. For steam turbine applications this behaviour results in high pressure stages which are extremely sensitive to roughness and low pressure stages which are very tolerant to roughness.

For the same reason the high pressure blading of gas turbine compressors is particularly sensitive to roughness. In experiments upon a three-stage compressor Bammert and Woelk[11.43] reported upon characteristics obtained with smooth blades and three different emery grain sizes. The deterioration in characteristic shape was quite marked. The characteristic steepened and considerable pressure ratio, flow and stall margin were all lost. A 13% deterioration in efficiency was obtained between the smooth blade and the roughest blade having 130 μm grain size on a mean chord length of 39.9mm.

The turbine blading of gas and steam turbines is particularly susceptible to roughness caused by erosion, corrosion and contamination. Bammert and Sandstede[11.44] have conducted experiments on a four stage turbine having surfaces coated with differing grades of emery; the efficiency degradation was up to 14%. Forster[11.41] has made similar measurements. Available commercial data from tests on shot-blasted surfaces having lower degrees of roughness give a considerably lower penalty. Tabakoff et al [11.45] have investigated theoretically and experimentally the effects of foreign particle ingestion on turbine blading; increasing particle concentration causes a reduction in blade loading.

11.3 Interactive Parameters

Two variables which involve subtle interactions of geometrical variables and complex flow phenomena are tip clearance and axial clearance between blade rows. The flow phenomena are viscous and unsteady and are thus not yet fully calculable. Accordingly the majority of available data are experimental.

11.3.1 Tip clearance effects. In section 3.5.2 the subject of the tip clearance effects on axial flow turbomachine performance was introduced and it was shown how a relationship between the annulus wall boundary layer displacement thickness and the tip clearance could be established by theoretical considerations.

Ainley's expression (3.22) for tip clearance loss gives a dependence of loss on C_L^2. This is also the relationship one would obtain by computation of the local induced velocities. But if the tip clearance gap is treated as an orifice with the through-flow driven by pressure difference across the gap then the loss coefficient has been shown by Rains[11.46] to be proportional to $(t_c/h)\cdot(C_L)^{3/2}$.

After consideration of these and other factors Lakshminarayana[11.47] has produced a semi-empirical expression for the decrease in efficiency due to tip clearance:-

$$\Delta\eta = \frac{0.7\, t_c \Psi}{h \cos \beta_m} \left\{ 1 + 10\sqrt{\frac{\Phi}{\Psi} \cdot \frac{t_c A}{h \cos \beta_m}} \right\} \qquad (11.4)$$

Upon comparison with some of the experimental data presented in Fig. 11.12 Lakshminarayana found equation (11.4) to predict efficiency degradation by tip clearance with reasonable accuracy.

It will be recalled from section 3.5.2 that tests on the eight-stage compressor 'Alice' [3.79] gave a 4% increase in efficiency as the tip clearance was decreased from 3% to 1% of blade height. Further information from series of tests on axial flow compressors, fans and pumps is presented in Fig. 11.12. In all cases the data have been presented for conditions around the peak efficiency point or higher on the characteristic. Since 'zero clearance' data are not available extrapolations to zero clearance have been made in order to establish the benchmark level. In most cases the data trends were sufficiently linear for such extrapolation to present no difficulties. Where any doubt was present equation (11.4) was used to assist in obtaining a reasonable extrapolation.

The rule of thumb that every 1% increase in tip clearance results in a 2% degradation in efficiency (consistent with the result from compressor 'Alice') is indicated by a dashed line in Fig. 11.12. Upon reviewing the data the designer might well conclude that the above simple rule fits most data reasonably well, provided that aspect ratios of the order two are not exceeded. However there are factors calling for a more sophisticated approach, one of the most important being the influence of aspect ratio variation.

The results of Mellor and Strong[3.74] appear to give too high an efficiency decrement. They do, nevertheless, indicate that as the aspect ratio is increased substantially beyond two the effect of tip clearance on efficiency becomes much stronger. Equation (11.4) would give an efficiency decrement which increases with the square root of aspect ratio. The data of ref.3.74 indicate a much stronger trend. Further data are required from high aspect ratio machines to resolve this question. It should also be noted that the Reynolds' Number can be expected to have a significant effect on efficiency decrements and that insufficient experimental information is available on this parameter.

Data on the effect of tip clearance on stall-free operating range are less readily available. Some data are presented in Fig. 11.12 in which the ordinate plotted is the

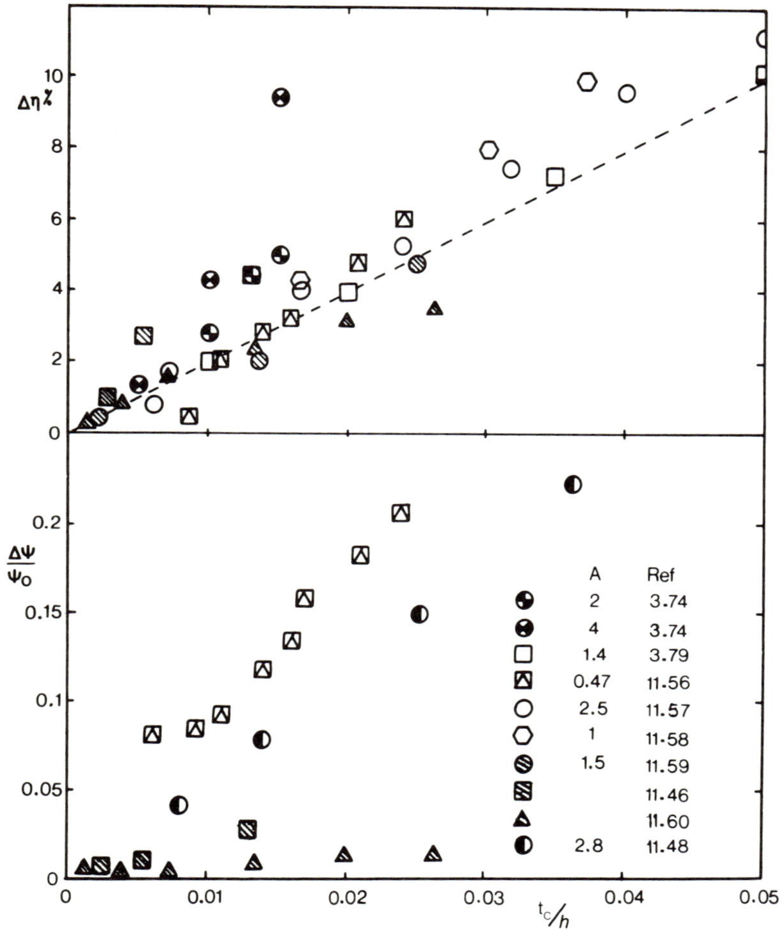

Fig. 11.12. Variation of efficiency and peak pressure-rise co-efficient with tip-clearance level for work-absorbing machines.

decrease in peak pressure rise coefficient normalised by the design pressure rise coefficient. Where the design value is not available data are normalised by the value of Ψ at the peak efficiency condition for each test.

It is quite clear that the different machines surveyed have a widely differing degree of sensitivity, in terms of peak pressure rise decrease, to increasing tip clearance.

A correlation by Hesselgreaves[11.48], in which the independent variable (t_c/h) is multiplied by a sensitivity factor, f, appears to effect an improved correlation of available data. The factor, f, is proportional to $(C_L)^{3/2}$ indicating a relationship to the Rains model mentioned previously.

The principal weakness in any attempt to produce a correlation for peak pressure rise is probably that the shape of the characteristic curve, and hence the machine's behaviour as stall is approached, is not taken into account. Any criterion for the inception of stall based purely on peak pressure rise will only have applicability to a limited range of machinery. Fan manufacturers often only consider the flow reduction attainable prior to stall and such considerations also have limitations. A more general criterion for stall margin must take into account some such parameter as flow coefficient divided by pressure rise coefficient or some other measure of distance from the operating line.

Information on the effect of tip clearance on degradation of axial turbine efficiency is presented in Fig. 11.13. The first impression is that the data are less coherent than those from work-absorbing machines. This impression is superficial because the range of configurations covered is much wider for the turbine data.

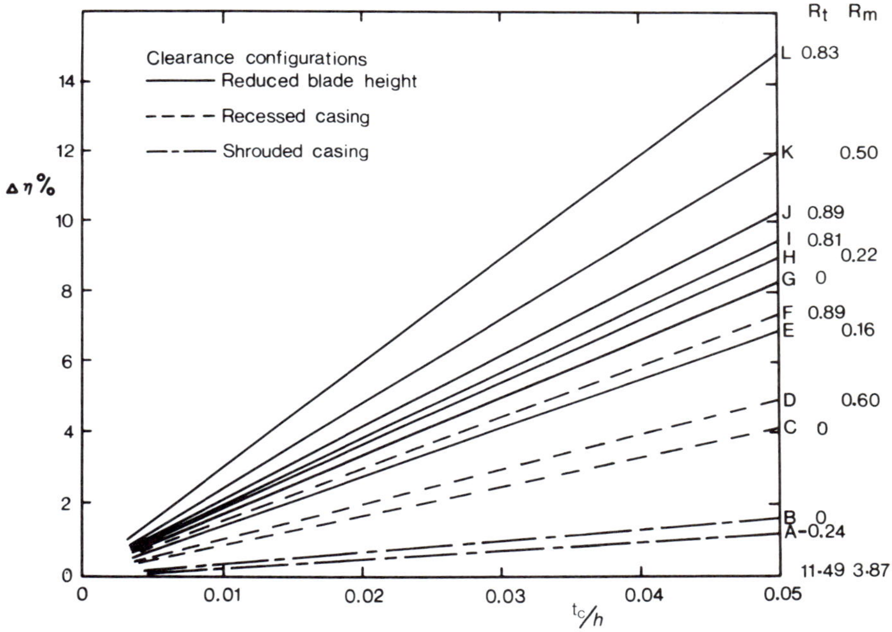

Fig. 11.13. Effect of tip clearance on efficiency for turbines.

Tests have been represented by straight lines rather than discrete points because results from a single turbine, as tip clearance is varied, usually exhibit marked linearity and little scatter.

Whereas all data in Fig. 11.13 were obtained by reducing blade height within a fixed casing, this was not so for the turbine data. Lines A and B of Fig. 11.13 were for cases having a tip-shrouded rotor. For cases C, D and F the clearance was obtained by cutting back the casing whilst retaining a constant blade height. The remaining cases, represented by full lines, are more comparable with those of Fig. 11.12 and were obtained by reducing the blade height within a fixed casing.

An additional parameter which might be expected to exert a substantial effect on tip clearance results is the degree of reaction. It might be expected that impulse blading, having a lower pressure difference between surfaces, would offer a lower performance loss due to tip clearance than reaction blading. This effect is observable in Fig. 11.13 to some extent. A complicating factor in interpretation of Fig. 11.13 is that whereas the Soviet data of ref. 3.87 quote the degree of reaction at the mid height, the U.S. data of ref. 11.49 quote the corresponding value at the blade tip. A further complication is that the data of ref. 3.87 pertain to blading having differing degrees of casing flare (varying between 4.5° and 12.4°) whereas the configurations of ref. 11.49 all had zero flare. When the above factors are all taken into account the turbine data do not seem to be too incompatible with those obtained from compressors.

11.3.2 Axial spacing between blade rows. As the spacing between rotor and stator rows is varied systematically in an axial direction pronounced variations in performance are observed. These variations may include changes in efficiency, stall margin, sound propagation and vibration excitation. The effects are always unsteady in nature, often viscous and generally difficult to measure correctly. Their analysis and prediction is still in an embryonic stage of development.

A number of the unsteady effects in turbomachines have been discussed by various reviewers (e.g. refs. 8.5, 8.84 and 11.50) and include the effects discussed in Chapter 8. Of particular interest to designers of steam turbines is the periodic excitation of a rotor blade passing through the shock and expansion regions emanating from an upstream nozzle row. This source produces an acoustic disturbance at rotor-blade passing frequency. A different source of excitation is the possibility of acoustic resonance which has been suggested as a problem area potentially leading to destruction in gas circulators for nuclear power stations[11.51] or steam turbines[8.107]. These effects are complex interactive ones which would be particularly dependent on axial spacing.

The work of Fleeter et al[8.126] was oriented towards quantification of the effects of axial spacing on the unsteady aerodynamics of blade rows, especially with regard to vibration sensitivity. Rotor wake profiles and unsteady stator pressure distributions have been measured for differing amounts of axial clearance.

Even for low-speed flows, few systematic data are available. It can be said that in the simplest case of interaction between a rotor row and a stator row (whether compressor or turbine) there are two predominant effects. These are the effects of potential interaction and of wake interaction.

If systematic predictions of axial spacing effects are to be undertaken it will be necessary to incorporate both of these effects. As indicated in Chapter 8 the mere linear superposition of the perturbations would lead to an inappropriate, and probably excessive, response. Potential interaction effects may be derived from quasi-steady and unsteady theories; this will be a reasonable extension of existing capabilities. The application of viscous flow theory to wake interaction is less tractable and progress in wake interaction will be made most rapidly by the use of suitable empirical data.

Semi-empirical equations for the decay of near and far wakes were presented in ref. 7.64 and appear to give a good correlation. Figure 11.14, taken from ref. 7.64, presents available data for maximum velocity defects on the basis of a $C_D^{\frac{1}{4}}$ correlation for the wakes of isolated aerofoils, cascades and rotors. Schlichting[2.25] has advocated the division of the maximum velocity defect by a $C_D^{\frac{1}{2}}$ term. In ref. 7.64 it is argued that the proposal of Silverstein et al[11.52] for a $C_D^{\frac{1}{4}}$ relationship should be used to give a tighter fit to the data.

The magnitude and nature of gusts encountered by downstream blading should be reliably predictable using the data of Fig. 11.14.

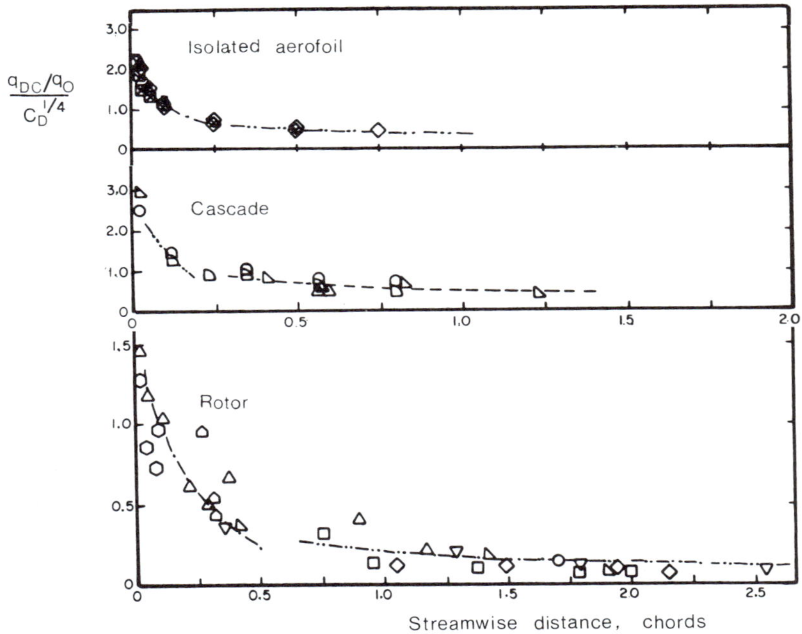

Fig. 11.14. Data and trends for maximum velocity defect; sources are given by Reynolds, Lakshminarayana and Ravindranath[7.64].

Inspection of the results of ref. 3.75 shows that there are significant variations in efficiency and peak pressure rise when axial clearance between compressor rows is varied. These differences correlate better with blade spacing than with chord and better with either of those parameters than with blade height. This suggests that potential interaction effects might be quite significant in the flow behaviour. The effect on efficiency for compressors appears to be that efficiency is highest when the blade rows are closest together and falls off as the blade rows are moved apart.

This is in accordance with the earlier findings of Möhle[11.53] who found that efficiency improvements were registered for a single stage compressor as the axial spacing was decreased from values of about half a blade chord.

Turbine designers usually argue that both potential interaction and wake interaction effects are deleterious for efficiency. For very small axial clearances the potential interaction effect is the strongest. However as axial clearance is increased viscous effects begin to predominate. It is therefore argued that some turbines might show an optimum axial clearance[3.87]. The optimum efficiency is obtained when axial clearance is increased to a value for which the increase in wake losses is no longer dominated by a decrease in potential interaction losses.

The data presented in Fig. 11.15 tend to confirm the validity of the above argument. Some of the configurations tested exhibit optimum efficiency for axial clearances of the order of 15% blade height. Unfortunately no attempt has been made to use either chord-length or throat as a normalising parameter which might have facilitated a tighter correlation.

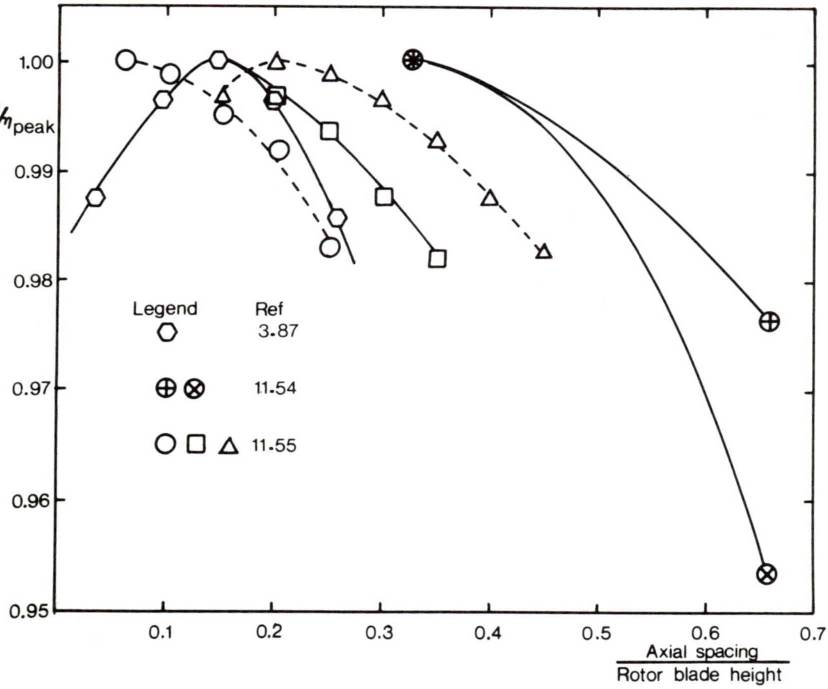

Fig. 11.15. Effect of axial spacing on normalised efficiency for axial flow turbines.

Ref. 3.87 quotes an example of a high reaction turbine in which the ratio of axial clearance to rotor blade height was varied from 8.3% to 25.5%. A corresponding monatonic decrease in efficiency of between 1 and 1.5% was observed.

For lower degrees of reaction, approaching impulse conditions, a definite optimum clearance is generally observed. The data from ref. 3.87 in Fig. 11.15 are from a stage having degrees of reaction of 50% or less at the casing and as low as 7% at the hub. The stage of ref. 11.54 was close to impulse in design and was transonic. An additional factor elucidated by Gogolev[11.55] was that the absolute stator discharge angle must be lower than 75° for an optimum clearance to be observed and that the hub/tip ratio must be greater than 0.66.

It must be concluded from the foregoing discussion that the state of empirical information and correlation on axial clearance effects is woefully deficient. There are important advantages at stake in efficiency, stall behaviour, noise and vibration. The state of theoretical development is not yet sufficiently advanced for predictions of performance penalties to be possible. The whole basis of correlation of data from compressor and turbine testing leaves much to chance. Accuracy will continue to be poor where true time averages of downstream traverse data are not used. Before further progress can be made an improved understanding of the physics of clearance flows and interaction effects is required.

11.4 Conclusions on the Design Application of Cascade Information

The information assembled in this chapter has mostly consisted of correlations, often quite early ones, of various parameters.

In contrast the major efforts over the last few years have tended to neglect the correlation approach and to concentrate on the generation of powerful analysis techniques for cascades.

The objective of each of these approaches is the provision of essential design techniques and information. Satisfactory design solutions for the complex and aerodynamically-sensitive blading of modern high-speed, high-loading turbomachinery must call upon both approaches, each of which is therefore necessary. The correlation data will be needed for preliminary surveys, rapid design analysis work and for filling in those detailed interactions which are not yet calculable. The powerful computational techniques are needed if any design is to be credible; they provide the framework and structure for the detailed aerodynamic design. It is desirable that the gap between theory and correlation should be bridged in constructive and far-sighted ways. Serious design packages will need to absorb the available information from correlations. The computational techniques may be used to generate more correlation information but their most likely use is as a one-off design tool suitable for iterative interaction using modern computer graphic capabilities. Design progress will therefore continue to be a pragmatic mixture of the best available computational techniques supplemented by the best available correlation data.

These design strategies do not imply a reduction in the importance of cascade testing. On the contrary the author considers that the demand on cascade tunnels will tend to increase. The testing will be different in kind with the emphasis switching from routine testing to the more careful elucidation of aerodynamic behaviour. There are exceptions to this change of emphasis. Many blading configurations are relatively unexplored and in some of these cases systematic design information would be best generated by cascade testing. An example of this is the unmapped territory of widely-spaced blading for propellers and fans. These configurations have traditionally fallen between isolated aerofoil work and cascade work and systematic cascade testing is urgently needed.

Nevertheless in general cascade testing will tend to be used for a more diverse range of purposes. Not the least important of these is the educational value of cascade testing. The characteristic behaviour of differing types of blading under differing conditions is more readily appreciated by observation of a cascade test than by observation of a machine's performance. This is especially so for high speed flow with schlieren visualisation capabilities.

The cascade tunnel has an essential role as 'calibrator' of new numerical methods. No such method should be used for design purposes until its predictions have been thoroughly tested by comparison with appropriate cascade tests. This is not to say that the cascade test always gives the correct answer. There are many difficult problems still to be overcome in cascade testing, especially at high speeds.

It is quite essential to establish the validity of physical models prior to their incorporation into the design process. This is often best done through appropriate cascade testing. There are a number of areas in which prediction capabilities are presently limited by the available physical understanding. An example of this is the strong shock-boundary layer interactions which occur in turbomachinery blading. Data are required on interaction-region behaviour in the presence of high levels of free-stream turbulence and surface curvature.

The status of viscous-flow modelling is still quite unsatisfactory. Information is inadequate on laminar separation bubbles, transition, turbulent separation and base flows. The concave surfaces of turbine blades still present problems for heat transfer predictions. Three-dimensional viscous effects and radial migrations are hardly predicted at all.

Separated flows are definitely responsible for the failure of prediction methods to perform satisfactorily for compressor blades operating under off-design conditions. These are also the biggest impediment to the application of cascade theories in the design of radial-flow machines. Much detailed experimental investigation, and the painstaking incorporation of results into prediction procedures, are required if the accuracy of flow calculations is to be improved. Experimental work on cascades could play an important role in any solution to the problem of stall propagation.

Probably the most important parameter for blading is its loss coefficient and it is in predicting losses that the present techniques are least satisfactory. As exciting new purpose-designed blade shapes are introduced compilations of the accumulated test experience of a company will become less reliable in predicting profile losses. The diffusion factor approach may also prove to be less useful. For accurate predictions to be made it will be essential to return to cascade tests and flow calculations for guidance; where these have given unsatisfactory loss predictions the deficiencies will need to be rectified. This conclusion applies even more strongly to secondary flow losses where existing predictions provide a most inadequate basis for design. Flow velocities and angles are predicted very well but a major breakthrough in the prediction of secondary losses is urgently needed.

In summary the author sees a healthy and interesting future for both theoretical and experimental work on cascades. Sophisticated calculation methods will form the backbone of the blading design process and if these can be viscous and three-dimensional then significant benefits will accrue. The calculation methods will be used in interactive modes using suitable graphic displays, thus freeing the aerodynamic designer to concentrate on the more creative aspects of the work. The methods will only produce accurate results if improved information on the physics of the flow is incorporated and cascade testing has a vital role to play in the provision of this information.

REFERENCES

11.1 ECKERT, B. Recent results on high pressure axial blowers. NACA TM 1128 (1947)

11.2 ANDREWS, S.J. Test related to the effect of profile shape and camber line on compressor cascade performance. ARC R&M 2743 (1950)

11.3 WALLIS, R.A. The development of blade sections for axial flow fans. Mech. & Chem. Eng. Trans. I. E. Aust., Vol. MC8. No. 2. pg. 111 (1972)

11.4 JACOBS, E.N., WARD, K.E. and PINKERTON, R.M. The characteristics of 78 related airfoil sections from tests in the variable density wind tunnel. NACA Rep. 460 (1933)

11.5 CARTER, A.D.S. Blade profiles for axial flow fans, pumps, compressors etc. Proc. I. Mech. E. 175, 15 (1961)

11.6 WRIGHT, L.C. Approximate effect of leading-edge thickness, incidence angle and inlet Mach Number in inlet losses for high solidity cascades of low-cambered blades. NACA TN 3327 (1954)

11.7 KLAPPROTH, J.F. Approximate relative-total-pressure losses of an infinite cascade of supersonic blades with finite leading edge thickness. NACA RM E 9421 (1950)

11.8 REDDY, J.L. and KLAPPROTH, J.F. Advanced fan development for high bypass engines. AIAA 68-563 (1968)

11.9 YORK, R.E. and WOODARD, H.S. Supersonic compressor cascades - an analysis of the entrance region flow field containing detached shock waves. Trans. ASME, J. Eng. Power 98, 2 pg. 247 (1976)
11.10 HERRIG, L.J., EMERY, J.C. and ERWIN, J.R. Effect of section thickness and trailing-edge radius on the performance of NACA 65-series compressor blades in cascade at low speeds. NACA RM L51J16 (1951)
11.11 INOUE, M. and KANEKA, K. Effect of trailing edge thickness on the cascade performance on circular-arc blades. Proc. 2nd Int. JSME Symp. Fluid Mech. and Fluidics. Tokyo (1972)
11.12 FARN, C.L.S. and WHIRLOW, D.K. Design estimate of cascade performance. Computers and Fluids 3, 2-3, 225 (1975)
11.13 REEMAN, J. and SIMONIS, E.A. The effect of trailing edge thickness on blade loss. RAE T.N. Eng. 116 (1943)
11.14 PRUST, H.W.Jr. and HELON, R.M. Flow conditions around the exit and downstream of certain stator blading with various trailing-edge thicknesses and geometries. NASA TM X-2659 (1972)
11.15 JOHNSTON, I.H., DRANSFIELD, D.C. and FULLBROOK, A.J. Experiments concerning the effect of trailing edge thickness on blade loss and turbine stage efficiency. ARC R&M 3459 (1967)
11.16 BAMMERT, K. and FIEDLER, K. Loss due to trailing edge and friction in turbine blade cascades. Forsch. Ing. Wes. 32,5 pg. 133 (1966)
11.17 MARKOV, N.H. Calculation of the aerodynamic characteristics of turbine blading. Trans. by Assoc. Tech. Serv. (1958)
11.18 ZARYANKIN, A.E. Edge losses in turbine cascades. Teploenergetika 13 (1) 38-42 pg. 49 (1966)
11.19 JASPAL, R. Investigation of the trailing edge losses of a symmetrical isolated aerofoil at transonic speeds. M. Sc. thesis, London Univ. (1970)
11.20 FENTRESS, W.K. and DESAI, K.A. Low speed cascade tests. Westinghouse Report. WANL-TME-1970 (1970)
11.21 KACKER, S.C. and OKAPUU, U. A mean line prediction method for axial flow turbine efficiency. ASME Paper 81-GT-58 (1981)
11.22 CARTER, A.D.S. The axial compressor. "Gas Turbine Principles and Practice." (ed. Sir H. Roxbee-Cox) G. Newnes, London (1955)
11.23 HOWELL, A.R. The present basis of axial flow compressor design: Part I - Cascade theory and performance. ARC R&M 2095 (1942)
11.24 HEILMANN, W., STARKEN, H. and WEYER, H. Cascade wind tunnel tests on blades designed for transonic and supersonic compressors. AGARD CP 34 (1968)
11.25 COHEN, H., ROGERS, G.F.C. and SARAVANAMUTTOO, H.I.H. Gas turbine theory. 2nd ed., Longman (1974)
11.26 HOWELL, A.R. The aerodynamics of the gas turbine. J. Roy. Aero. Soc. 70, pg. 329 (1948)
11.27 ZWEIFEL, D. The spacing of turbo-machine blading especially with large angular deflection. Brown Boveri Rev. 32 pg. 436 (1945)
11.28 DUNCOMBE, E. Aerodynamic design of axial flow turbines. 'Aerodynamics of Turbines and Compressors' Sir W.R. Hawthorne, ed. Princeton (1964)
11.29 GOSTELOW, J.P. Design and performance evaluation of four transonic compressor rotors. Trans. ASME, J. Eng. Power 93, 1 pg. 33 (1971)
11.30 DUNAVENT, J.C. and ERWIN, J.R. Investigation of a related series of turbine blade profiles in cascades. NACA TN 3802 (1956)
11.31 HORLOCK, J.H., SHAW, R., POLLARD, D. and LEWKOWICZ, A. Reynolds' Number effects in cascades and axial flow compressors. Trans. ASME, J. Eng. Power, pg. 236 (July, 1964)
11.32 AMERICAN SOCIETY OF MECHANICAL ENGINEERS. Symposium on the effects of Reynolds' Number on turbomachinery. (1963)
11.33 BRIDLE, E.A. Some high speed tests on turbine cascades. ARC R&M 2697 (1949)
11.34 DRING, R.P. and HEISER, W.H. Turbine aerodynamics. Chapter 18 of "The Aerothermodynamics of Aircraft Gas Turbine Engines". Oates, G.C. ed. AFAPL-TR-78-52 (USA) (1978)

11.35 CRAIG, H.R.M. and COX, H.J.A. Performance estimation of axial flow turbines. Proc. I. Mech.E. 185, 32/71 pg. 407 (1970-71)

11.36 DENTON, J.D. A method of predicting the profile loss of turbine cascades. ARC 33,222 (1971)

11.37 FULLBROOK, D.J. The effect of Reynolds' Number and air turbulence level on the performance of a turbine rotor blade in a two-dimensional cascade. NGTE Aero. Note 707 (1962)

11.38 KIOCK, R. Turbulence downstream of stationary and rotating cascades. ASME Paper 73-GT-80 (1973)

11.39 SHAW, R., LEWKOWICZ, A.K. and GOSTELOW, J.P. Measurement of turbulence in the Liverpool University turbomachinery wind tunnels and compressor. ARC CP 847 (1966)

11.40 EVANS, R.L. Stream turbulence effects on the turbulent boundary layer in a compressor cascade. ARC 34,587 (1973)

11.41 FORSTER, V.T. Performance loss of modern steam-turbine plant due to surface roughness. Proc. I. Mech. E. 181 pg. 391 (1966-67)

11.42 BAMMERT, K. and MILSCH, R. Boundary layers on rough compressor blades. ASME Paper 72-GT-48 (1972)

11.43 BAMMERT, K. and WOELK, G.U. The influence of the blading surface roughness on the aerodynamic behaviour and characteristic of an axial compressor. Trans. ASME, J. Eng. Power 102, 2 pg. 383 (1980)

11.44 BAMMERT, K. and SANDSTEDE, H. Measurements concerning the influence of surface roughness and profile changes on the performance of gas turbines. Trans. ASME, J. Eng. Power 94 pg. 207 (1972)

11.45 TABAKOFF, W., HOSNY, W. and HAMED, A. Effect of solid particles on turbine performance. Trans. ASME 98 A, 1, pg. 47 (1976)

11.46 RAINS, D.A. Tip clearance effects in axial flow compressors and pumps. Cal. Tech., Mech. Eng. Lab. Rep. 5 (1954)

11.47 LAKSHMINARAYANA, B. Methods of predicting the tip clearance effects in axial flow turbomachinery. Trans. ASME, J. Basic Eng. (1970)

11.48 HESSELGREAVES, J.E. A correlation of tip-clearance/efficiency measurements on mixed-flow and axial-flow turbomachines. Proc. 3rd. Conf. on Fluid Mech. & Fluid Machines, Budapest (1969)

11.49 HAAS, J.E. and KOFSKEY, M.G. Effect of rotor tip clearance and configuration on overall performance of a 12.77-centimetre tip diameter axial-flow turbine. ASME Paper 79-GT-42 (1979)

11.50 HORLOCK, J.H. Unsteady flow in turbomachines. I. E. Aust. Conf. Hyd. and Fluid Mech. (1968)

11.51 RIZK, W. and SEYMOUR, D.F. Investigations into the failure of gas circulators and circuit components at Hinkley Point Nuclear Power Station. Proc. I. Mech. E. 179, 1 (1964-65)

11.52 SILVERSTEIN, A., KATZOFF, W. and BULLIVANT, W.K. Downwash and wake behind plain airfoils. NACA Rep. 651 (1939)

11.53 MOHLE, H. Untersuchungen über den Einfluss des Abstandes zwischen Lauf-und Leitrad auf das Betriebsverhalten von einstufigen axialen Strömungsmachinen. Konstruktion im Maschinen-Apparate-und Gerätebau, 13, 213-222 (1961)

11.54 KANE, M.J. Experimental investigation of the effect of rotor to stator axial spacing on the performance of a single stage transonic axial turbine. MSME thesis, Nav. Postgrad. Sc., Monterey (1978)

11.55 GOGOLEV, I.G. et al. The influence of blade clearance on the characteristics of a turbine stage. Teploenergetika pg. 70 (1973)

11.56 SPENCER, E.A. The performance of an axial-flow pump. Proc. I. Mech. E. 170, 25 pg. 874 (1956)

11.57 RUDEN, P. Investigation of single stage axial fans. NACA TM 1062 (1944)

11.58 KOLESHNIKOV, A.V. Effect of gap between the impeller and the casing on the characteristics of an axial flow fan. (in Russian) Prom. Aero. 17, pg. 20 (1960)

11.59 WILLIAMS, A.D. The effect of tip clearance flow on the performance of axial flow compressors. AE thesis, Cal. Tech. (1960)
11.60 MUHLEMANN, N.E. Experimental investigations of a single-stage axial blower. (in German) Mitt. Inst. Aero. E.T.H. Zurich 12 (1946)

Index